AF597965

Methods in Molecular Biology™

Series Editor
John M. Walker
School of Life Sciences
University of Hertfordshire
Hatfield, Hertfordshire, AL10 9AB, UK

For further volumes:
http://www.springer.com/series/7651

Phosphatase Modulators

Edited by

José Luis Millán

Sanford Children's Health Research Center, Sanford-Burnham Medical Research Institute, La Jolla, CA, USA

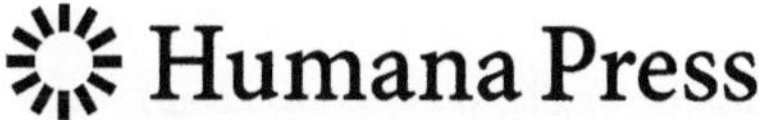

Editor
José Luis Millán
Sanford Children's Health Research Center
Sanford-Burnham Medical Research Institute
La Jolla, CA, USA

ISSN 1064-3745 ISSN 1940-6029 (electronic)
ISBN 978-1-62703-561-3 ISBN 978-1-62703-562-0 (eBook)
DOI 10.1007/978-1-62703-562-0
Springer New York Heidelberg Dordrecht London

Library of Congress Control Number: 2013942869

Printed on acid-free paper

Humana Press is a brand of Springer
Springer is part of Springer Science+Business Media (www.springer.com)

Preface

For decades, medicinal chemistry efforts, including target validation, small-molecule screening, structure–activity relationships, and lead optimization, were terms and activities largely restricted to scientists working at large pharmaceutical companies with well-established drug discovery programs. However, that situation changed with the implementation of the roadmap initiative by the National Institutes of Health in the USA, which supported the development of screening centers at academic institutions, known as the Molecular Libraries Probe Production Centers Network (MLPCN). Suddenly, basic scientists could be involved in the design of assays to screen for modulators of their favorite targets and develop probes to help them investigate the function of their pet protein or pathway. They could also now aspire to become involved in the development of pharmaceuticals for clinical use.

One of these comprehensive screening centers is the outstanding Conrad Prebys Center for Chemical Genomics (CPCCG) at the Sanford-Burnham Medical Research Institute (SBMRI), an Institution that I joined almost four decades ago and where I have driven my research on the structure and function of mammalian alkaline phosphatases ever since. Tissue-nonspecific alkaline phosphatase (TNAP) was one of the early targets used for screening at CPCCG and one for which we are developing pharmaceuticals for the treatment of medial vascular calcification. I am pleased that many of the scientists from CPCCG have agreed to contribute chapters to this book.

My intention in planning this volume has been to provide a "how to" manual for basic scientists interested in initiating a drug discovery effort. Thus, while this book contains the traditional "method" chapters and some typical reviews on the structure and known functions of phosphatases, other contributions, while formally "reviews," are meant to discuss approaches and alternatives useful in making "go/no-go" decisions in high-throughput screening (HTS) and lead optimization campaigns. In that light, Michael Jackson opens the book (Chapter 1) with a clear presentation of what constitutes a probe versus a drug. Ed Sergienko discusses the challenges associated with "Phosphatase HTS assay design and selection" (Chapter 2). In Chapter 3, Professors Buchet, Millán, and Magne introduce what is known about the function of alkaline phosphatases, as a prelude to a related "review" and five method chapters. Using TNAP as an example, Thomas Chung beautifully elucidates the efforts that lead to "Robotic implementation of assays" (Chapter 4), while chemists Teriete, Pinkerton, and Cosford discuss how to go from hits to leads (Chapter 5). Chapter 6 presents a precise protocol on how to use endogenous TNAP activity in neat plasma as a biomarker in vivo, while Chapters 7 and 8 describe protocols for the isolation and study of mineralization in isolated matrix vesicles and cells. Chapter 9 explains the development of a modulator of intestinal alkaline phosphatase, and Chapter 10 describes new activity assays for NPP1, an enzyme with structural similarity to TNAP that is also involved in the control of skeletal and soft-tissue calcification. Chapters 11 and 12, contributed by Professor Vihko, summarize what is known about acid phosphatases and early efforts to probe the function of this class of enzymes.

The book finishes with five chapters dedicated to protein phosphatases. In Chapter 13, Lutz Tautz presents a comprehensive review of the structure and function of protein tyrosine phosphatases (PTP) to introduce efforts to screen for modulators of PTPs (Chapter 14) and then to use those modulators in an exploration of T cell receptor signaling (Chapter 15). Finally, Professor Bollen discusses interactor-guided dephosphorylation of protein phosphatase-1 (Chapter 16), while Professor Janssens discusses the structure, regulation, and pharmacological modulation of protein phosphatase 2A (Chapter 17).

I hope that these presentations will help the reader gain an understanding of the many facets and complexities associated with undertaking a drug discovery effort, and will serve as a roadmap to initiating those efforts. In closing, I would like to express my sincere gratitude to my administrative assistant, Amy Zimmon, for keeping track of all the materials submitted to this book project and for editing all of the manuscripts for consistency.

La Jolla, CA, USA *José Luis Millán*

Contents

Contributors

CÉSAR L. ARAUJO • *Department of Clinical Chemistry, Institute of Clinical Medicine, University of Helsinki, Helsinki, Finland; Helsinki University Hospital Laboratory, Helsinki, Finland*

EKATERINA V. BOBKOVA • *Conrad Prebys Center for Chemical Genomics, Sanford-Burnham Medical Research Institute, La Jolla, CA, USA*

SHANNAH BOENS • *Laboratory of Biosignaling & Therapeutics, Department of Cellular and Molecular Medicine, University of Leuven, Leuven, Belgium*

MATHIEU BOLLEN • *Laboratory of Biosignaling & Therapeutics, Department of Cellular and Molecular Medicine, University of Leuven, Leuven, Belgium*

RENÉ BUCHET • *Equipe Organisation et Dynamique des Membranes Biologiques, UMR-CNRS 5246, Institut de Chimie et Biochimie Moléculaires et Supramoléculaires, Université Claude Bernard–Lyon 1, Université de Lyon, Villeurbanne, France*

THOMAS D.Y. CHUNG • *Conrad Prebys Center for Chemical Genomics, Sanford-Burnham Medical Research Institute, La Jolla, CA, USA*

NICHOLAS D.P. COSFORD • *NCI-Designated Cancer Center, Sanford-Burnham Medical Research Institute, La Jolla, CA, USA*

DAVID A. CRITTON • *Molecular Discovery Technologies, Bristol-Myers Squibb, Lawrenceville, NJ, USA*

LAURENCE DELACROIX • *Immunology and Infectious Diseases, GIGA-Signal Transduction, University of Liège, Liège, Belgium*

STEFAN GROTEGUT • *Conrad Prebys Center for Chemical Genomics, Sanford-Burnham Medical Research Institute, La Jolla, CA, USA*

DORIEN HAESEN • *Laboratory of Protein Phosphorylation and Proteomics, Department of Cellular and Molecular Medicine, University of Leuven, Leuven, Belgium*

ANNAKAISA M. HERRALA • *Department of Clinical Chemistry, Institute of Clinical Medicine, University of Helsinki and Helsinki University Hospital Laboratory, Helsinki, Finland*

ELITSA IVANOVA • *Laboratory of Protein Phosphorylation and Proteomics, Department of Cellular and Molecular Medicine, University of Leuven, Leuven, Belgium*

MICHAEL R. JACKSON • *Conrad Prebys Center for Chemical Genomics, Sanford-Burnham Medical Research Institute, La Jolla, CA, USA*

VEERLE JANSSENS • *Laboratory of Protein Phosphorylation and Proteomics, Department of Cellular and Molecular Medicine, University of Leuven, Leuven, Belgium*

TINA KIFFER-MOREIRA • *Sanford Children's Health Research Center, Sanford-Burnham Medical Research Institute, La Jolla, CA, USA*

CAROLINE LAMBRECHT • *Laboratory of Protein Phosphorylation and Proteomics, Department of Cellular and Molecular Medicine, University of Leuven, Leuven, Belgium*

WALLACE H. LIU • *Infectious and Inflammatory Disease Center, Sanford-Burnham Medical Research Institute, La Jolla, CA, USA*

CHEN-TING MA • *Conrad Prebys Center for Chemical Genomics, Sanford-Burnham Medical Research Institute, La Jolla, CA, USA*
DAVID MAGNE • *Equipe Organisation et Dynamique des Membranes Biologiques, UMR-CNRS 5246, Institut de Chimie et Biochimie Moléculaires et Supramoléculaires, Université Claude Bernard–Lyon 1, Université de Lyon, Villeurbanne, France*
SAÏDA MEBAREK • *Equipe Organisation et Dynamique des Membranes Biologiques, UMR-CNRS 5246, Institut de Chimie et Biochimie Moléculaires et Supramoléculaires, Université Claude Bernard–Lyon 1, Université de Lyon, Villeurbanne, France*
JOSÉ LUIS MILLÁN • *Sanford Children's Health Research Center, Sanford-Burnham Medical Research Institute, La Jolla, CA, USA*
SONOKO NARISAWA • *Sanford Children's Health Research Center, Sanford-Burnham Medical Research Institute, La Jolla, CA, USA*
SLAWOMIR PIKULA • *Department of Biochemistry, Nencki Institute of Experimental Biology, Polish Academy of Sciences, Warsaw, Poland*
ANTHONY B. PINKERTON • *NCI-Designated Cancer Center, Sanford-Burnham Medical Research Institute, La Jolla, CA, USA*
ILEANA B. QUINTERO • *Department of Clinical Chemistry, Institute of Clinical Medicine, University of Helsinki and Helsinki University Hospital Laboratory, Helsinki, Finland*
SOUAD RAHMOUNI • *Immunology and Infectious Diseases, GIGA-Signal Transduction, University of Liège, Liège, Belgium*
WARD SENTS • *Laboratory of Protein Phosphorylation and Proteomics, Department of Cellular and Molecular Medicine, University of Leuven, Leuven, Belgium*
EDUARD A. SERGIENKO • *Conrad Prebys Center for Chemical Genomics, Sanford-Burnham Medical Research Institute, La Jolla, CA, USA*
QING SUN • *Conrad Prebys Center for Chemical Genomics, Sanford-Burnham Medical Research Institute, La Jolla, CA, USA*
KATHELIJNE SZEKÉR • *Laboratory of Biosignaling & Therapeutics, Department of Cellular and Molecular Medicine, University of Leuven, Leuven, Belgium*
LUTZ TAUTZ • *Infectious and Inflammatory Disease Center, Sanford-Burnham Medical Research Institute, La Jolla, CA, USA*
PETER TERIETE • *NCI-Designated Cancer Center, Sanford-Burnham Medical Research Institute, La Jolla, CA, USA*
ALEYDE VAN EYNDE • *Laboratory of Biosignaling & Therapeutics, Department of Cellular and Molecular Medicine, University of Leuven, Leuven, Belgium*
PIRKKO T. VIHKO • *Department of Clinical Chemistry, Institute of Clinical Medicine, University of Helsinki, Helsinki, Finland; Helsinki University Hospital Laboratory, Helsinki, Finland*

Chapter 1

Chemical Probe Development Versus Drug Development

Michael R. Jackson

Abstract

Phosphatases as a class of proteins have recently attracted significant attention from the pharmaceutical industry. As our knowledge of this diverse family of proteins has grown, the relationship between phosphatases and human disease has clearly been established, with model systems proving much validation for the potential of some members of this family to be candidate drug targets. This, coupled with the fact that there have been a flood of successful drug development efforts over the past 10 years targeting protein kinases, has led some to propose that phosphatases as a class of enzymes might be equally as rich a source of drug targets as kinases. However to date there remain relatively few molecules targeting protein phosphatases in clinical development. This is less a reflection of their importance in key processes associated with disease, but rather seems to reflect inherent issues with developing drugs for many members of this family. This seems especially so for intracellular phosphatases where the development of selective, potent cell penetrant molecules with good drug-like properties has proven a formidable challenge.

This chapter provides a brief outline of the two major processes that have resulted in the existing armament of chemical modulators of protein phosphatases, namely, chemical probe development and drug development. These two processes initially seem to be rather similar and while they do overlap, the stated goals of the two approaches at project initiation are distinct.

Key words Phosphatases, Chemical probe development, Drug development, Automated assays, Robotics, Phosphatase assays, Assay development

1 Chemical Probe Development

It is only in the past 5 years or so that chemical probe development has really taken off. This has largely been driven by the NIH roadmap initiative which engaged a network (Molecular Libraries Probe Production Centers Network: http://mli.nih.gov/mli/mlpcn/mlpcn/) of academic-based centers that provided state-of-the-art high-throughput screening (HTS) capabilities of a large diverse small-molecule library (the Molecular Libraries Small Molecule Repository (MLSMR) >350,000 compounds) as well as much expertise in the activities of assay development and downstream hit to probe (http://mli.nih.gov/mli/). Prior to this initiative, HTS was almost exclusively conducted within pharmaceutical

José Luis Millán (ed.), *Phosphatase Modulators*, Methods in Molecular Biology, vol. 1053,
DOI 10.1007/978-1-62703-562-0_1, © Springer Science+Business Media, LLC 2013

and biotechnology companies where the goal was to discover and develop drugs. The goal of the MLPCN was somewhat different; it was to develop chemical probes as research tools for use in the study of protein and cell functions as well as the biological processes relevant to physiology and disease. The process to derive these probes relied on MLPCN centers to optimize biochemical, cellular, and model organism-based assays submitted by the biomedical research community and to perform automated HTS, and in some cases high-content (HCS) phenotypic screens followed by medicinal chemistry optimization on confirmed hits to produce chemical probes.

At the outset of projects, assay providers (biomedical researchers typically from the public sector) with deep knowledge of specific targets, pathways, and biological processes provided a rationale for the relevance of the biology to be screened and defined a profile of properties that a chemical compound would need to attain to be of value to probe biology. The expectation was that such probes, which would be freely available for distribution to academic scientists, would then be used to further advance our understanding of biology at the biochemical, cellular, and possibly organismal level with the recognition that these new tools would allow scientists to explore the functions of genes and signaling pathways in human health and disease. The diversity of biology that has been screened by the MLPCN network is extremely wide; assay formats cover the full spectrum of platforms from straightforward biochemical assays that measure enzyme activity through to highly complex image-based high-content screens of cells for compounds that affect cellular function, to screens of whole organisms, e.g., *C. elegans*. Information on the assays and screens generated by the MLPCN is made fully available through PubChem, a comprehensive database of chemical structures and their biological activities developed by NCBI. Details of the chemical probes that have been derived by this program (currently over 300) are available on the NCBI bookshelf (http://www.ncbi.nlm.nih.gov/books/NBK47352/).

While drug discovery was not a stated objective at the outset for the majority of chemical probe development projects, the NIH has stated that they anticipate that some of these projects will ultimately facilitate the development of new drugs by providing early-stage chemical compounds to researchers in the public and private sectors for validation of new drug targets, which could then move into the drug-development pipeline. Indeed, with additional optimization some of the chemical probes developed within the network are now advancing into in vivo studies (e.g., animal models of disease) and in one case into clinical testing, where they are allowing targets to be chemically validated for a specific disease. In this way the probe discovery path is crossing over into the traditional drug discovery and development pathway.

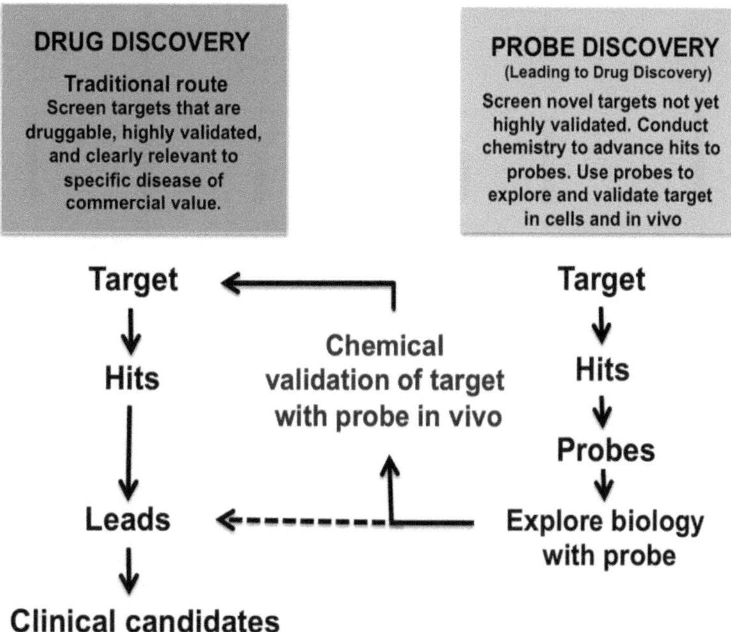

Fig. 1 Depicts a schematic of the starting point and initial steps of the traditional process of early drug discovery compared to those of probe discovery. Arrows indicate typical project flow and also how these two activities can be linked. Specifically how chemical probes can validate a novel target for a specific disease indication and thereby kick-start a drug discovery program to initiate at a lead-stage gate

As shown in Fig. 1, chemical probes with sufficiently optimized properties are being used to explore targets in vitro (cell models of disease) and in vivo with a goal to generate the critical data that tips a target over a threshold of validation required by pharmaceutical companies to initiate a drug discovery project (*see* section below on Drug Development). The exciting aspect of this approach is that if chemical validation of targets can be attained with a probe compound, then the project enters the drug discovery process at lead status which reduces the overall risk and timeline.

2 Drug Development

The traditional process employed at pharmaceutical companies to discover novel drugs that will treat diseases with high unmet medical need is deliberate and by now highly refined. This process typically starts with identifying which diseases or indications merit such an expensive and risky endeavor. Many aspects are taken into consideration including the medical need, the availability and satisfaction of current therapy, the competition, and of course the commercial opportunity, i.e., the potential market. Once the disease

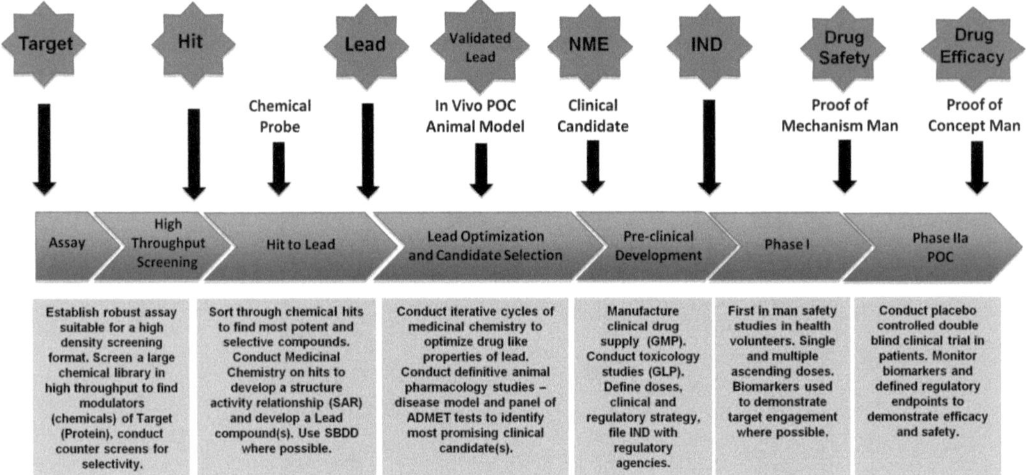

Fig. 2 Depicts the steps and stage gates that are traditionally followed for small-molecule drug discovery and early development. Short descriptions of the activities conducted at each step are provided along with the milestones achieved. Abbreviations used in the figure are as follows: *SBDD* structure-based drug design, *ADMET* absorption, distribution, metabolism, excretion, and toxicity, *GMP* good manufacturing practice, *GLP* good laboratory practice, *IND* investigational new drug, *NME* new molecular entity

focus is defined, the landscape of potential targets is assessed. Significant effort is deployed prior to the start of any drug discovery project to assemble a body of evidence that validates the target as having a high degree of relevance to the disease. Typically the key information for target validation is provided by human genetic linkage studies supported by data from genetically modified mice (gene knock-in or -out) along with studies using small interfering RNA (SiRNA), antibodies, and where possible chemical compound tools. The collective evidence from these lines of research is required to provide compelling connectivity between the target and the disease. In addition to considerations on target validation, other aspects are also considered at this stage. Particularly important at the outset of a project is the tractability of the target to drug discovery, recognizing that some protein families have proven significantly more straightforward to drug than others. Indeed all the steps downstream of target identification and validation are evaluated up front prior to ultimate section of a drug discovery and development program. These steps are relatively well defined. A linear process with distinct activities being carried out at each stage is shown in Fig. 2.

Given that the overall process from target selection through to clinical studies in patients and ultimate approval of the drug based on large clinical studies is of inherent high risk, project leaders spend much time up front laying out as many of the issues that are likely to be encountered as possible so that attrition can be minimized. Hence defining what endpoints or biomarkers will be used to demonstrate proof of mechanism and then proof of concept in both animal studies and patients becomes important. Similarly defining which patients will benefit from the drug, how the drug will be differentiated from other treatment options, how safe the drug needs to be, and how the drug will be administered come together to define the so-called target product profile (TPP) that outlines the key attributes of the ultimate marketed drug. It is this profile that is constantly at the center of the decision-making process as a project advances from target validation through to the clinic. The drug discovery and development process is heavily influenced by considerations of the expense versus the probability of completing not just the next stage in the overall process but all the stages in the process to get a competitive product to the market. The high degree of up-front target validation and relatively narrow focus of many companies to diseases of very high value have resulted in much competition among companies around a relatively small number of targets. As such, chemical probes may be ideally suited to expand the breadth of targets under consideration by providing chemical validation in animal models of disease.

In the case of phosphatases, there are a number of members of this family that have passed the validation hurdle, e.g., PTPN22 for autoimmune disease, PTP1B for type II diabetes, STEP for neuropsychiatric disorders, and drug discovery efforts have advanced. Once the molecular target to be modulated is defined then the process of drug development is relatively straightforward; for phosphatases this typically involved generating purified recombinant protein and establishing enzyme assays in which the activity of the protein can be measured in a simple format suitable for ultra high-throughput screening. While the assay development and screening activity are very much overlapping activities between the drug development and chemical probe development processes, they may diverge as early as the decision of which compounds from the hit set are selected to conduct hit-to-probe efforts compared to hit to lead. For a drug discovery researcher who is seeking an orally available drug defined in the TPP, evaluation occurs as early as the hit stage as to how closely a compound fits Lipinski's rule of 5, i.e., an orally active drug typically has no more than one violation of the following criteria: no more than five hydrogen bond donors, no more than ten hydrogen bond acceptors, a molecular weight under 500 Da, and an octanol–water partition coefficient log P of less than 5.

If a chemical hit (series) identified from screening has clear potential chemical liability that would limit the ability to develop the compound and related analogs then it is unlikely to be selected with high priority to move into the hit-to-lead phase. This is less of an issue for probe development, for example oral bioavailability is rarely a criteria for a chemical probe (even if desirable), similarly potential in vivo toxicity liabilities of a hit are not of concern for initiating a probe development program. While medicinal chemists and pharmacologists on a drug discovery or probe discovery team start by assessing potency and selectivity, other properties of the hits are considered much earlier in drug discovery process such that compounds that are prioritized to advance may not be the most potent nor selective but rather have good drug-like properties and potential for generation of many analogs are prioritized for hit to lead efforts. Ultimately the advancement of drug discovery and development projects depends on satisfying an ever-growing list of properties that need to be attained, within the drug discovery and preclinical phase of drug development this list includes potency (activity), selectivity, pertinent physicochemical properties plus a large panel of in vitro and in vivo assessments of absorption, distribution, metabolism, excretion, and toxicity (ADMET) and efficacy in relevant models. Unfortunately many exciting drug discovery programs struggle to make it past the hit-to-lead phase, for example, generating cell penetrant selective compounds that target intracellular protein phosphatases has proven an issue whether the goal was to develop a chemical probe or a drug. Whether this reflects inherent issues with targeting this class of proteins or limitations on the types of compounds that are identified by the common assay platform deployed in screening is not clear. More success has been had with targeting alkaline phosphatases; the reported development of orally available potent selective inhibitors of tissue-nonspecific alkaline phosphatase (TNAP) that demonstrate in vivo activity in animal models represents a successful example where probe development has crossed over into the drug development path.

Chapter 2

Phosphatase High-Throughput Screening Assay Design and Selection

Eduard A. Sergienko

Abstract

Phosphatases are a heterogeneous group of enzymes catalyzing dephosphorylation of diverse substrates ranging from small organic molecules to large phosphorylated multiprotein complexes. A wide variety of biochemical approaches for measuring phosphatase activity exists. Spectrophotometric methods utilizing artificial chromogenic, fluorogenic, and luminogenic substrates and taking advantage of the optical properties of dephosphorylated products are broadly used by research community. Another major assay type is based on quantitation of the second product of any phosphatase reactions, inorganic phosphate, using a variety of phosphate detection methods. Although, in theory, compatible with any phosphatase substrate, these assays often are unable to provide acceptable high-throughput screening adaptations of native phosphatase reactions. Conversely, phosphatase assays with artificial substrates frequently are incapable to mirror the intricacies of substrate binding and catalysis of the native reaction and, as a result, unable to deliver biologically relevant phosphatase modulators. Utilization of comprehensive phosphatase assay panels, employing honed biochemical assays and cell-based model systems, in conjunction with novel approaches for screening phosphatases may aid in identification of potent, selective, and biologically active phosphatase modulators.

Key words Phosphatase detection, Phosphate quantitation, Artificial and native phosphatase substrates, Protein phosphatases, Alkaline phosphatases, Tissue nonspecific phosphatase, High-throughput screening, Assay development

1 Introduction

High-throughput screening (HTS) became the primary approach for identification of small-molecule chemical leads in drug discovery. Its wide acceptance throughout the drug discovery industry and academic institutions involved in chemical biology is a tribute to its power and efficiency; with HTS, hundreds of thousands of compound samples could be tested in a single day. At the heart of the screening process are the HTS assays that link specific changes of biological targets induced by screened molecules with a detectable output signal measurable in a high-density format (Fig. 1). Examples of biological properties successfully utilized for building HTS assays include enzymatic activity, small-molecule ligand- and

José Luis Millán (ed.), *Phosphatase Modulators*, Methods in Molecular Biology, vol. 1053, DOI 10.1007/978-1-62703-562-0_2, © Springer Science+Business Media, LLC 2013

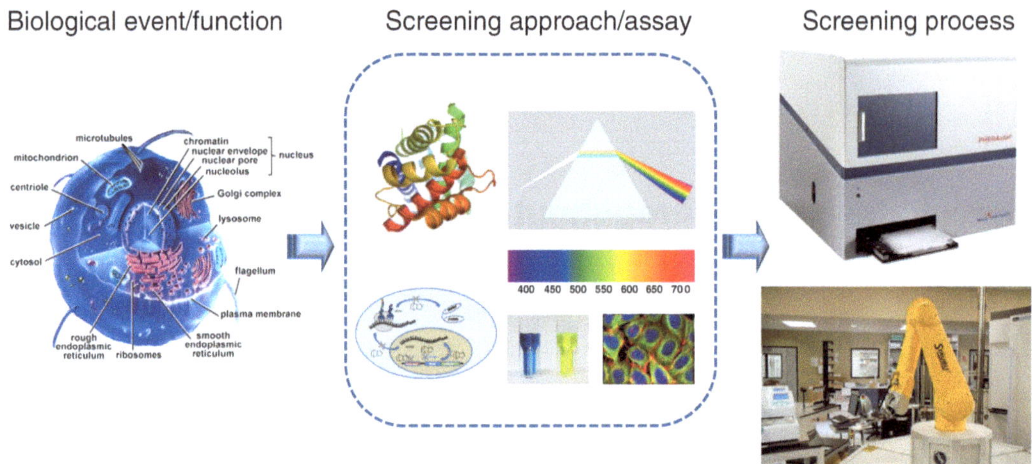

Fig. 1 HTS assays connect target biology and screening technology. 3D cell image is reproduced from http://www.ebi.ac.uk/microarray/biology_intro.html and ref therein

protein partner-binding, and protein expression level. Assays based on activity are by far the most popular way of screening enzymes, directly targeting their biological function.

Phosphatase enzymes are part of EC 3 group of hydrolases. They catalyze hydrolytic removal of a phosphate from biological substrate molecules, producing inorganic orthophosphate and an unsubstituted alcohol group. Most phosphatases have narrow substrate specificities and are able to recognize and efficiently hydrolyze a particular phosphorylated molecule or a group of analogous molecules in vivo. Nevertheless, some of them, such as tissue-nonspecific alkaline phosphatase (TNAP), appear to have broad substrate specificity and are able to hydrolyze diverse sets of substrates. Irrespective of their in vivo substrate specificity, most of the phosphatases are also able to hydrolyze phosphorylated small-molecule artificial substrates when tested in vitro, supplying an easy handle for their screening.

Majority of HTS is performed in microplates; most frequently used formats contain 96, 384, and 1,536 wells per plate (Fig. 2). Consequently, HTS assays require detection approaches that could be easily executed in a plate format. Although the number of novel detection approaches adapted in HTS is constantly growing, assays based on spectrophotometric properties are by far the most prevalent, reflecting their technical and technological simplicity and low cost. Absorbance, fluorescence, luminescence, and derivations thereof are common in HTS assays. Broad selection of relatively inexpensive monochromator-, optical filter-, or diode array-based plate readers and suitable generic plates provides a foundation to spectroscopic HTS assays.

Small molecules broadly vary in physical or chemical properties that could interfere with HTS assays. Compound interference may make inactive compounds appear active (false positives) or, to the

Fig. 2 Common plate formats. Reproduced with permission from the image author Maggie Bartlett, NHGRI

contrary, it may hide actual activity of positive compounds (false negatives). Although, compound chemical reactivity and solubility are known to interfere with some types of targets or assays, optical properties of compounds have a broader reach affecting detection signal in practically any spectroscopic assay. The severity of optical interference is strongly influenced by the spectral characteristics of the employed detection system; generally, the shorter the detection wavelengths utilized in the assay, the higher the number of interfering compounds [1].

Most of HTS assays contain multiple components, such as enzyme and their substrates in biochemical assays. Some of these components define detection signal strength, while others determine assay sensitivity to compounds with particular properties [2]. For example, enzyme concentration is usually lower than the concentration of other assay components and is directly proportional to the enzymatic activity, i.e., signal strength. On the other hand, substrate concentration, although capable of influencing the signal strength, is normally utilized to define the mechanism of action (MOA) of the identified hits. Utilization of the substrate concentration equal to its K_m value ensures comparable sensitivity to competitive, noncompetitive, and uncompetitive hits, while using saturating concentrations of the substrate would make the assay less sensitive to competitive compounds.

In the following sections, we describe the most common approaches utilized for screening phosphatases, paying attention to their advantages and disadvantages. In addition, using case studies of real-life phosphatase projects we will illustrate common challenges observed in phosphatase assays and the helpful approaches for overcoming them.

2 Diversity of Detection Techniques and Assay Formats Available for Phosphatase Screening

There are multiple assay types that exist for each target class; especially, this holds true for phosphatases, a class of enzymes that has been identified and studied since the dawn of biochemistry. All earlier assays were based on detecting phosphate through analytical chemistry approaches. These assays utilized a propensity of inorganic phosphate to form a complex with ammonium molybdate in a strongly acidic environment, detecting the resultant phosphomolybdate complex through colorimetric reactions. Later on, protein-based phosphate detection approaches were devised; they relied on enzymatic and protein-binding assays sensing the phosphate concentration. In parallel with the phosphate detection approaches, artificial substrates were explored and designed to enable visualization of dephosphorylation in real time, without the need for detection reagents. Altogether, this broad selection of phosphatase detection approaches provides a reusable toolbox for studying any new phosphatase target of interest.

2.1 Phosphate Detection Approaches

The classical phosphate detection method originally developed by Bell and Doisy [3] and further optimized by Fiske and SubbaRow [4] employed quantitation of phosphomolybdate complex through reduction with 1-amino-2-naphthol-4-sulfonate in strong acids resulting in blue color. Fifty years later, a much more sensitive approach was proposed; it employed color change of malachite green associated with its acid–base equilibrium shift upon binding to phosphomolybdate complex in strong acids [5]. Its undisputed advantages are high sensitivity and long wavelengths utilized for absorbance measurement. Although these analytical chemistry approaches were initially designed for work with organic phosphates, for example, determination of phosphate content in protein preparations after acid hydrolysis [6, 7], they later were applied and are broadly used for quantification of inorganic phosphate released in reactions of phosphatases [8–10].

Over the years, the malachite green-based approach became the most popular phosphate detection approach applied to phosphatases despite its known disadvantages, such as aggregation and precipitation of phosphomolybdate complex and the presence of strong acids in the detection reagent. Pronounced hydrolysis of some organic compounds, such as sugar phosphates and phosphoamino acids, specifically phosphotyrosine [11], in the presence of strong acids may result in elevated release of phosphate from these substrates in the absence of phosphatases. Recently commercialized modification of the malachite green-based phosphate detection approach appears to handle the problem of hydrolysis by integrating a pH-adjustment step, which has a stabilizing effect on both phosphomolybdate complex and acid-labile compounds. This phosphate detection kit by Innova Biosciences achieves high sensitivity and broad dynamic range and is suitable in screening assays with acid-labile substrates [12, 13].

Several additional phosphate detection procedures were designed to deal with substrate and signal stability through the use of protein-based approaches. One of the procedures employs bacterial enzyme purine-nucleoside phosphorylase and an artificial chromogenic substrate, 2-amino-6-mercapto-7-methylpurine ribonucleoside [14]. Phosphorolysis of the substrate leads to the release of nucleotide and concomitant increase of absorbance at 360 nm proportional to the phosphate concentration. Although this assay provides a convenient procedure for real-time detection of phosphate generation with diverse phosphate-generating enzymes [15, 16], it is not suitable for HTS, suffering from low signal-to-background ratio and significant compound optical interference.

Another coupled-enzyme phosphate detection assay involves a complex sequence of enzymatic reactions leading to a generation of a fluorescent signal. In the initial step of the detection, maltose phosphorylase consumes phosphate to produce glucose-1-phosphate and glucose; in the second step, glucose molecule is converted to gluconolactone and hydrogen peroxide by glucose oxidase [17]. In the final step, HRP consumes hydrogen peroxide to convert fluorogenic substrate Ampex Red into fluorescent product resorufin [18]. Thus overall, the fluorescent signal is proportional to the amount of the inorganic phosphate produced by a phosphatase. Although capable of detecting low concentration of phosphate, this assay suffers from pronounced spontaneous oxidation of Ampex Red with ambient oxygen and concomitant low signal-to-background ratio.

Webb and colleagues designed and implemented another ingenious phosphate detection approach; they utilized the ability of *Escherichia coli* phosphate-binding protein (PBP) for tight phosphate binding and concomitant conformational changes to generate a phosphate sensor [19]. They cloned and purified a variant of the protein that contained a single cysteine and specifically labeled it with a coumarin-type fluorophore. The resulting protein demonstrates fivefold fluorescence increase upon phosphate binding (excitation at 425 nm and emission at 465 nm). The fluorescence increase correlates with PBP–phosphate complex detecting low-micromolar concentrations of phosphate. This and other mentioned protein-based phosphate detection approaches operate at near-neutral pH values overcoming the drawbacks of the molybdate-based approaches. However, they were not able to gain the popularity of the malachite green-based phosphate detection due to their own limitations, such as low wavelengths and significant compound optical interference, as in the assay with PBP.

2.2 Phosphatase Assays Based on Alcohol Moiety Detection

Majority of phosphatase assays employed in HTS today rely on the ability of these enzymes to catalyze dephosphorylation of artificial chromogenic or fluorogenic substrates. Their popularity is due to simplicity, high sensitivity, and robustness of these assays. The most popular chromogenic substrate is para-nitrophenyl phosphate (pNPP) (Fig. 3a). It is stable in aqueous solutions at different pH values and is readily hydrolyzed by majority of phosphatases [20–22]. Hydrolysis of pNPP results in generation of para-nitrophenol, which has $pK_a = 7.16$ and once deprotonated becomes yellow-colored para-nitrophenolate ion with absorbance at 405 nm. It can be monitored continuously at pH values above 7; assays with lower pH values are performed in the end-point format, requiring alkalinization of solution prior to measuring absorbance.

Two other chromogenic substrates employed in phosphatase assays, phenolphthalein monophosphate (PPMP) and thymolphthalein monophosphate (TPMP) shown in Fig. 3b, c, although did not find as broad a popularity as pNPP, offer distinct advantages over the latter [23, 24]. Their products have two- to threefold greater extinction coefficients than the product of pNPP. Dephosphorylation of colorless PPMP and TPMP leads to generation of phenolphthalein and thymolphthalein, respectively. Similarly to para-nitrophenol, the acid forms of the products are colorless, and require deprotonation for color development, characterized with pK_a values around 9.5–10, making them especially attractive for use in alkaline phosphatase assays [25, 26]. Additional advantage of PPMP and TPMP is that the products of their dephosphorylation absorb light at 552 and 595 nm, respectively, offering improved resistance to compound optical interference.

Colorimetric assays dominated the field of spectrophotometric detection in the past and still are widely used now; however, fluorescent and especially luminescent approaches significantly gained in popularity over the past decade due to their superb sensitivity

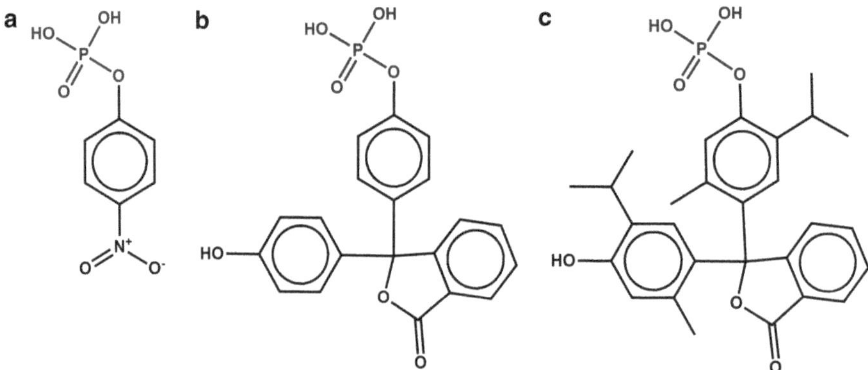

Fig. 3 Structures of chromogenic phosphatase substrates: (**a**) Para-nitrophenyl phosphate; (**b**) phenolphthalein monophosphate; (**c**) thymolphthalein monophosphate

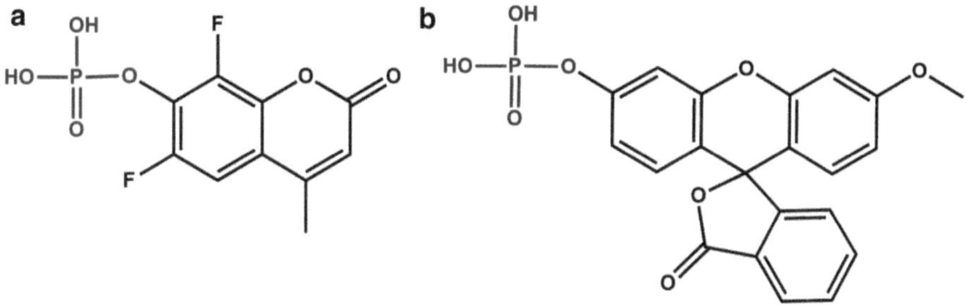

Fig. 4 Structures of fluorogenic phosphatase substrates: (a) 6,8-difluoro-4-methylumelliferyl phosphate; (b) 3-*O*-methylfluorescein phosphate

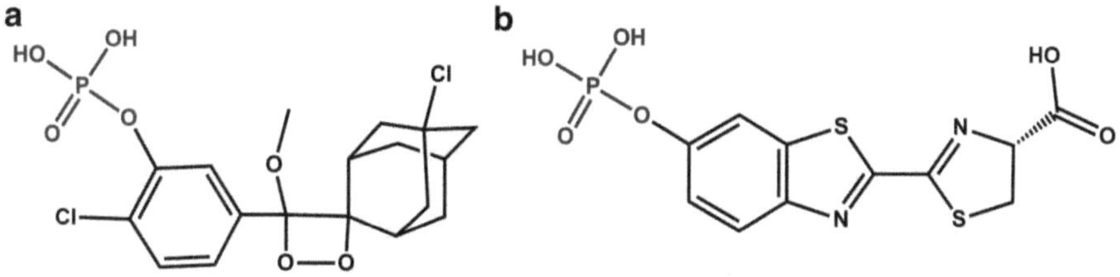

Fig. 5 Structures of luminogenic phosphatase substrates: (a) CDP-star; (b) D-luciferin 6′-*O*-phosphate

and extended dynamic range. Several fluorescent substrates are available for screening phosphatases. One of them, 6,8-difluoro-4-methylumbelliferyl phosphate (DiFMUP; Fig. 4a), generates a product with fluorescence excitation and emission maxima 358 and 450 nm, respectively [27]. Another popular phosphatase substrate, 3-*O*-methyl fluorescein phosphate (OMFP; Fig. 4b), generates fluorescent product with excitation and emission maxima at 485 and 520 nm, respectively [28]. Higher wavelengths utilized in OMFP assay provide better protection from compound optical interference. The products generated from both substrates, similarly to the products of pNPP, PPMP, and TPMP, fluoresce only in deprotonated form. Since both DiFMU and OMF have pK_a values ~4.6–4.7 [29], both fluorescent substrates are suitable for continuous assays at pH values above 5. These substrates are especially popular in studies of phosphotyrosine and dual-specificity protein phosphatases [29, 30]; nevertheless, both of them were successfully utilized for other classes of phosphatases [27, 28].

Luminescent assays are characterized with extremely low backgrounds consistent with the transient nature of photons, as opposed to cumulative nature of the detectable species in colorimetric and fluorescent assays. There are several luminogenic phosphatase substrates available; however, most of them, such as dioxetane-based CDP-star (Fig. 5a), CSPD, and PPD, require alkaline pH and thus easily applicable only for alkaline phosphatases. These substrates generate light through chemiluminescence, in which high-energy intermediate produced after the substrate dephosphorylation decays to a low energy level with a concomitant production of light [31]. Another interesting and potentially promising approach is using luciferin phosphate (Fig. 4d) as a substrate for phosphatases [32]. Its potential advantage is that dephosphorylation leads to generation of stable compound D-luciferin. The light is generated in a bioluminescent reaction that requires the detection enzyme, firefly luciferase. This compulsory dependence on the detection enzyme should allow decoupling the main reaction from the light production and thus performing the phosphatase reaction in any buffer of choice, not necessarily at pH 7.8, the pH-optimum of the luciferase.

2.3 Phosphatase Substrate Types

Phosphatase assays can be subdivided according to a type of substrate utilized in the reaction. The two main types of phosphatase substrates are native substrates and artificial substrates; most commonly, the former require utilization of aforementioned phosphate detection approaches, while the latter generate measurable spectroscopic changes upon dephosphorylation. The binding of native substrates usually involves multiple points of specific interaction with the enzyme molecule, frequently translated into low values of K_m. Reflecting functional diversity of the phosphatase class, the native substrates vary in nature and sizes and can be as small as a molecule of pyrophosphate or as large as phosphorylated multiprotein complexes. Consistent with their sizes, the binding of substrates involves additional sites that may be located in the near proximity or far away from the active site. In contrast, the binding of artificial substrates is driven majorly through interactions of the phosphate group with the active site cleft; as a consequence, their K_m values are usually much higher than those of native substrates.

As mentioned above, small-molecule artificial substrates are very attractive for using in HTS phosphatase assays. Detection of the phosphatase reaction progress through spectroscopic changes of the alcohol moiety usually provides a convenient, one-step procedure. In addition, these assays utilize commercial substrates, readily available in large quantities and at low cost. However, since the binding mode for these substrates is different from that of the native substrates, the mechanism and the rate-limiting steps of their hydrolysis are frequently dissimilar too [33, 34]. This phenomenon has important ramifications for identification of biologically relevant modulators of enzymes; the change in rate-limiting steps is likely to result in major changes in potency ranking of modulators. Thus, phosphatase modulators found through assays with artificial substrates might not actually work against native substrates.

Clearly, the assays with the native substrates would probably be a logical choice for any phosphatase screening; however, these assays are frequently much more challenging and not always are the most effective choice for a primary assay. To start, they usually require utilization of multistep inorganic phosphate detection approaches. More importantly, they rely on custom-made substrates that could be difficult to obtain; for example, full-length substrates for protein phosphatases require either co-expression with a suitable kinase or an in vitro phosphorylation of a purified recombinant protein. Finally, low K_m values for native phosphatase substrates, e.g., submicromolar K_m values of protein phosphatases [34, 35], are frequently much lower than the quantitation limits of phosphate detection approaches, necessitating HTS to be performed at saturating substrate concentrations.

Unfortunately, both artificial and the native substrates, paired with the aforementioned phosphate detection approaches, are only fit for use in biochemical phosphatase assays; the number of diverse phosphatases and the concentration of inorganic phosphate in the cells make these approaches unsuitable in cell-based systems. One could hope that new detection approaches devised in the future would allow performing phosphatase assays in the cells. An interesting, and potentially relevant to phosphatases, approach was proposed for monitoring posttranslational modifications, e.g., protein phosphorylation, in cell-based assays. In this approach, a protein substrate fused with green fluorescent protein (GFP) is transfected in the cells. Substrate phosphorylation level in lysed cells is detected with anti-phosphoprotein-specific antibodies labeled with lanthanide terbium that generate time-resolved fluorescence resonance energy transfer (TR-FRET) signal when brought in proximity to GFP-phosphoprotein [36]. This approach was successfully used for testing protein kinases and could potentially be extended to protein phosphatases. Similar biochemical assay could be envisioned and would resolve the aforementioned issue of high substrate affinity, since TR-FRET detection works best in nM-concentration range. Technically, common protein tags paired with commercially available anti-tag antibodies labeled with TR-FRET reagents could be utilized instead of the GFP fusion.

An intermediate type of substrates, alas specific to protein phosphatases, is phosphopeptide substrates that encompass a consensus sequence flanking the phosphorylated amino acid residue. Similarly to the assays with other substrates, release of phosphate from phosphopeptides could be paired with the standard phosphate detection approaches. Two recently proposed ingenious approaches provide a convenient way of testing synthetic peptide-based protein phosphatase substrates. In one of the approaches, fluorogenic phosphocoumaryl amino propionic acid mimicking phosphotyrosine is incorporated in a peptide substrate [37]. In the other approach, fluorogenic peptide contains a quenched umbelliferone attached to amino group of either phosphoserine or phosphothreonine via carbamate bond. Fluorescent umbelliferone is released after dephosphorylation as a result of cyclization reaction involving carbamate and hydroxy-group of serine/threonine [38].

Some properties of phosphopeptides make them preferable to the native phosphoprotein substrates for use with phosphate detection systems. The values of K_m for peptide substrates are two to three orders of magnitude larger than for protein substrates and allow setting assays with an appropriate substrate concentration using standard phosphate detection approaches. In addition, short synthetic peptides are inexpensive and easy to obtain. Nonetheless, although phosphopeptide substrates are clearly useful in exploring interactions in the immediate vicinity of the phosphatase active site, they are unable to probe distant (allosteric) sites.

These allosteric sites appear to be critical for biological function of some protein phosphatases. For example, binding of protein phosphatases hematopoietic protein tyrosine phosphatase (HePTP) and MKP3 with their native substrate phosphoERK2 relies on two distant-from-one-another binding sites, the active site and kinase interaction motif (KIM), with the major affinity coming from the latter one [34, 35, 39]. In addition, the allosteric sites possess a greater structural diversity between subfamilies of phosphatases and thus are expected to be a more attractive target compared to the active (orthosteric) sites of phosphatases.

3 Challenges of Phosphatase Assays and Screening Projects

Some examples of technical challenges encountered in phosphatase assays were mentioned earlier in the chapter; they usually originate from the limitations of screening approaches and available instrumentation. For example, acid-labile phosphatase substrates are incompatible with classical malachite green-based phosphate detection systems. Another example is low sensitivity of all current phosphate detection systems necessitates using native protein phosphatase substrates at concentrations well above their K_m values. Although design and implementation of novel detection approaches, techniques, and/or instrumentation might be the most direct response to the technical challenges, this path is usually too time-consuming and expensive to be taken for the benefit of a single project. Utilization of the "next-best" existing assay is more practical and could be sufficient for a successful completion of the project.

Another group of challenges arise from the lack of comprehensive understanding of the biology of a protein target. Indeed, limited knowledge on the regulation and biological function of a phosphatase of interest makes the task of finding a biologically relevant HTS assay challenging. Both technical and epistemic issues are resolved in a similar manner, through testing multiple assays, selecting the most appropriate one for primary HTS while using the other ones for hit validation. Usually, a hit validation panel of assays is designed and utilized to ensure that the hits are real, are selective, and have the optimal mechanism of action. These assay panels are especially critical in the projects with poorly characterized targets or for which optimal primary screening assays are not available; the latter include assays utilizing undesirable substrates or concentration thereof.

We will illustrate some of the challenges and their resolution using select case studies of phosphatase projects performed at the Conrad Prebys Center for Chemical Genomics (CPCCG) at Sanford-Burnham Medical Research Institute. The described projects were performed within Molecular Libraries Program (MLP), a part of NIH Roadmap for Medical Research, so one of the benefits of these examples is that the details and the experimental data for all described assays are available through NCBI database PubChem (http://pubchem.ncbi.nlm.nih.gov/). Within MLP, a large national network of screening centers was assembled to perform HTS on a shared compound collection against a plethora of biochemical and cell-based assays. This undertaking resulted in unparalleled, comprehensive, and constantly growing open-access data on efficacy, potency, and selectivity of a large number of compounds in an enormous panel of diverse assays.

Multiple projects performed at CPCCG were aimed to identify inhibitors of protein phosphatases. Although aforementioned compound optical interference is prevalent in most types of phosphatase assays, it is easily handled through the use of an assay with a different detection system or through testing spectrophotometric properties of compounds in the absence of enzymatic reaction. It is much more difficult to deal with other types of interference, for example, modification of active site of protein phosphatases resulting in enzyme inactivation. Most protein phosphatases contain an active site cysteine residue critical for their catalytic reaction. Thiol groups of cysteines, especially the ones involved in enzymatic reactions, are easily oxidized by ambient oxygen, reactive oxygen species, like hydrogen peroxide, or modified by various agents. Biochemical HTS assays for cysteine-based enzymes are generally performed in the presence of reducing agents, most frequently dithiothreitol (DTT), to prevent enzyme inactivation.

One of the protein phosphatases screened at CPCCG was MAP kinase phosphatase 3 (MKP3). The activity assay employed fluorogenic substrate, OMFP, and was performed in an end-point format in the presence of DTT. To counteract potential optical interference of compounds, the hits were reconfirmed in the OMFP-based assay performed in kinetic mode. Visual inspection of the progress curves revealed time-dependent character of inhibition for several scaffolds (Fig. 6). The experimental curves for these compounds conformed to an equation describing a transition between 100 and 0 % activity at any compound concentration, suggesting MKP3 inactivation. Further experimentation revealed that hydrogen peroxide was accumulated in the presence of these compounds in the assay; moreover,

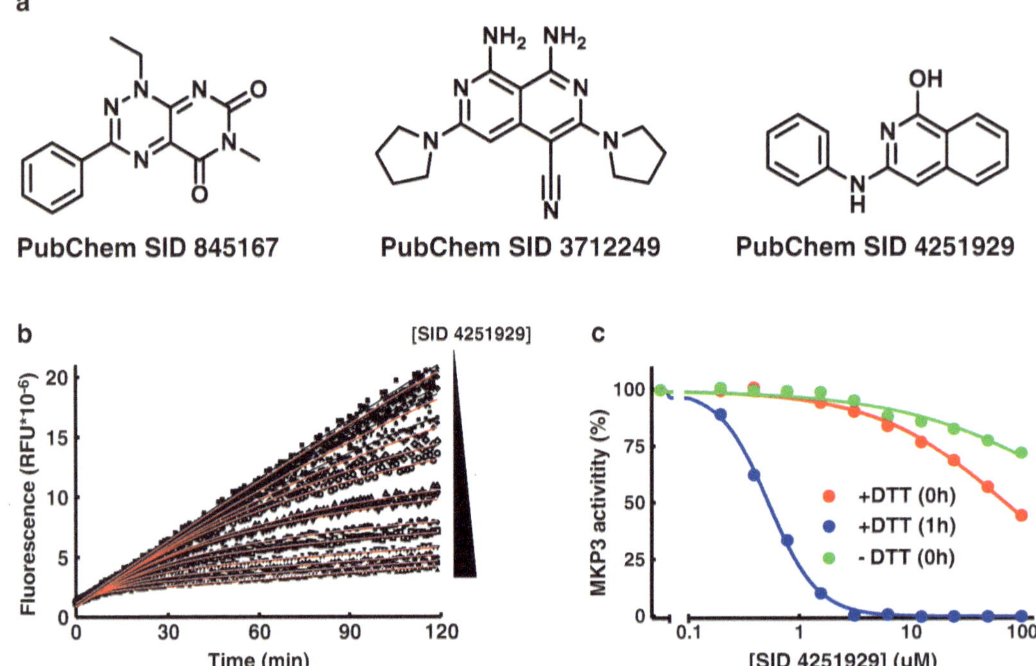

Fig. 6 Select artifacts identified in MKP3 project. (**a**) Structures of exemplary MKP3 inactivators. (**b**) Progress curves of one of the inactivators. *Black symbols* represent experimental data, *red curves* correspond to fitted curves of inactivation. (**c**) Apparent compound inhibition curves in different MKP3 assays. *Blue* and *red symbols* correspond to the initial rates of the MKP3 reaction performed with and without preincubation with compounds, respectively. *Green symbols* correspond to MKP3 assay performed without DTT

the process required the presence of DTT, and was completely abrogated in the absence of the reducing agent.

A panel of enzymatic assays was designed and established to help in identification of assay artifacts (Fig. 7). This panel included two kinetic assays for MKP3 that differed only in that one of them had DTT in the buffer while the other one had no reducing agents. The panel also included a counter screen assay with HePTP described in more detail below. An assay for out-of-class enzyme, glyceraldehyde-3-phosphate dehydrogenase (GAPDH), was also included in the panel. The rationale behind this assay was that a reactive cysteine residue present in GAPDH active site behaves very similarly to the cysteine of protein phosphatases. Hits active against GAPDH were taken as nonspecific and likely to result from cysteine modification. This panel of assays was instrumental in identifying and culling out the chemical artifacts.

Concomitant to MKP3 project, a screening project with HePTP was performed at CPCCG. This project utilized HePTP assay with pNPP substrate and malachite green-based phosphate detection. The HePTP and MKP3 assays were included in the same panel and utilized as counter screens for one another helping in identification of compounds selective against either phosphatase. Two of HePTP hit scaffolds demonstrated a behavior opposite to

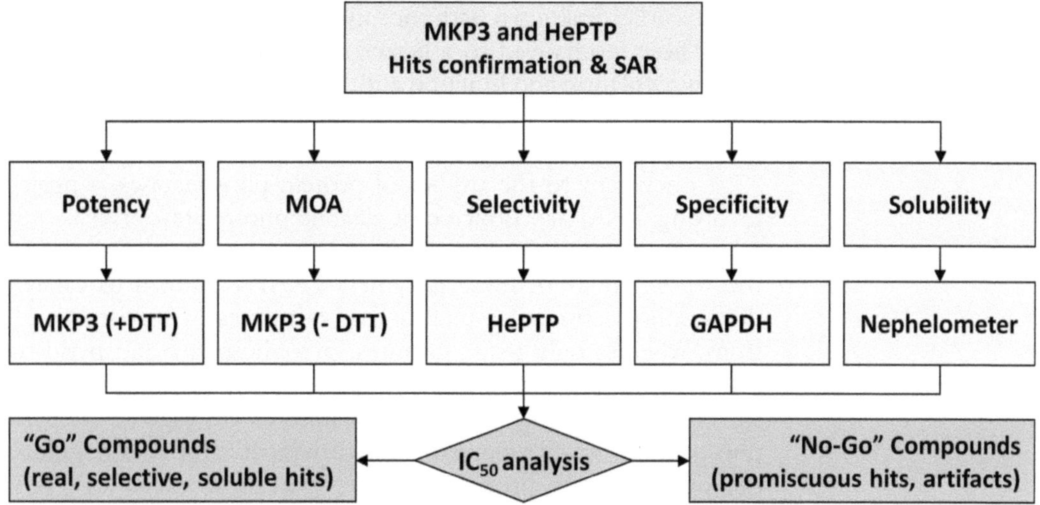

Fig. 7 Assay panel utilized for protein phosphatase screening projects at CPCCG

that of H_2O_2-generating compounds; they were clearly more potent in the absence of DTT than in its presence. It appears that these compounds rely on not only the binding to HePTP active site but also partial oxidation of the active site cysteine. Surprisingly, these same two scaffolds were the only ones that also demonstrated cell-based inhibition of ERK2 dephosphorylation, the native HePTP activity in the cells, measured by western blot approach. Structure–activity relationship (SAR) studies focusing on these scaffolds resulted in identification of chemical analogs selective against HePTP [22].

Several points of general interest could be deduced from the above case studies. Expanded panels including diverse assays for multiple closely related or sharing unique property targets are very effective in both rounding up artifacts and identification of real selective hits. In addition, same-target assays with altered conditions are very powerful in identification of a specific mechanism of inhibition. Although a biochemical HePTP assay with the native protein substrate, phosphoERK2 (pERK2), was considered at the start of the project, it was found too expensive and impractical. The above-mentioned outcome of the project suggests that, in fact, the primary screening with the native substrate was perhaps less critical than a utilization of a broad panel of assays and direct confirmation of the hits in the HePTP cell-based assay.

Another important point from the HePTP case study illustrates epistemic challenges and their resolution with the use of small-molecule chemical probes. Although reversible oxidation of the phosphatase active site cysteines was hypothesized as one of the cellular pathways regulating protein phosphatase activity prior to our studies [40, 41], this mechanism was not experimentally demonstrated for HePTP. Identified small-molecule hits not only enabled biologically relevant inhibition of HePTP but also pointed out an earlier-unknown path for modulating HePTP activity in the cell. These results lead to a better understanding of HePTP biological regulation and function and, potentially, build a foundation for a future discovery of therapeutically relevant modulators with specific MOA.

Concurrent to the studies of protein phosphatases, a project involving TNAP, an isozyme of alkaline phosphatases (AP), commenced at CPCCG leading to a broad progeny of projects covering the whole family of these enzymes. TNAP is known to catalyze dephosphorylation of a plethora of substrates in in vitro conditions, whereas only two substrates, pyrophosphate and pyridoxal phosphate, were unequivocally demonstrated to serve as TNAP substrates in vivo [42–44]. TNAP-catalyzed cleavage of the pyrophosphate, a known suppressor of hydroxyapatite formation, induces the mineralization process in bone tissue. The excess of TNAP activity was indicated as a causal factor in pathological ectopic calcification within blood vessels and soft tissues [10, 45], suggesting that inhibition of TNAP pyrophosphatase activity would remedy the abnormal calcification. Intriguingly, unlike other AP isozymes, TNAP is a ubiquitous enzyme with elevated levels in not only bone tissue but also liver, kidneys, and brain; thus, it is conceivable that its biologically relevant substrates and functions could be diverse and regulated by specific environments.

At the start of the project, existing malachite green-based reagents were tested and found incompatible with monitoring TNAP pyrophosphatase reaction. Indeed, pyrophosphate hydrolysis in the presence of highly acidic reagents resulted in extremely high background values dwarfing the TNAP-catalyzed phosphate release. Development of an HTS assay for TNAP pyrophosphatase activity became possible only several years later, after the commercialization of the aforementioned PiColorLock reagent by Innova Biosciences.

Another assay, utilized for determination of AP in clinical samples, employs high concentrations of both pNPP substrate and aminoalcohol buffer at pH values near 10. Although performance of this assay is compatible with HTS, requirement for saturating concentrations of an aminoalcohol buffer makes it unattractive. Indeed, an aminoalcohol (depicted as AOH in Fig. 8) participates in in vitro AP reaction as facultative phosphate-acceptor substrate taking part of the water in the reaction and significantly accelerating its rate; its high concentration in clinical tests is intended to compensate for low sensitivity of a colorimetric reaction. Intriguingly, the existence, identity, and concentration of the natural phosphate-acceptor substrate of TNAP are yet unknown.

A luminescent reaction utilized for visualization of antibodies labeled with AP in immunoblotting techniques appeared very attractive for application to solution assays of TNAP. The reaction utilizes CDP-star substrate and requires alkaline pH in the assay.

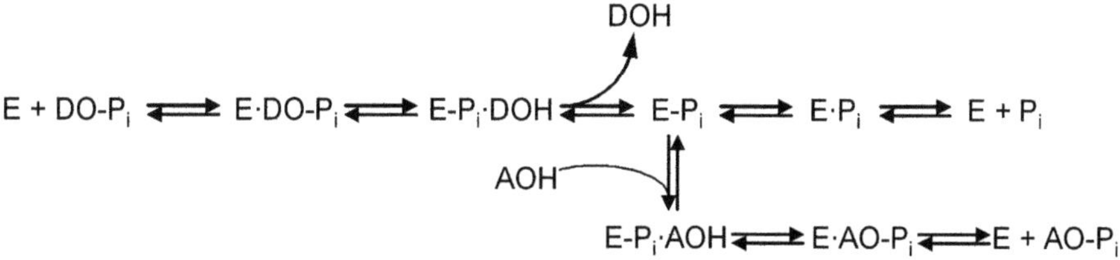

Fig. 8 Catalytic mechanism of alkaline phosphatase reaction [44]. The initial alkaline phosphatase (E)-catalyzed reaction consists of a substrate (DO-Pi) binding step, phosphate-moiety transfer to Ser-93 (in the TNAP sequence of its active site), and product alcohol (DOH) release. In the second part of the reaction, phosphate is released through hydrolysis of the covalent intermediate (E-Pi) and non-covalent complex (E·Pi) of inorganic phosphate in the active site. In the presence of alcohol molecules (AOH), phosphate is also released via a transphosphorylation reaction

This reaction when directly compared with pNPP-based assay demonstrated 1,000-fold higher sensitivity and 100-fold larger dynamic range, eliminating the necessity for including aminoalcohol. Prior to HTS, we optimized the assay to have the concentrations of both CDP-star and aminoalcohol DEA at their respective K_m values [31].

Screening of the NIH MLP compound collection identified four major and several minor TNAP inhibitor scaffolds [21]. They demonstrated diverse MOA against both CDP-star and DEA substrates. Compounds competitive and noncompetitive against DEA, as well as competitive and uncompetitive against CDP-star, were identified. Diversity in properties of inhibitors was hypothesized to be of great benefit for finding a suitable chemical probe for studies unraveling potential TNAP functions in diverse environments present in different tissues [21]. Gratifyingly, at least one of the scaffolds was active in a later-established biochemical TNAP pyrophosphatase assay, as well as in a cell-based mineralization assay with primary mouse osteoblasts [46]. An additional colorimetric assay for monitoring TNAP activity in situ, directly in the blood plasma samples from various mammalian species, was developed enabling accelerated medicinal chemistry optimization of the scaffold and helping to advance the project into preclinical stage.

Several additional projects evolved from the initial TNAP project. Thanks to the broad dynamic range of the luminescent assay, several TNAP activator scaffolds were observed during HTS, resulting in an independent project with a potential therapeutic indication for osteomalacia and osteoporosis. Other human APs, intestinal (IAP) and placental (PLAP) isozymes, were initially utilized to establish counter screen assays based on CDP-star. One of the TNAP scaffolds demonstrated inhibition of IAP, and was further optimized to produce a selective chemical probe for human IAP. Later, the assays for IAP and PLAP were utilized for full-scale screening of these isozymes, eventually leading to identification of selective inhibitors for both isozymes. Interestingly, human IAP inhibitors were inactive against mouse ortholog enzyme, inhibitors for which were successfully identified in a separate project. Since the biological functions of IAP and PLAP are not well understood, availability of small-molecule chemical probes will enable further studies of these enzymes.

Some points derived from the TNAP project confirm the observations made in the protein phosphatase projects. First, extended assay panels provide very useful, and frequently serendipitous, information about not only the target but also other proteins in the panels. Second, the use of a primary HTS assay with the native substrate, although desirable, may not be necessary for a successful outcome of the project. Utilization of a robust primary HTS assay and extensive profiling of hits in a panel of diverse assays at the early stages of lead optimization could be sufficient. An important point is that since the mechanism of enzymatic reactions with different substrates may differ significantly [33, 47], the assay utilized for primary HTS needs to be unbiased with respect to diverse MOA [22]. Finally, when a biologically relevant MOA of modulators is unknown, biochemical assays with native substrates and cell-based models are critical in establishing biological relevance of compounds identified in the assays with artificial substrates.

4 Promising Future Directions in Phosphatase Screening

Phosphatase activity assays remain the most attractive option for identification of biologically relevant modulators for this class of enzymes; utilization of native substrates and cell-based models is critical to achieving this goal. Although, some native substrates may not be compatible with the existing phosphate detection approaches, binding assays may provide a viable alternative capitalizing on high-affinity binding of these ligands. For example, tight complexes between protein phosphatases and their substrates, such as phosphoErk2 and MKP3 or HePTP, could be utilized in a displacement assay detecting the compounds targeting enzyme–substrate interface; understandably, minor modification of either the substrate or the enzyme is needed to prevent the substrate turnover. Compounds identified in the described displacement assays would be expected to specifically inhibit the phosphatase reaction with these particular substrates.

A very promising future direction for phosphatase screening is provided by recently developed label-free binding approaches, such as differential scanning fluorimetry [48] and optical biosensor technology [49] that are gaining popularity in HTS. These approaches directly detect the compound binding to the protein molecule. Unlike activity or displacement assays, they are mechanism agnostic detecting the binding to any available site, irrespective of its potential function and our knowledge thereof. If combined with phenotypic cell-based assays these assays could lead to uncovering a biological role for a new binding site and, therefore, a yet-unknown function for the phosphatase of interest, thus helping to resolve epistemic challenge for the target [50]. Alternatively, the identified hits could be profiled in assays performed in the presence of known ligands as well as in a panel of mechanism-based assays, aligning the compounds with potential binding sites.

Advances in design of chemical libraries offer alternatives to conventional HTS collections that are usually redundant and cover only a minor portion of the overall chemical space [51]. Recent advances in synthetic chemistry led to development of diversity-oriented synthesis (DOS) that furnishes compact diverse compound collections [52]. Having small-size diverse libraries offers a great advantage; it enables utilization of low-throughput and prohibitively expensive assays to which many biologically relevant assays belong. Another approach relies on small-size compounds, known as fragments, for building libraries; it also provides high structural diversity with a small number of compounds. Utilization of virtual library screening against the target of interest appears to help in design of a successful fragment collection [53]. Although fragment libraries are applicable to any screening approach, they are particularly powerful in combination with NMR-based screening [54, 55].

Phosphatases are an important and practically untapped drug target class; its significance and potential may be on par with those of the kinases. However, majority of past efforts in finding phosphatase modulators led nowhere, plagued with artifacts and non-specific compounds targeting the active site, leading to notion of phosphatases as a difficult and "undruggable" class of enzymes. It is however possible that these challenges simply suggest that the utilized screening approaches, heavily relying on screening of conventional libraries in activity assays with artificial substrates, are not effective. Hopefully, utilization of alternative screening approaches and diversified compound collections would help to turn the page in discovery and design of effective and selective phosphatase modulators.

References

1. Simeonov A, Jadhav A, Thomas CJ et al (2008) Fluorescence spectroscopic profiling of compound libraries. J Med Chem 51:2363–2371
2. Sergienko E (2012) Chapter 12. Basics of HTS assay design and optimization. Chemical genomics. Cambridge University Press, New York, pp 159–172
3. Bell RD, Doisy EA (1920) Rapid colorimetric methods for the determination of phosphorus in urine and blood. J Biol Chem 44:55–67
4. Fiske CH, Subbarow Y (1925) The colorimetric determination of phosphorus. J Biol Chem 66: 375–400
5. Hess HH, Derr JE (1975) Assay of inorganic and organic phosphorus in the 0.1-5 nanomole range. Anal Biochem 63:607–613
6. Sergienko EA, Kharitonenkov AI, Bulargina TV et al (1992) D-glyceraldehyde-3-phosphate dehydrogenase purified from rabbit muscle contains phosphotyrosine. FEBS Lett 304:21–23

7. Sergienko EA, Ermakova AA, Muronets VI et al (1993) Phosphorylation of glyceraldehyde-3-phosphate dehydrogenase. Biokhimiia 58: 636–647
8. Fisher DK, Higgins TJ (1994) A sensitive, high-volume, colorimetric assay for protein phosphatases. Pharm Res 11:759–763
9. Tautz L, Mustelin T (2007) Strategies for developing protein tyrosine phosphatase inhibitors. Methods 42:250–260
10. Narisawa S, Harmey D, Yadav MC et al (2007) Novel inhibitors of alkaline phosphatase suppress vascular smooth muscle cell calcification. J Bone Miner Res 22:1700–1710
11. Plimmer RH (1941) Esters of phosphoric acid: phosphoryl hydroxyamino-acids. Biochem J 35:461–469
12. Low HH, Lowe J (2006) A bacterial dynamin-like protein. Nature 444:766–769
13. Freschauf GK, Karimi-Busheri F, Ulaczyk-Lesanko A et al (2009) Identification of a small molecule inhibitor of the human DNA repair enzyme polynucleotide kinase/phosphatase. Cancer Res 69:7739–7746
14. Webb MR (1992) A continuous spectrophotometric assay for inorganic phosphate and for measuring phosphate release kinetics in biological systems. Proc Natl Acad Sci U S A 89: 4884–4887
15. Sergienko EA, Srivastava DK (1994) A continuous spectrophotometric method for the determination of glycogen phosphorylase-catalyzed reaction in the direction of glycogen synthesis. Anal Biochem 221:348–355
16. Cheng Q, Wang ZX, Killilea SD (1995) A continuous spectrophotometric assay for protein phosphatases. Anal Biochem 226:68–73
17. Huwel S, Haalck L, Conrath N et al (1997) Maltose phosphorylase from Lactobacillus brevis: purification, characterization, and application in a biosensor for ortho-phosphate. Enzyme Microb Technol 21:413–420
18. Zhou M, Diwu Z, Panchuk-Voloshina N et al (1997) A stable nonfluorescent derivative of resorufin for the fluorometric determination of trace hydrogen peroxide: applications in detecting the activity of phagocyte NADPH oxidase and other oxidases. Anal Biochem 253:162–168
19. Brune M, Hunter JL, Corrie JE et al (1994) Direct, real-time measurement of rapid inorganic phosphate release using a novel fluorescent probe and its application to actomyosin subfragment 1 ATPase. Biochemistry 33:8262–8271
20. Tietz NW, Burtis CA, Duncan P et al (1983) A reference method for measurement of alkaline phosphatase activity in human serum. Clin Chem 29:751–761
21. Sergienko E, Su Y, Chan X et al (2009) Identification and characterization of novel tissue-nonspecific alkaline phosphatase inhibitors with diverse modes of action. J Biomol Screen 14:824–837
22. Sergienko E, Xu J, Liu WH et al (2012) Inhibition of hematopoietic protein tyrosine phosphatase augments and prolongs ERK1/2 and p38 activation. ACS Chem Biol 7:367–377
23. Sparkers MC, Crist ML, Sparkes RS (1975) High sensitivity of phenolphthalein monophosphate in detecting acid phosphatase isoenzymes. Anal Biochem 64:316–318
24. Ewen LM, Spitzer RW (1976) Improved determination of prostatic acid phosphatase (sodium thymolphthalein monophosphate substrate). Clin Chem 22:627–632
25. Babson AL, Greeley SJ, Coleman CM et al (1966) Phenolphthalein monophosphate as a substrate for serum alkaline phosphatase. Clin Chem 12:482–490
26. Coleman CM (1966) The synthesis of thymolphthalein monophosphate, a new substrate for alkaline phosphatase. Clin Chim Acta 13: 401–403
27. Gee KR, Sun WC, Bhalgat MK et al (1999) Fluorogenic substrates based on fluorinated umbelliferones for continuous assays of phosphatases and beta-galactosidases. Anal Biochem 273:41–48
28. Hill HD, Summer GK, Waters MD (1968) An automated fluorometric assay for alkaline phosphatase using 3-O-methylfluorescein phosphate. Anal Biochem 24:9–17
29. Wang WQ, Bembenek J, Gee KR et al (2004) Kinetic and mechanistic studies of a cell cycle protein phosphatase Cdc14. J Biol Chem 279: 30459–30468
30. McCain DF, Catrina IE, Hengge AC et al (2002) The catalytic mechanism of Cdc25A phosphatase. J Biol Chem 277:11190–11200
31. Sergienko EA, Millan JL (2010) High-throughput screening of tissue-nonspecific alkaline phosphatase for identification of effectors with diverse modes of action. Nat Protoc 5:1431–1439
32. Mountfort DO, Kennedy G, Garthwaite I et al (1999) Evaluation of the fluorometric protein phosphatase inhibition assay in the determination of okadaic acid in mussels. Toxicon 37:909–922
33. Sparks JW, Brautigan DL (1986) Molecular basis for substrate specificity of protein kinases and phosphatases. Int J Biochem 18:497–504
34. Huang Z, Zhou B, Zhang ZY (2004) Molecular determinants of substrate recognition in hematopoietic protein-tyrosine phosphatase. J Biol Chem 279:52150–52159
35. Zhou B, Zhang J, Liu S et al (2006) Mapping ERK2-MKP3 binding interfaces by hydrogen/deuterium exchange mass spectrometry. J Biol Chem 281:38834–38844
36. Robers MB, Horton RA, Bercher MR et al (2008) High-throughput cellular assays for

regulated posttranslational modifications. Anal Biochem 372:189–197

37. Mitra S, Barrios AM (2005) Highly sensitive peptide-based probes for protein tyrosine phosphatase activity utilizing a fluorogenic mimic of phosphotyrosine. Bioorg Med Chem Lett 15: 5142–5145
38. Xue F, Seto CT (2010) Fluorogenic peptide substrates for serine and threonine phosphatases. Org Lett 12:1936–1939
39. Mustelin T, Tautz L, Page R (2005) Structure of the hematopoietic tyrosine phosphatase (HePTP) catalytic domain: structure of a KIM phosphatase with phosphate bound at the active site. J Mol Biol 354:150–163
40. den Hertog J, Groen A, van der Wijk T (2005) Redox regulation of protein-tyrosine phosphatases. Arch Biochem Biophys 434:11–15
41. Heneberg P, Draber P (2005) Regulation of cys-based protein tyrosine phosphatases via reactive oxygen and nitrogen species in mast cells and basophils. Curr Med Chem 12: 1859–1871
42. Narisawa S, Wennberg C, Millan JL (2001) Abnormal vitamin B6 metabolism in alkaline phosphatase knock-out mice causes multiple abnormalities, but not the impaired bone mineralization. J Pathol 193:125–133
43. Hessle L, Johnson KA, Anderson HC et al (2002) Tissue-nonspecific alkaline phosphatase and plasma cell membrane glycoprotein-1 are central antagonistic regulators of bone mineralization. Proc Natl Acad Sci U S A 99: 9445–9449
44. Millán JL (2006) Mammalian alkaline phosphatases: from biology to applications in medicine and biotechnology. Wiley-VCH Verlag GmbH, Weinheim, Germany
45. Lomashvili KA, Garg P, Narisawa S et al (2008) Upregulation of alkaline phosphatase and pyrophosphate hydrolysis: potential mechanism for uremic vascular calcification. Kidney Int 73: 1024–1030
46. Villa-Bellosta R, Wang X, Millan JL et al (2011) Extracellular pyrophosphate metabolism and calcification in vascular smooth muscle. Am J Physiol Heart Circ Physiol 301:H61–H68
47. Chia JY, Gajewski JE, Xiao Y et al (2010) Unique biochemical properties of the protein tyrosine phosphatase activity of PTEN-demonstration of different active site structural requirements for phosphopeptide and phospholipid phosphatase activities of PTEN. Biochim Biophys Acta 1804:1785–1795
48. Niesen FH, Berglund H, Vedadi M (2007) The use of differential scanning fluorimetry to detect ligand interactions that promote protein stability. Nat Protoc 2:2212–2221
49. Du Y, Xu J, Fu H et al (2012) Chapter 18. Label-free biosensor technologies in small molecule modulator discovery. Chemical Genomics. Cambridge University Press, New York, pp 245–258
50. Sergienko EA, Heynen-Genel S (2013) Experimental approaches for rapid identification, profiling and characterization of specific biological effects of DOS compounds. Diversity-oriented synthesis: basics and applications in organic synthesis, drug discovery, and chemical biology. Wiley-Blackwell, New Jersey, In Press
51. Jacoby E, Mozzarelli A (2009) Chemogenomic strategies to expand the bioactive chemical space. Curr Med Chem 16:4374–4381
52. Nielsen TE, Schreiber SL (2008) Towards the optimal screening collection: a synthesis strategy. Angew Chem Int Ed Engl 47:48–56
53. Chen Y, Shoichet BK (2009) Molecular docking and ligand specificity in fragment-based inhibitor discovery. Nat Chem Biol 5: 358–364
54. Johnson S, Barile E, Farina B et al (2011) Targeting metalloproteins by fragment-based lead discovery. Chem Biol Drug Des 78: 211–223
55. Friberg A, Vigil D, Zhao B et al (2013) Discovery of potent myeloid cell leukemia 1 (mcl-1) inhibitors using fragment-based methods and structure-based design. J Med Chem 56:15–30

Chapter 3

Multisystemic Functions of Alkaline Phosphatases

René Buchet, José Luis Millán, and David Magne

Abstract

Human and mouse alkaline phosphatases (AP) are encoded by a multigene family expressed ubiquitously in multiple tissues. Gene knockout (KO) findings have helped define some of the precise exocytic functions of individual isozymes in bone, teeth, the central nervous system, and in the gut. For instance, deficiency in tissue-nonspecific alkaline phosphatase (TNAP) in mice (*Alpl*$^{-/-}$ mice) and humans leads to hypophosphatasia (HPP), an inborn error of metabolism characterized by epileptic seizures in the most severe cases, caused by abnormal metabolism of pyridoxal-5′-phosphate (the predominant form of vitamin B6) and by hypomineralization of the skeleton and teeth featuring rickets and early loss of teeth in children or osteomalacia and dental problems in adults caused by accumulation of inorganic pyrophosphate (PP_i). Enzyme replacement therapy with mineral-targeting TNAP prevented all the manifestations of HPP in mice, and clinical trials with this protein therapeutic are showing promising results in rescuing life-threatening HPP in infants. Conversely, TNAP induction in the vasculature during generalized arterial calcification of infancy (GACI), type II diabetes, obesity, and aging can cause medial vascular calcification. TNAP inhibitors, discussed extensively in this book, are in development to prevent pathological arterial calcification.

The brush border enzyme intestinal alkaline phosphatase (IAP) plays an important role in fatty acid (FA) absorption, in protecting gut barrier function, and in determining the composition of the gut microbiota via its ability to dephosphorylate lipopolysaccharide (LPS). Knockout mice (*Akp3*$^{-/-}$) deficient in duodenal-specific IAP (dIAP) become obese, and develop hyperlipidemia and hepatic steatosis when fed a high-fat diet (HFD). These changes are accompanied by upregulation in the jejunal-ileal expression of the *Akp6* IAP isozyme (global IAP, or gIAP) and concomitant upregulation of FAT/CD36, a phosphorylated fatty acid translocase thought to play a role in facilitating the transport of long-chain fatty acids into cells. gIAP, but not dIAP, is able to modulate the phosphorylation status of FAT/CD36. dIAP, even though it is expressed in the duodenum, is shed into the gut lumen and is active in LPS dephosphorylation throughout the gut lumen and in the feces. *Akp3*$^{-/-}$ mice display gut dysbiosis and are more prone to dextran sodium sulfate-induced colitis than wild-type mice. Of relevance, oral administration of recombinant calf IAP prevents the dysbiosis and protects the gut from chronic colitis. Analogous to the role of IAP in the gut, TNAP expression in the liver may have a proactive role from bacterial endotoxin insult. Finally, more recent studies suggest that neuronal death in Alzheimer's disease may also be associated with TNAP function on certain brain-specific phosphoproteins. This review recounts the established roles of TNAP and IAP and briefly discusses new areas of investigation related to multisystemic functions of these isozymes.

Key words TNAP, IAP, Mineralization, Epilepsy, Vascular calcification, LPS, Inflammation, Detoxification

José Luis Millán (ed.), *Phosphatase Modulators*, Methods in Molecular Biology, vol. 1053, DOI 10.1007/978-1-62703-562-0_3, © Springer Science+Business Media, LLC 2013

1 Introduction

1.1 Mouse and Human Genes Encoding Alkaline Phosphatases

Alkaline phosphatases (AP) are present widely in nature, from bacteria to man [reviewed in ref. 1]. In man, AP isozymes are encoded by four different genes [1]. Three, *ALPI*, *ALPP*, and *ALPPL2*, are quite tissue specific (TSAP) in their expression pattern restricted primary to the intestine, placenta, and germ cells, respectively. The fourth, *ALPL*, is designated as tissue-nonspecific alkaline phosphatase (TNAP), since it is expressed in bone, liver, kidney, and other tissues. Mice have five loci encoding AP genes, *Alpl* (a.k.a. *Akp2*) encoding TNAP, *Akp3* encoding a duodenal-specific IAP isozyme (dIAP), *Akp5* encoding an embryonic AP (EAP), *Akp6* that encodes a global IAP (gIAP) isozyme, and an inactive pseudogene (Akp-ps1).

The *ALPL* gene encoding human TNAP is localized on chromosome 1p. Its structure was first reported by Weiss and collaborators [2]. It exceeds 50 kb and contains 12 exons, the first exon being part of the 5′-untranslated region (UTR) of the *TNAP* mRNA. Soon after this discovery, Kishi et al. reported the presence of another exon, 3.4 kb upstream of exon 2 [3]. They showed that the 5′-UTR consists of either exon 1A or exon 1B obtained by alternative transcription initiation [3, 4]. Although species-related differences may exist, transcription of the upstream exon 1A appears preferentially driven by a promoter active in osteoblasts, whereas transcription may be preferentially initiated with exon 1B by a distinct promoter active in liver and kidney [4, 5]. However TNAP's expression is ubiquitous.

The human tissue-specific alkaline phosphatase (TSAP) genes, *ALPP*, *ALPPL2*, and *ALPI*, are clustered on human chromosome 2, bands q34–q37 [6, 7], and are closely related to one another (Table 1). Their structure is almost identical, consisting of 11 exons interrupted by small introns (74–425 bp) at analogous positions, all compressed in less than 5 kb of genomic DNA [8–10]. Figure 1 shows the structure of the *ALPPL2* gene as an example, in relation to the *ALPL* gene. The similarity in structure between all three TSAP genes suggests a divergent evolution for these genes since the chromosome mapping results show that the three related loci—*ALPP*, *ALPPL2*, and *ALPI*—are located in this order from centromere to telomere in the same region of the long arm of chromosome 2 (2q34–q37) [6, 7]. Each gene comprises 11 exons and 10 small introns contained within 4.5 kb of DNA. Specific regions of the introns, as well as in the 3′-UTR of exon XI, show major differences in sequence and these regions have proven useful in the development of gene-specific probes. The *PLAP* gene also contains an Alu repeat sequence inserted in exon XI [9]. This Alu repeat sequence creates a new polyadenylation signal that may be responsible for alternative usage and consequently alternative size *PLAP* mRNA molecules approximately 300 bp shorter. Alternative *PLAP* mRNA

Table 1
Summary of the gene nomenclature, accession numbers, common names, tissue distribution, and function, if known, for the human and mouse alkaline phosphatase isozymes

	Accession numbers	Protein names	Common names, synonyms	Tissue distribution	Function
Human genes					
ALPL	NM_000478	TNAP	Tissue-nonspecific alkaline phosphatase, TNSALP; "liver–bone–kidney-type" AP	Developing nervous system, skeletal tissues, liver, kidney	Bone and dental mineralization
ALPP	NM_001632	PLAP	Placental alkaline phosphatase, PLALP	Syncytiotrophoblast, a variety of tumors	Unknown
ALPPL2	NM_031313	GCAP	Germ cell alkaline phosphatase, GCALP	Testis, malignant trophoblasts, testicular cancer	Unknown
ALPI	NM_001631	IAP	Intestinal alkaline phosphatase, IALP	Gut, influenced by fat feeding and ABO status	Intestinal absorption, gut barrier function, detoxyfication
Mouse genes					
Alpl (aka Akp2)	NM_007431	TNAP	Tissue-nonspecific alkaline phosphatase, TNSALP; "liver–bone–kidney-type" AP	Developing nervous system, skeletal tissues, kidney	Bone, dentin, cementum, enamel mineralization
Akp3	NM_007432	dIAP	Duodenal-specific intestinal alkaline phosphatase (dIALP)	Gut	Fat absorption, gut barrier function, detoxyfication
Akp5	NM_007433	EAP	Embryonic alkaline phosphatase	Preimplantation embryo, testis, gut	Early embryogenesis
Akp6	NM_007433	qIAP	Global intestinal alkaline phosphatase (gIALP)	Gut	Unknown

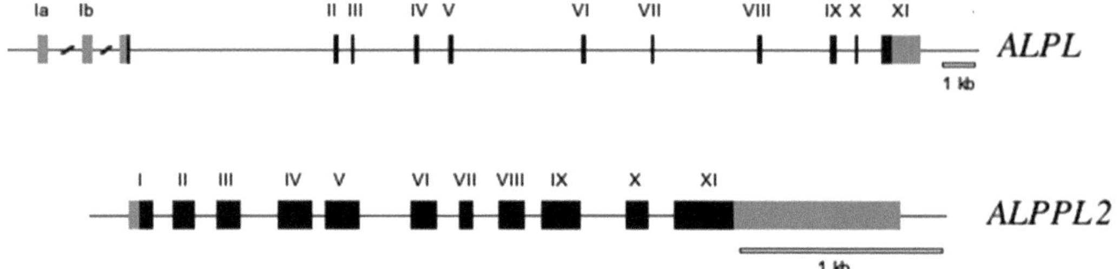

Fig. 1 Genomic organization of the human *ALPL* and *ALPPL2* genes [from ref. 1]. *Solid black boxes* indicate coding exons, while *gray boxes* mark the 5′ and 3′ untranslated regions (UTR)

molecules have already been observed in choriocarcinoma cells [11], colonic adenocarcinoma cells [12], and Hela cells [13]. The intron–exon junctions in the coding region are, however, remarkably similar for all four genes. Knoll et al. have proposed that insertion/deletion events in the promoters of these three isozyme genes may partially explain evolution of promoter activity [9]. In mice, besides the *Alpl* gene, which maps to chromosome 4 [14], four additional AP loci are found clustered on chromosome 1: the *Akp3* locus that encodes the IAP isozyme [15]; the *Akp5* gene that encodes the embryonic AP (EAP) isozyme [15]; the *Akp-ps1* non-transcribed, intron-containing, pseudogene [15]; and the *Akp6* gene [16].

1.2 AP Ubiquitous Expression in Mice and Humans

In mouse, the *Alpl* gene encoding TNAP is widely expressed during development. It is already expressed in one-cell-stage embryos [17], where it has been proposed to ensure that ES cells are maintained in an undifferentiated state [1]. *Alpl* expression can be observed in different stages in several tissues, such as the neural tube and the brain [18, 19], in endothelial cells, in renal tubules in kidney, and in biliary canaliculi in the liver [20] and is also expressed at high levels in bone, cartilage, and teeth [20]. Additionally, TNAP has been evidenced in different cell types of the brain [21, 22]. Mouse neurons seem to express the so-called liver (exon 1B-containing) *Alpl* transcript, but in humans, neuronal cells likely express the so-called bone (exon 1A-containing) *Alpl* transcript [23]. Another difference between mice and humans is the laminar distribution in humans and a more diffuse distribution in mice [24].

Mouse EAP [15], encoded by the *Akp5* gene, is expressed in the two-cell- to blastocyst-stage embryos at levels tenfold higher than *Akp2* in these embryonic stages [25]. EAP is not expressed after gastrulation but reappears in the adult testis, the thymus, and also the intestine [25]. In the human fetus, it is the germ cell AP (GCAP) isozyme, and not TNAP as in mouse, that is expressed in migrating PGCs and in the fetal gonads [26], but it remains to be established if GCAP is also expressed in human ES cells and in the preimplantation human embryo. GCAP is also found in trace amounts in the testis, lung, cervix, and thymus [27, 28].

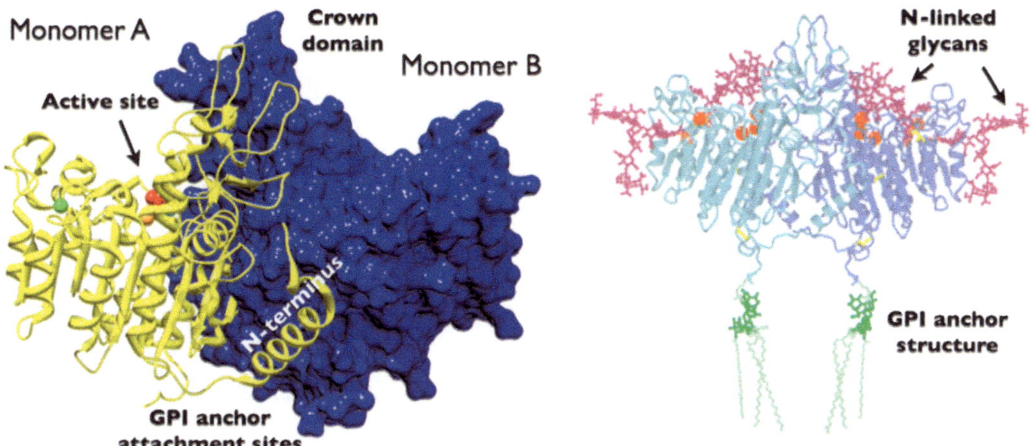

Fig. 2 Three-dimensional structure of PLAP. *Left panel*: Monomer A is shown in *ribbon representation* and in *yellow*, while monomer B is shown in *surface representation* in *blue*. The relative location of the active site, N-terminus, and GPI anchoring site are indicated. *Right panel*: Modeled structure of the GPI anchor attached to the 3D structure of PLAP. N-linked complex glycan was also modeled at N122 and at N249. This figure was kindly produced and contributed by Dr. Mark R. Wormald, Oxford Glycobiology Institute, University of Oxford, Oxford, UK

Both humans and monkeys also express GCAP in type I pneumocytes in the lung. PLAP is expressed in large amounts in the syncytiotrophoblast cells of the placenta from about the eighth week of gestation throughout pregnancy [29] where it shows a clustered distribution [30]. Intestinal AP (IAP), as the name implies, is expressed in the small intestine of many species. Mice co-express *Akp3*, *Akp5*, and *Akp6* in the gut, although *Akp3* is expressed just before weaning [16].

1.3 AP Protein Structures

PLAP, GCAP, and IAP are 90–98 % homologous, while TNAP is about 50 % identical to the other three [1]. All APs are anchored to the plasma membrane via a glycosyl phosphatidyl inositol (GPI) anchor. During synthesis of the enzyme, monomeric precursors dimerize and each monomer binds one magnesium, two zinc, and one calcium ion as cofactors. An analysis of the structural-functional relationship of residues conserved between the *Escherichia coli* AP and the PLAP structure revealed a conserved function for those residues that stabilize the active-site zinc and magnesium metal ions, whereas the nonhomologous disulfide bonds differ in their structural significance and non-conserved residues take part in determining the heat stability and uncompetitive inhibition properties of mammalian alkaline phosphatases [31–33]. The N-terminus of one monomer wraps around the second monomer conferring allosteric properties to the AP dimer [34, 35], while the flexible top domain of APs, designated as the "crown domain" stabilizes the positioning of AP inhibitors and determines isozyme-specific properties, such as heat stability and collagen binding [36] (Fig. 2).

The presence of carbohydrate chains does not seem to matter for the catalytic activities of the TSAP isozymes, but is of importance for TNAP. The bone and liver TNAP isoforms exhibit the same 507-amino acid sequence but differ in their posttranslational glycosylation modifications. The cDNA sequence of *ALPL* suggests the presence of five N-linked glycosylation sites [2]. In addition, TNAP seems to be O-glycosylated in bone but not in liver. Indeed, the difference between the bone, liver, and kidney isoforms of TNAP is caused by differences in glycosylation, and these posttranslational modifications affect the catalytic activities of these isoforms [37].

2 Pathophysiological Function of TNAP

TNAP is anchored at the cell membrane and also at the membrane of matrix vesicles (MVs) released from osteoblasts and chondrocytes [38] (Fig. 3). Importantly in the context of biomineralization, the membrane of MVs is markedly enriched with (GPI)-anchored TNAP compared with the cell membrane from which they originate [39]. TNAP can be released in the circulation, where it is found as a homodimer [40], upon the action of two circulating phospholipases, GPI-phospholipase C and GPI-phospholipase D [41–43]. In healthy adults, the bone and liver isoforms represent about 95 % of the serum AP activity with similar contribution of each form [44]. Three bone isoforms, B1, B2, and B/I, account for 16, 37, and 4 %, respectively. B1 and B2 differ by

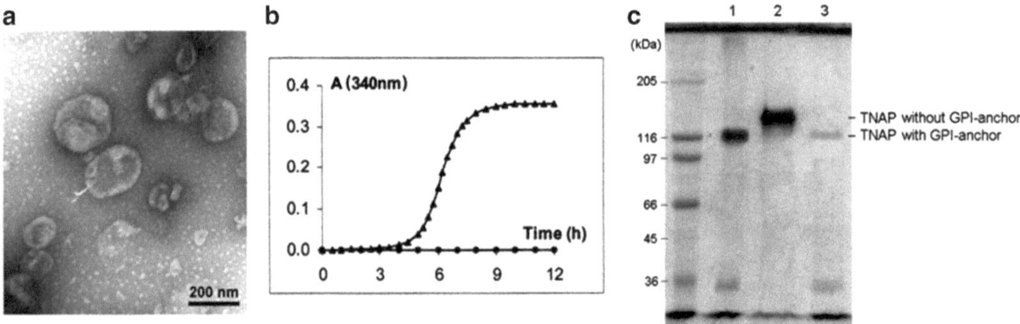

Fig. 3 Characterization of MVs isolated from femurs of 17-day-old chicken embryos [from ref. 38]. (**a**) Morphology of MVs observed under electron microscopy. The size range of spherical to oval MVs was 100–200 nm in diameter. (**b**) Mineralization induced by MVs in buffer containing 2 mM Ca, 1.42 mM P_i, and 10 g of protein/ml MVs; *filled circle*: control of mineralization in SCL containing 2 mM Ca and 1.42 mM P_i, but without MV. (**c**) Treatment of MVs by PI-PLC as evidenced by gel electrophoresis; the gel was stained with BCIP-NBT, only the bands of active TNAP are revealed (*line 1*: MVs; *line 2*: supernatants of MVs after treatment with PI-PLC; *line 3*: pellets of MVs after treatment with PI-PLC)

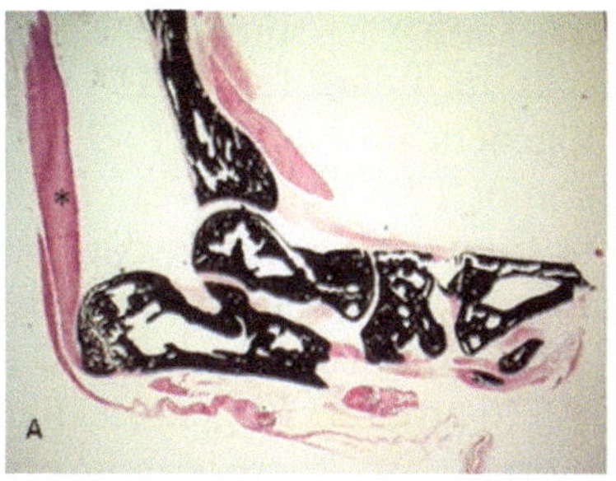

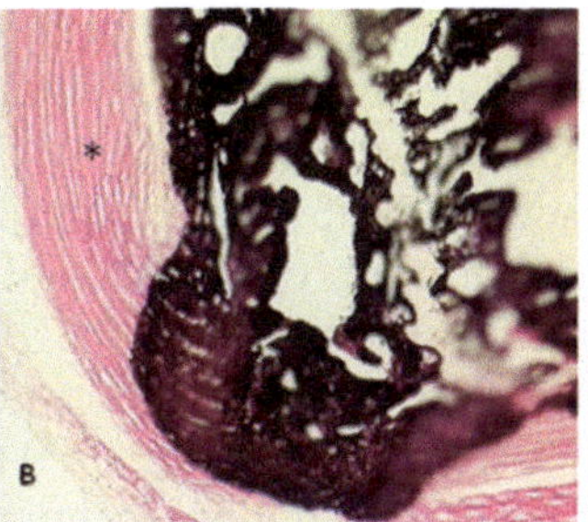

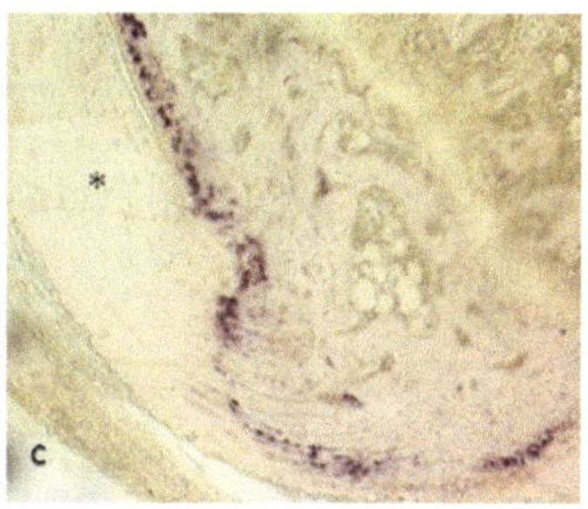

Fig. 4 TNAP activity colocalizes with the mineralization front (developing mouse Achilles tendon enthesis) [from ref. 50]. Ankles from 12-week-old C57BL6 mice were embedded, and sections were stained with von kossa staining to reveal mineralized bones in *black* (**a**). (**b**) shows at higher magnification the insertion of the developing enthesis (bony insertion) of the Achilles tendon (*asterisk*) in the calcaneum. (**c**) is a staining of TNAP activity indicating that TNAP activity colocalizes perfectly with the mineralization front. Moreover, TNAP activity is absent in already mineralized areas

their N-glycosylation pattern, which modulates their catalytic properties. The B/I is not a "pure" bone isoform since it is composed of 70 % TNAP and 30 % IALP [44]. APs catalyze the hydrolysis of monoesters of phosphoric acids and have broad substrate specificity in vitro. Purified osseous plate TNAP for instance is able to hydrolyze ATP, ADP, AMP, PP_i, glucose-1-phosphate, glucose-6-phosphate, fructose-6-phosphate, and β-glycerophosphate. However, only a few compounds have been considered as natural in vivo substrates: PP_i, pyridoxal-5′-phosphate (PLP), and phosphoethanolamine (PEA) [reviewed in ref. 1]. The ability of TNAP to hydrolyze PP_i is discussed extensively in the next section. PLP has been shown to be a physiological substrate for the TNAP present in leukocytes and in the central nervous system [45, 46]. PEA may be part of the GPI linkage apparatus and may be a degradation product from this tether for cell-surface proteins [47]. Its levels are increased in patients with HPP and in TNAP-deficient mice [48]. However and in contrary to what was previously thought, PEA does not seem to be a substrate for TNAP. In TNAP-deficient mice, the levels of PLP in the liver are significantly reduced. PLP is a cofactor for the enzyme O-phosphoethanolamine phospho-lyase, which transforms phosphoethanolamine into acetaldehyde, P_i, and NH_3 [49]. This may explain the increased levels in phosphoethanolamine observed in patients with hypophosphatasia.

2.1 TNAP's Necessary Role in Bone and Tooth Mineralization

TNAP expression is precisely regulated to coincide with the initiation of skeletal and dental mineralization. For instance, TNAP activity colocalizes perfectly with the mineralization front in the developing Achilles tendon in mouse (Fig. 4) [50]. Mineralizing cells express TNAP before the onset of mineralization and TNAP expression slowly drops as the mineralization progresses [1].

A wide variety of signals have been shown to stimulate TNAP expression in human mineralizing cells, including 1,25$(OH)_2$ vitamin D3, retinoic acid, bone morphogenetic protein (BMP)-2, and Wnt factors [51, 52]. The stimulation of TNAP activity can result from stabilization of mRNA and/or from increased transcription. Increased transcription of the *ALPL* gene seems inversely correlated with promoter methylation [53]. In human chromosome 1, a CpG island extends from −579 to +836 bp of *ALPL* gene, including the promoter region. In human osteoblasts, there is an inverse relationship between the methylation status of this island and *ALPL* transcript abundance; and interestingly, demethylation strongly enhances TNAP expression and activity [53]. In bone, TNAP is localized on the entire cell surface of preosteoblasts, as well as the basolateral cell membrane of osteoblasts. It is also localized on hypertrophic cells in cartilage. TNAP is moreover also anchored at the membrane of MVs released by osteoblasts and growth plate chondrocytes [39, 54]. Conflicting results have been published as to whether TNAP must be anchored to participate to mineralization or whether TNAP released from cell membranes might also be active [55–57]. The fact that infusions of alkaline phosphatase correct hypophosphatasemia but do not improve radiographic symptoms of hypophosphatasia (HPP) first seemed to suggest that TNAP has to be tethered to cell membranes to fulfill its functions [58–60]. Indeed, the treatment of *Alpl*$^{-/-}$ mice with recombinant TNAP to which ten Asp residues have been added to increase its affinity to apatite crystals prevented both the skeletal manifestations and the epilepsy in this mouse model of infantile HPP [61]. Ongoing clinical trials of enzyme replacement therapy using mineral-targeting TNAP is proving life-saving for HPP patients with severe, life-threatening disease [62]. These data suggest that the GPI-membrane anchor is not required for TNAP activity, but may help localize TNAP in proximity to the mineralization front. Mutations encoding GPI anchor biosynthesis, class V (PIGV), a member of the GPI-anchor biosynthesis pathway, lead to Mabry syndrome, a disease with seizures, hyperphosphatasemia, neurologic deficit, facial abnormalities, and brachytelephalangy [63].

HPP is a heritable form of rickets or osteomalacia characterized by subnormal activity of TNAP. Loss-of-function mutations in the *ALPL* gene leading to HPP were first documented in 1988 [64], and since then about 260 mutations have been reported [65]. The most severe form of HPP manifests in utero with dramatic hypomineralization, and causes death at, or soon after, birth. The least severe form, odonto HPP, manifests as dental problems in children or in adults. Perinatal/infantile cases of HPP have high mortality with 50 % of the patients succumbing to respiratory failure caused by undermineralization of the ribs. Electron microscopy of perinatal HPP revealed a normal distribution of MVs containing apatite crystals [66]. However, Anderson and colleagues showed that both in human and mouse HPP specimens, these crystals fail to grow after MV rupture [67], arguing that TNAP is not involved in the very first steps of crystal formation inside MVs, but rather in the growth and multiplication of crystals in the collagen matrix. Recent data have implicated the function of PHOSPHO1, a phosphatase with substrate preference for phosphoethanolamine and phosphocholine, in the initiation of MV-mediated mineralization, while it is clear that the role of TNAP is to control the concentrations of extracellular PP_i, and thus controlling extravesicular propagation of mineralization [68, 69]. In the case of TNAP deficiency, the activity of TNAP is inversely correlated with osteoid accumulation [70]. In the murine moldel of infantile HPP (*Alpl*$^{-/-}$ mice), skeletal disease first appears radiographically at about 6–10 days of life [70]. These differences in hypomineralization onset as compared with human HPP may be due to differences in PP_i levels in the absence of TNAP in both species possibly due to compensatory activity by other phosphatases. For instance, ectonucleotide pyrophosphatase phosphodiesterase 1 (ENPP1), whose active site is structurally similar to the active site of APs [71], has been described as an enzyme that could act as a backup phosphatase in the absence of TNAP, at least in the axial skeleton [68, 72].

Historically, the function of TNAP was thought to be to provide inorganic phosphate (P_i) for tissue mineralization. However, the circulating levels of P_i are usually normal or elevated during HPP [65]. Secondly, mice fed with a high-phosphorus diet for 2 months develop a significant increase in serum P_i levels but do not develop any detectable pathological mineralization [73]. Finally, P_i itself is known to inhibit TNAP activity through a competitive mechanism [74], and therefore the level of P_i in vivo will impact the ability of TNAP to hydrolyze PP_i. Compelling evidence indicates that the role of TNAP is primarily to remove PP_i, which is a constitutive inhibitor of apatite crystal deposition. However, TNAP is also an extremely efficient ATPase [72], and the possibility that TNAP is also involved in the perivesicular production of P_i from ATP has recently been discussed in the context of the interplay between PHOSPHO1, TNAP, and NPP1 [68, 69]. The nucleotide triphosphate pyrophosphohydrolase activity of NPP1 and the transmembrane PP_i-channeling progressive ankylosis protein (ANK) are responsible for supplying PP_i to the extracellular compartment. Indeed PP_i levels are elevated in *Alpl*$^{-/-}$ mice, and affecting the function of either NPP1 or ANK has beneficial effects on HPP by reducing the levels of PP_i in *Alpl*$^{-/-}$ mice [75, 76]. It appears therefore that rather than just P_i, it is the P_i/PP_i ratio that is the critical determinant of appropriate controlled mineralization. We reported that the formation of apatite by MVs isolated from growth plate chondrocytes is optimal when the P_i/PP_i molar ratio is above 140, but is totally inhibited when the ratio

a

Initial $[P_i]/[PP_i]$	Final $[P_i]/[PP_i]$	Induction time (%)	Minerals formed, IR spectra
∞	∞	12.5 ± 1.4	HA, Spectrum I
142 ± 47	198.3 ± 65.6	33.4 ± 5.6	HA, Spectrum II
71 ± 14.2	138.7 ± 27.7	50.1 ± 3.5	HA + other
28.4 ± 5.7	102.9 ± 20.7	100 ± 5.0	CPPD + other, Spectrum III
14.2 ± 1.4	24.5 ± 2.4	68.1 ± 4.2	CPPD + other, Spectrum IV
2.8 ± 0.3	6.2 ± 0.7	51.4 ± 4.2	CPPD, Spectrum V
1.4 ± 0.1	2.8 ± 0.3	30.1 ± 2.8	CPPD, Spectrum VI

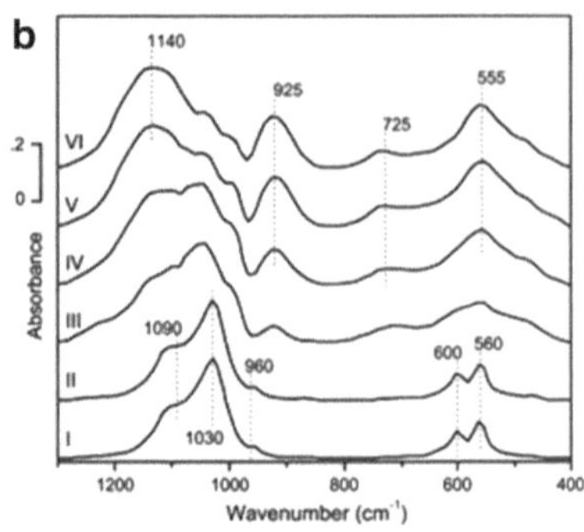

Fig. 5 Characterization of crystals formed by isolated MVs in the presence of varying $[P_i]/[PP_i]$ ratio [from ref. 77]. (**a**) Minerals formed by MVs in buffer containing 2 mM Ca, 1.42–3.42 mM P_i, and varying amounts of PP_i. (**b**) Infrared spectra of minerals produced by MVs revealing the presence of hydroxyapatite (HA) or calcium pyrophosphate dihydrate (CPPD) crystals

decreases below 70 [77]. The retardation of any mineral formation is maximal at P_i/PP_i ratio around 30, and crystals of calcium PP_i dihydrate (CPPD) are exclusively produced by MVs when the ratio is below 6 (Fig. 5) [77]. This likely explains why some patients with adult HPP suffer from CPPD deposition, occasionally with attacks of pseudogout [65]. It is now well established that odontoblasts mineralize the dentin via the production of MVs and regulation of P_i/PP_i ratio, as demonstrated through the analysis of *Alpl*$^{-/-}$ hypophosphatasia mice and the success of enzyme replacement using mineral-targeting TNAP to correct the dentin phenotype [61, 78]. Similarly, the acellular cementum is under strict regulation by the P_i/PP_i ratio [79], and this ratio is normalized, and acellular cementum corrected, in HPP mice treated with enzyme replacement [80]. Enamel defects are also present in HPP mice and these defects are preventable by enzyme replacement with mineral-targeting TNAP [81], arguing that the P_i/PP_i ratio is also involved in the regulation of enamel matrix mineralization.

The fact that mineralization is inhibited by PP_i and that TNAP promotes apatite crystal formation by hydrolyzing PP_i questions why mineralization does not also occur in other soft tissues, which also express TNAP. Murshed et al. elegantly answered this question. Using different loss- and gain-of-function approaches in vitro and in vivo, they demonstrated that mineralization requires the colocalization of TNAP and a fibrillar collagen, such as type I (in bone and dentin) and type II (in growth plate cartilage) [73]. Collagen fibrils are necessary to provide a scaffold for propagation of apatite crystals. TNAP may play a role in the quantity of crystals formed but also in their association pattern with collagen fibrils.

Indeed, Tesch et al. showed that in TNAP-deficient mice, the crystal axis was not ordered parallel with the long axis of collagen fibrils as it is in wild-type animals [82].

2.2 TNAP's Involvement in Neurotransmitter Synthesis

Neuronal abnormalities in HPP are likely due to deregulated metabolism of vitamin B6, and in particular in the PLP vitamer [83], which appears to be a physiological substrate for TNAP in serum and/or in leukocytes [46, 47]. After ingestion, phosphorylated vitamin B6 vitamers are dephosphorylated by IAP, and the resulting vitamers, as well as ingested pyridoxine, pyridoxal (PL), and pyridoxamine, are then rapidly absorbed by a carrier-mediated diffusion process [84]. In intestinal cells, vitamers are converted via pyridoxal kinase and pyridoxamine 5′-phosphate oxidase to PLP. PLP is dephosphorylated at the serosal surface and pyridoxal then reaches the portal circulation. In TNAP-deficient mice, absence of TNAP activity results in intracellular deficiency of B-vitamers and decreased ability to sustain vitamin B6-dependent cellular homeostasis, including glutamate decarboxylase activity and the synthesis of neurotransmitters such as serotonin or gamma-aminobutyric acid (GABA) in the brain. Thus, $Alpl^{-/-}$ mice die from seizures before weaning [48, 85, 86]. The most likely cause of death in $Alpl^{-/-}$ animals appears to be apnea, which occurs in conjunction with epileptic seizures [1, 85]. In mutant mice administered PL, no seizures are observed and concentrations of PLP and GABA in the brain return within the normal range [48, 86]. Elevated plasma PLP levels may also be the cause of epileptic seizures experienced by patients with HPP [87]. When seizures are present, they may respond to treatment with vitamin B6 [88]. Conversely, in hyperphosphatasia, neurodevelopmental abnormalities are likely due to decreased levels of PLP [83].

2.3 TNAP's Suspected Role in Neuronal Differentiation

In addition to its role in neurotransmitter synthesis, TNAP may also be involved in neural cell differentiation. During early mouse embryonic development, TNAP activity becomes apparent in the neuroepithelium already at E8.5 where it is highly expressed at the plasma membrane of the neuroepithelial stem cells of the neural tube [19]. Most notably, the enzyme remains highly expressed during further embryonic development in the ventricular neurogenic regions and remains associated with neural precursor cells in the adult mouse brain [22]. Ablation of TNAP function in the mouse results in a significant decrease in the white matter of the spinal cord, and absence of myelinated axons in the cerebral cortex [89]. In vitro, inhibition of TNAP with RNAi in neural stem cells results in reduced cell proliferation and differentiation into neurons or oligodendrocytes [90]. In cultured hippocampal neurons, TNAP is co-localized at axonal growth cones with ionotropic ATP receptors (P2X7 receptor). In these cells, TNAP is required for axonal growth since it hydrolyzes ATP and prevents ATP inhibitory effects through P2X7 receptor binding [91]. TNAP has also been proposed to participate in neurotransmission [21]. Very recently, it was shown that active TNAP physically interacts with the cellular prion protein in serotonergic and noradrenergic differentiated neuronal cells [92]. In these cells, TNAP inhibition resulted in a significant decrease in serotonin and catecholamine synthesis. These data define TNAP as a player in neurotransmitter metabolism. TNAP activity is regulated by sensory experience [21]. Since serotonin-containing fibers are abundant in sensory regions of the brain, Kellermann et al. suggested that TNAP regulates serotonin or dopamine synthesis and participates in cortical function and neuronal plasticity by regulating neurotransmitter synthesis [92].

2.4 Function of TNAP in the Liver

TNAP is expressed in human hepatocytes, and bile acids increase its activity [93] and secretion in the bile [94]. TNAP in rat hepatocytes is predominantly localized in the bile canalicular domain of the plasma membrane [95, 96], but can be addressed to the basolateral membrane in the presence of high levels of bile acid [97]. In contrast, mouse hepatocytes do not express TNAP [98]. In humans, liver TNAP may be expressed both at the sinusoidal and biliary pole of the hepatocyte. This explains why a significant proportion of TNAP activity in the circulation of healthy individuals originates from the liver. TNAP serum levels are of major clinical relevance as a marker of cholestasis. AP levels increase due to retrograde reflux of biliary alkaline phosphatase, enhanced hepatic synthesis and enzyme release into the serum, and induction of the intestinal alkaline phosphatase form [94, 99, 100].

It has been suggested that TNAP inhibits bile secretion [101]. Alkaline phosphatase intraluminal injection inhibits the basal activity of the Cl^-/HCO_3^- exchanger [101]. Interestingly, cystic fibrosis transmembrane receptor (CFTR) is activated by TNAP inhibitors [102, 103], but the underlying mechanisms remain obscure. Becq et al. suggested that TNAP dephosphorylates the CFTR that had been activated by PKA [102, 103]. This hypothesis may however seem unlikely given that TNAP is normally produced as a GPI-anchored protein, even if experiments suggested that TNAP is present in significant amounts in cytoplasmic vesicles in rat hepatocytes [96]. It has also been proposed that TNAP acts in the extracellular compartment by hydrolyzing ATP and modulating purinergic receptor signaling [101].

The liver is the major lipopolysaccharide (LPS)-removing organ. The majority of systemic LPS is removed from the bloodstream by Kupffer cells and excreted in the bile by hepatocytes [104]. One important function of TNAP in the liver may be to dephosphorylate endotoxins. The toxic part of LPS is located in the lipid A part of the molecule, which is the most well-conserved moiety among LPS serotypes from a variety of gram-negative bacteria. Two phosphate groups coupled to two glucosamine in the lipid A part largely determine the toxicity of LPS [105]. Removal of a single phosphate group results in the formation of a nontoxic monophosphoryl lipid A moiety [106, 107]. Administration of LPS in rats stimulates the expression of TNAP by hepatocytes [108] and TNAP is able to dephosphorylate LPS [109]. Poelstra et al. proposed that TNAP function is to dephosphorylate endotoxin, whose levels are elevated in cholestasis [109]. The phagocytic activity of Kupffer cells may be suppressed during cholestasis and the normal enterohepatic route for endotoxin removal via the biliary system is blocked during this disease. The enhanced TNAP activity during cholestasis may reflect a physiological response of the liver upon bile duct obstruction [109].

2.5 Function of TNAP in the Kidney

In human kidney, TNAP is expressed along the proximal tubule in segments S1, S2, and S3 [110]. IAP is also expressed in the kidney but is restricted to the pars recta (S3) of the proximal tubule [110]. The expression of TNAP in the kidney is counterintuitive since TNAP potently hydrolyzes PP_i, and PP_i is a well-known inhibitor of mineralization in urine [111]. Indeed, individuals prone to form calcium stones appear to show reduced urinary PP_i excretion [112], and in a group of 107 patients with recurrent calcium stones, PP_i:creatinine ratios were reduced compared with control subjects [113]. Actually, it would seem that PP_i is generated in the segments of the nephron downstream of TNAP-expressing proximal segments. However, as discussed above, TNAP can be shed from cells on the surface of nanosized vesicles and thus can conceivably act downstream from its site of synthesis, as has been shown to be the case for IAP in the gut [114]. Whereas PP_i levels in urine are around 10 μM [111], PP_i is present in plasma at a concentration of 1–6 μM [115], mainly arising from liver metabolism [116]. Intravenous $^{32}PP_i$ is rapidly hydrolyzed in plasma, with PP_i also being filtered at the glomerulus and subject to further hydrolysis within the kidney; only less than 5 % of intravenous $^{32}PP_i$ appears in urine. These data indicate that by far the largest source of PP_i in the kidney is local generation [117]. In the kidney, PP_i may be generated from ATP by nucleoside triphosphate diphosphohydrolase (NTPD) [117]. However, the largest source of PP_i generation in the kidney may be intracellular. ANK is expressed in the kidney, and since ank/ank mice exhibit nephrocalcinosis, it is likely that PP_i is transported through ANK to prevent mineralization within the urinary system [118]. Expression of ANK happens in the segments of the nephron downstream from the alkaline phosphatase-expressing segments [119]. Since ANK is located both at the apical and basolateral membranes, it may participate to inhibit mineralization within both the renal interstitium and the tubule lumen. While further studies are needed, to date the function of TNAP in the kidney appears to be linked to PP_i metabolism and LPS detoxification [120].

2.6 Pathophysiological role of TNAP in Vascular Calcification

Paradoxically, whereas aging is associated with a decrease in bone formation [121], it is also linked to vascular calcification [122]. Vascular calcification can form by intramembranous ossification or by endochondral bone formation, and eventually lead to the appearance of mature bone, with bone marrow, in the vasculature [123]. Vascular calcification develops during aging independently of cardiovascular diseases, but is exacerbated by several age-related diseases, such as atherosclerosis and diabetes type 2. Atherosclerosis increases calcification of the intima [124]. Over 70 % atherosclerotic plaques observed in the aging population are calcified. Calcification can lead to plaque rupture and acute cardiovascular events [125]. Vascular smooth muscle cells (VSMCs) play a central role during atherosclerotic calcification by expressing several markers of chondrocytes and/or osteoblasts such as TNAP, bone sialoprotein, osteocalcin, and the transcription factor RUNX2 [126]. On the other hand, diabetes type 2 is associated with calcification of the media (also named Mönckeberg's sclerosis) [reviewed in ref. 123]. Medial calcification appears to be an indicator of the severity and or duration of diabetes type 2, relates to the degree of glycemic control [127], and has emerged as the most significant predictor of lower extremity amputation and cardiovascular mortality risk. Like in atherosclerosis, calcification associated with diabetes is accompanied by expression by VSMCs of chondrocyte and/or osteoblast markers such as collagen type II, TNAP, and osteocalcin [128]. Several papers to date have documented that not only TNAP upregulation is associated with medial vascular calcification, but also it is a necessary and sufficient pathophysiological event. TNAP upregulation has been documented in the aortas of uremic rats, a model of chronic kidney disease [129], as well as in *Enpp1*$^{-/-}$ mice, a model of generalized arterial calcification of infancy (GACI: OMIM # 208000) [130]. Ankylosis mice (*ank/ank* mice) also show a vascular calcification phenotype comparable to *Enpp1*$^{-/-}$ mice [130]. In both the *Enpp1*$^{-/-}$ and *ank/ank* models, the development of vascular calcification could be attributed to the combined effects of increased degradation of PP_i—caused by upregulated TNAP expression—and the reduced production (in *Enpp1*$^{-/-}$ mice) or transport (in *ank/ank* mice) of PP_i. Another rare disease, arterial calcification due to deficiency of CD73 (ACDC; OMIM #211800), a 5′-nucleotidase encoded by the *NT5E* gene, has been shown to involve TNAP upregulation as the likely mechanism of pathogenesis [131]. Keutel syndrome (OMIM # 245150), caused by deficiency in matrix gamma-carboxyglutamic acid protein (MGP), also manifests medial vascular calcification [132]. MGP is a local inhibitor of vascular calcification that inhibits vascular calcification only in the absence of PP_i [133]. Medial vascular calcification in *Mgp*$^{-/-}$ mice, a murine model of Keutel syndrome, also appears to be mediated via upregulation of TNAP expression at sites of calcification [134]. Given that TNAP is required for mineralization, and since it is induced in VSMCs during vascular calcification, it may also be necessary for vascular calcification [122, 129]. Therefore, the development of TNAP inhibitors to prevent vascular calcification represents a promising therapeutic strategy [130, 135, 136]. In fact, about half of the contributions in this issue of Methods in Molecular Biology, Humana Press, use TNAP to illustrate the path from target identification to the development of pharmaceuticals suitable for clinical trials.

Inflammation may be an important factor in the development of vascular calcification [reviewed in ref. 137]. Several studies have shown that levels of C-reactive protein (CRP) and inflammatory cytokines are associated with increased coronary artery calcification [138, 139]. Cytokine stimulation of TNAP by VSMCs probably plays an important role in calcification associated with diabetes, since the TNF-α inhibitor infliximab reduces the extent of medial calcification in *ldlr*$^{-/-}$ diabetic mice, without reducing obesity, hypercholesterolemia, and hyperglycemia [140]. Interestingly, osteogenesis associates with inflammation and atherosclerosis in *Apoe*$^{-/-}$ mice as revealed by molecular imaging in vivo [139]. Macrophages in atherosclerotic lesions secrete TNF-α, which stimulates TNAP expression by VSMCs [141, 142]. Moreover, in a positive loop, crystals activate macrophages to secrete more TNF-α [143]. An unresolved issue in atherosclerosis is whether preventing calcification of an atherogenic plaque would be beneficial or deleterious to the patient [143]. The availability of pharmacological inhibitors of TNAP will enable studies to answer this important clinical question.

2.7 Pathophysiological Association of TNAP Expression and Alzheimer's Disease

Finally, TNAP has recently been suspected to play a role in Alzheimer's disease (AD). AD is characterized by the loss of neurons and the presence of amyloid plaques and neurofibrillary tangles. The plaques are extracellular deposits of amyloid-β peptide, whereas the neurofibrillary tangles are intracellular aggregates of the microtubule-associated protein tau, which has become hyperphosphorylated [144]. In Alzheimer's disease, tau initially accumulates in the entorhinal cortex and hippocampus and then spreads to the surrounding areas. During this process, neuronal loss causes extracellular release of monomeric or aggregated tau, assembled in extracellular ghost tangles.

An increase in TNAP expression and activity has been observed in the hippocampus and serum of AD patients as compared with control patients [145, 146]. Importantly, TNAP activity in patients with AD is inversely correlated with cognitive functions [146], suggesting that TNAP may be involved in the degenerative process. Diaz-Hernández et al. recently reported that cell membrane-anchored TNAP is able to dephosphorylate extracellular tau released upon cell death in neuronal cells [145]. They observed that once dephosphorylated, tau behaves as an agonist of muscarinic M1 and M3 receptors, provoking a robust and sustained intracellular calcium increase finally triggering neuronal death [145]. Interestingly, TNAP expression and activity are up-regulated by extracellular dephosphorylated tau in human neuronal cells [145], suggesting a vicious circle during which TNAP is stimulated by the product it forms. Thus, the association between TNAP expression and Alzheimer's disease is intriguing and merits careful investigation.

3 Pathophysiological Function of IAP

3.1 Role of IAP in Fatty Acid Absorption

The brush border enzyme IAP is involved in a rate-limiting step of fatty acid transport in the gut. Lymph and serum levels of IAP increase after a fatty meal and the lymphatic triglyceride output is directly correlated with the lymphatic IAP output [147]. In turn, IAP levels are dramatically decreased upon starvation [148]. The mouse gut expresses three IAP isozymes, *Akp3* which is expressed specifically in the duodenum (duodenal IAP, or dIAP) after weaning, *Akp5* expressed at very low levels throughout the gut, and *Akp6* that is expressed at high levels throughout the gut (global IAP, or gIAP) [16]. IAP knockout mice (*Akp3*$^{-/-}$ mice) display an accelerated transport of fatty acids both upon forced oil feeding as well as with long-term (36 weeks) feeding with an 11 % fat mouse chow (Fig. 6) [149].

In follow-up experiments using a 30 % HFD [150], the authors showed that oral administration of the IAP inhibitor L-Phe markedly inhibited the postprandial TAG increase in male C57Bl/6 mice indicating that the decreased postprandial triglyceridemia was associated with the inhibition of IAP activity. A similar effect was also seen when the mice were injected with the chylomicron inhibitor Pluronic L81, which is known to inhibit lipid absorption [151] and to suppress the postprandial increase in serum triacylglycerol (TAG) [152]. After 10 weeks of feeding on a 30 % fat chow, hepatic steatosis was observed in *Akp3*$^{-/-}$ mice. These livers had significantly higher mean TAG and cholesterol contents than those of the littermate controls. Electron microscopic examination clearly shows intrahepatocyte lipid inclusions.

Interestingly, *Akp3*$^{-/-}$ mice under normal feeding conditions displayed increased expression of gIAP in the ileum. This increased expression is greatly exacerbated upon feeding *Akp3*$^{-/-}$ mice an HFD [16]. Furthermore, this increased jejunal-ileal expression of gIAP was accompanied by a concomitant elevation of FAT/CD36,

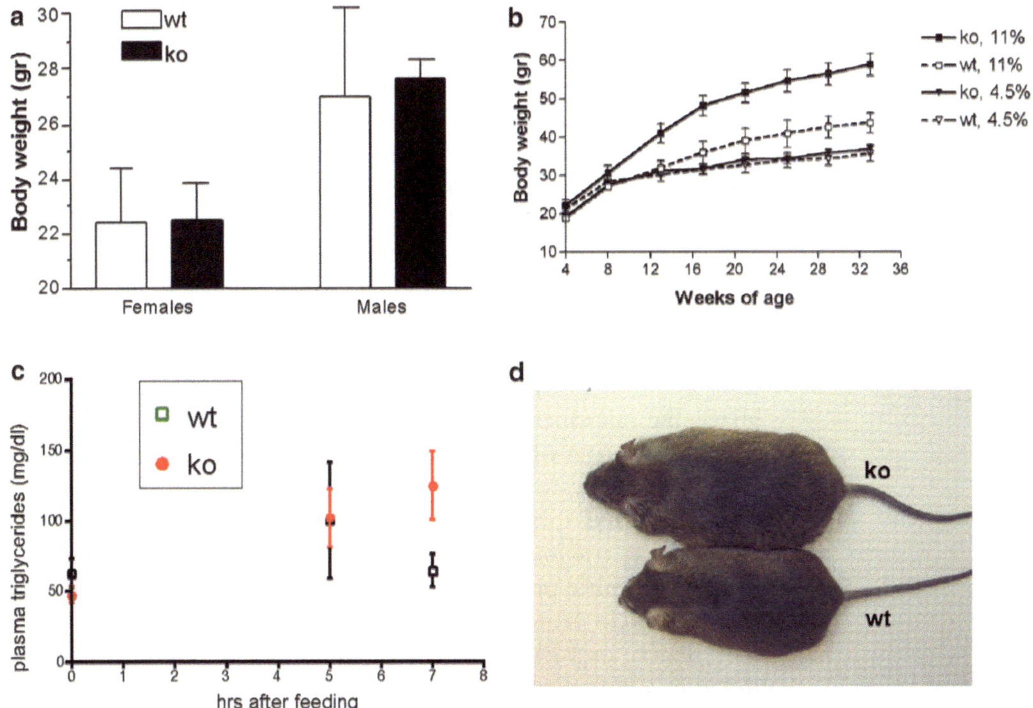

Fig. 6 Plasma triglycerides (TAG) and body weight of control and *Akp3*$^{-/-}$ (knockout; ko) mice under different feeding conditions [from ref. 149]. (**a**) Triglyceride levels in plasma from the mice subjected to forced oil feeding were measured at 0, 5, and 7 h following gavage. (**b**) Body weight of *Akp3*$^{-/-}$ and control wt mice at 8 weeks of age fed a regular diet containing 4.5 % fat. Average values were obtained from 7–11 mice in each group. (**c**) Body weight of control and *Akp3*$^{-/-}$ male mice under regular-fat diet (4.5 %) or high-fat diet (11 %) feeding regime initiated at 4 weeks of age. (**d**) Difference in body size between *Akp3*$^{-/-}$ and wt male mice at 30 weeks of age fed long term with a high-fat diet

a phosphorylated fatty acid translocase thought to play a role in facilitating the transport of long-chain fatty acids into cells. In addition, gIAP, but not dIAP, is able to dephosphorylate FAT/CD36 [153]. These experiments suggest that the upregulated jejunum-ileal expression of gIAP, via its ability to dephosphorylate and modulate FAT/CD36 function, is part of the mechanism of increased fatty acid absorption in *Akp3*$^{-/-}$ mice.

3.2 Role of Intestinal Alkaline Phosphatase in Gut Barrier Function

Several studies have shown that the IAP is capable of detoxifying bacterial LPS, likely through dephosphorylation of the lipid A moiety, the primary source of its endotoxic effects. Despite the fact that *Akp3* expression is restricted to the duodenum in mice, dIAP is shed into the gut lumen and is active in LPS dephosphorylation throughout the gut lumen and in the feces [114]. Furthermore, dIAP prevents bacterial invasion across the gut mucosal barrier using the superior mesenteric artery ligation followed by reperfusion [114]. Since, in the clinical setting, gut barrier dysfunction is also known to occur in response to a variety of remote systemic insults, the effects of IAP on gut mucosal integrity in the context of remote injury/trauma were explored, using a model of bilateral hind-limb ischemia (2 h) followed by 24 h of reperfusion. Histologic damage in hind-limb muscle and lung tissues were identical in WT and *Akp3*$^{-/-}$ mice, but the degree of gut mucosal injury was much more pronounced in the *Akp3*$^{-/-}$ mice and a greater degree of bacterial translocation to the mesenteric nodes was observed in the *Akp3*$^{-/-}$ animals subjected to the remote leg I/R injury. These data indicate that IAP is a component of the gut mucosal defense system. In addition, dIAP is involved in the maintenance of normal gut microbial homeostasis [154]. *Akp3*$^{-/-}$ mice had dramatically fewer and also different types of aerobic and anaerobic microbes in their stools compared with WT mice. Oral supplementation of IAP favored the growth of commensal bacteria, enhanced restoration of gut microbiota lost due to antibiotic treatment, and inhibited the growth of a pathogenic bacterium (*Salmonella typhimurium*). Thus, oral administration of IAP may have therapeutic potential against dysbiosis and pathogenic infections and help protect gut barrier function.

Indeed, oral IAP administration protected *Akp3*$^{-/-}$ mice from dextran sodium sulfate (DSS)-induced colitis [155]. WT and *Akp3*$^{-/-}$ mice received four cycles of 2 % DSS ad libitum for 7 days. Each cycle was followed by a 7-day DSS-free interval during which mice received either recombinant calf IAP (cIAP) or vehicle in the drinking water. Microscopic colitis scores of DSS-treated *Akp3*$^{-/-}$ mice were higher than DSS-treated WT mice. Oral cIAP treatment attenuated the disease in both groups. These data indicate that endogenous dIAP appears to play a role in protecting the host against chronic colitis and that orally administered cIAP exerts a protective effect in an experimental model of chronic colitis.

4 Concluding Remarks

As highlighted in the present review chapter, alkaline phosphatases play a large variety of functions in several very distinct tissues. For instance, an increasing amount of data today indicates that TNAP is expressed in many cell types and displays more functions than previously appreciated. TNAP is expressed during development and growth, and plays a fundamental role during skeletal and dental mineralization and in the development of the neural tube. During adulthood, TNAP may in some circumstances play protective roles. For instance, TNAP may be induced by LPS particularly in liver where it seems to participate in LPS detoxification. Conversely, ectopic TNAP induction causes medial vascular calcification adversely influencing other common conditions such as type II diabetes, obesity, and ageing in general and it is possible that TNAP expression in the aging brain may be associated with neuronal death and the development of Alzheimer's disease. At a moment when therapeutic strategies based on the replacement, alternatively inhibition, of TNAP are being explored a better understanding of TNAP functions appears crucial.

IAP also appears to play several critical functions. IAP plays an important role in fatty acid absorption. IAP is able to dephosphorylate the phosphorylated fatty acid translocase FAT/CD36, which is thought to play a role in facilitating the transport of long-chain fatty acids into cells. In addition, like TNAP, IAP is able to dephosphorylate LPS. The localization of the active site of IAP into the gut lumen allows IAP to protect gut barrier function and determine the composition of the gut microbiota. These functions of IAP, like those of TNAP, may have very important therapeutic applications.

References

1. Millan JL (2006) Mammalian alkaline phosphatases: from biology to applications in medicine and biotechnology. Wiley-VCH, Weinheim
2. Weiss MJ, Ray K, Henthorn PS et al (1988) Structure of the human liver/bone/kidney alkaline phosphatase gene. J Biol Chem 263:12002–12010
3. Kishi F, Matsuura S, Kajii T (1989) Nucleotide sequence of the human liver-type alkaline phosphatase cDNA. Nucleic Acids Res 17:2129
4. Matsuura S, Kishi F, Kajii T (1990) Characterization of a 5′-flanking region of the human liver/bone/kidney alkaline phosphatase gene: two kinds of mRNA from a single gene. Biochem Biophys Res Commun 168:993–1000
5. Studer M, Terao M, Gianni M et al (1991) Characterization of a second promoter for the mouse liver/bone/kidney-type alkaline phosphatase gene: cell and tissue specific expression. Biochem Biophys Res Commun 179:1352–1360
6. Martin D, Tucker DF, Gorman P et al (1987) The human placental alkaline phosphatase gene and related sequences map to chromosome 2 band q37. Ann Hum Genet 51: 145–152
7. Griffin CA, Smith M, Henthorn PS et al (1987) Human placental and intestinal alkaline phosphatase genes map to 2q34-q37. Am J Hum Genet 41:1025–1034
8. Henthorn PS, Raducha M, Edwards YH et al (1987) Nucleotide and amino acid sequences of human intestinal alkaline phosphatase: close homology to placental alkaline phosphatase. Proc Natl Acad Sci U S A 84:1234–1238
9. Knoll BJ, Rothblum KN, Longley M (1988) Nucleotide sequence of the human placental alkaline phosphatase gene. Evolution of the 5′ flanking region by deletion/substitution. J Biol Chem 263:12020–12027
10. Millán JL, Manes T (1988) Seminoma-derived Nagao isozyme is encoded by a germ-cell alkaline phosphatase gene. Proc Natl Acad Sci U S A 85:3024–3028
11. Ovitt CE, Strauss AW, Alpers DH et al (1986) Expression of different-sized placental alkaline phosphatase mRNAs in placenta and choriocarcinoma cells. Proc Natl Acad Sci U S A 83:3781–3785
12. Gum JR, Kam WK, Byrd JC et al (1987) Effects of sodium butyrate on human colonic adenocarcinoma cells. Induction of placental-like alkaline phosphatase. J Biol Chem 262:1092–1097
13. Chou JY, Takahashi S (1987) Control of placental alkaline phosphatase gene expression in HeLa cells: induction of synthesis by prednisolone and sodium butyrate. Biochemistry 26:3596–3602
14. Terao M, Pravtcheva D, Ruddle FH et al (1988) Mapping of gene encoding mouse placental alkaline phosphatase to chromosome 4. Somat Cell Mol Genet 14:211–215
15. Manes T, Glade K, Ziomek CA et al (1990) Genomic structure and comparison of mouse tissue-specific alkaline phosphatase genes. Genomics 8:541–554
16. Narisawa S, Hoylaerts MF, Doctor KS et al (2007) A novel phosphatase upregulated in Akp3 knockout mice. Am J Physiol Gastrointest Liver Physiol 293:G1068–G1077
17. Merchant-Larios H, Mendlovic F, Alvarez-Buylla A (1985) Characterization of alkaline phosphatase from primordial germ cells and ontogenesis of this enzyme in the mouse. Differentiation 29:145–151

18. MacGregor GR, Zambrowicz BP, Soriano P (1995) Tissue non-specific alkaline phosphatase is expressed in both embryonic and extraembryonic lineages during mouse embryogenesis but is not required for migration of primordial germ cells. Development 121:1487–1496
19. Narisawa S, Hasegawa H, Watanabe K et al (1994) Stage-specific expression of alkaline phosphatase during neural development in the mouse. Dev Dyn 201:227–235
20. Hoshi K, Amizuka N, Oda K et al (1997) Immunolocalization of tissue non-specific alkaline phosphatase in mice. Histochem Cell Biol 107:183–191
21. Fonta C, Négyessy L, Renaud L et al (2004) Areal and subcellular localization of the ubiquitous alkaline phosphatase in the primate cerebral cortex: evidence for a role in neurotransmission. Cereb Cortex 14:595–609
22. Langer D, Hammer K, Koszalka P et al (2008) Distribution of ectonucleotidases in the rodent brain revisited. Cell Tissue Res 334:199–217
23. Brun-Heath I, Ermonval M, Chabrol E et al (2011) Differential expression of the bone and the liver tissue non-specific alkaline phosphatase isoforms in brain tissues. Cell Tissue Res 343:521–536
24. Négyessy L, Xiao J, Kántor O et al (2011) Layer-specific activity of tissue non-specific alkaline phosphatase in the human neocortex. Neuroscience 172:406–418
25. Hahnel AC, Rappolee DA, Millan JL et al (1990) Two alkaline phosphatase genes are expressed during early development in the mouse embryo. Development 110:555–564
26. Hustin J, Collette J, Franchimont P (1987) Immunohistochemical demonstration of placental alkaline phosphatase in various states of testicular development and in germ cell tumours. Int J Androl 10:29–35
27. Chang CH, Angellis D, Fishman WH (1980) Presence of the rare D-variant heat-stable, placental-type alkaline phosphatase in normal human testis. Cancer Res 40:1506–1510
28. Goldstein DJ, Rogers C, Harris H (1982) A search for trace expression of placental-like alkaline phosphatase in non-malignant human tissues: demonstration of its occurrence in lung, cervix, testis and thymus. Clin Chim Acta 125:63–75
29. Fishman L, Miyayama H, Driscoll SG et al (1976) Developmental phase-specific alkaline phosphatase isoenzymes of human placenta and their occurrence in human cancer. Cancer Res 36:2268–2273
30. Jemmerson R, Klier FG, Fishman WH (1985) Clustered distribution of human placental alkaline phosphatase on the surface of both placental and cancer cells. Electron microscopic observations using gold-labeled antibodies. J Histochem Cytochem 33:1227–1234
31. Kozlenkov A, Manes T, Hoylaerts MF et al (2002) Function assignment to conserved residues in mammalian alkaline phosphatases. J Biol Chem 277:22992–22999
32. Kozlenkov A, Le Du MH, Cuniasse P et al (2004) Residues determining the binding specificity of uncompetitive inhibitors to tissue-nonspecific alkaline phosphatase. J Bone Miner Res 19:1862–1872
33. Le Du MH, Millan JL (2002) Structural evidence of functional divergence in human alkaline phosphatases. J Biol Chem 277:49808–49814
34. Hoylaerts MF, Manes T, Millán JL (1997) Mammalian alkaline phosphatases are allosteric enzymes. J Biol Chem 272:22781–22787
35. Hoylaerts MF, Ding L, Narisawa S et al (2006) Mammalian alkaline phosphatase catalysis requires active site structure stabilization via the N-terminal amino acid microenvironment. Biochemistry 45:9756–9766
36. Bossi M, Hoylaerts MF, Millán JL (1993) Modifications in a flexible surface loop modulate the isozyme-specific properties of mammalian alkaline phosphatases. J Biol Chem 268:25409–25416
37. Halling LC, Narisawa S, Millán JL et al (2009) Glycosylation differences contribute to distinct catalytic properties among bone alkaline phosphatase isoforms. Bone 45:987–993
38. Zhang L, Balcerzak M, Radisson J et al (2005) Phosphodiesterase activity of alkaline phosphatase in ATP-initiated Ca^{2+} and phosphate deposition in isolated chicken matrix vesicles. J Biol Chem 280:37289–37296
39. Morris DC, Masuhara K, Takaoka K et al (1992) Immunolocalization of alkaline phosphatase in osteoblasts and matrix vesicles of human fetal bone. Bone Miner 19:287–298
40. Hawrylak K, Stinson RA (1988) The solubilization of tetrameric alkaline phosphatase from human liver and its conversion into various forms by phosphatidylinositol phospholipase C or proteolysis. J Biol Chem 263:14368–14373
41. Anh DJ, Eden A, Farley JR (2001) Quantitation of soluble and skeletal alkaline phosphatase, and insoluble alkaline phosphatase anchor-hydrolase activities in human serum. Clin Chim Acta 311:137–148
42. Low MG, Prasad AR (1988) A phospholipase D specific for the phosphatidylinositol anchor of cell-surface proteins is abundant in plasma. Proc Natl Acad Sci U S A 85:980–984

43. Davitz MA, Hereld D, Shak S et al (1987) A glycan-phosphatidylinositol-specific phospholipase D in human serum. Science 238:81–84
44. Magnusson P, Degerblad M, Sääf M et al (1997) Different responses of bone alkaline phosphatase isoforms during recombinant insulin-like growth factor-I (IGF-I) and during growth hormone therapy in adults with growth hormone deficiency. J Bone Miner Res 12:210–220
45. Smith GP, Peters TJ (1981) Subcellular localization and properties of pyridoxal phosphate phosphatases of human polymorphonuclear leukocytes and their relationship to acid and alkaline phosphatase. Biochim Biophys Acta 661:287–294
46. Wilson PD, Smith GP, Peters TJ (1983) Pyridoxal 5′-phosphate: a possible physiological substrate for alkaline phosphatase in human neutrophils. Histochem J 15:257–264
47. Low MG, Saltiel AR (1988) Structural and functional roles of glycosyl-phosphatidylinositol in membranes. Science 239:268–275
48. Waymire KG, Mahuren JD, Jaje JM et al (1995) Mice lacking tissue non-specific alkaline phosphatase die from seizures due to defective metabolism of vitamin B-6. Nat Genet 11:45–51
49. Fleshood HL, Pitot HC (1970) The metabolism of O-phosphorylethanolamine in animal tissues. I. O-phosphorylethanolamine phospho-lyase: partial purification and characterization. J Biol Chem 245:4414–4420
50. Lencel P, Delplace S, Pilet P et al (2011) Cell-specific effects of TNF-α and IL-1β on alkaline phosphatase: implication for syndesmophyte formation and vascular calcification. Lab Invest 91:1434–1442
51. Krause U, Harris S, Green A et al (2010) Pharmaceutical modulation of canonical Wnt signaling in multipotent stromal cells for improved osteoinductive therapy. Proc Natl Acad Sci U S A 107:4147–4152
52. Briolay A, Lencel P, Bessueille L et al (2013) Autocrine stimulation of osteoblast activity by Wnt5a in response to TNF-α in human mesenchymal stem cells. Biochem Biophys Res Commun 430:1072–1077
53. Delgado-Calle J, Sañudo C, Sánchez-Verde L et al (2011) Epigenetic regulation of alkaline phosphatase in human cells of the osteoblastic lineage. Bone 49:830–838
54. Bernard GW (1978) Ultrastructural localization of alkaline phosphatase in initial intramembranous osteogenesis. Clin Orthop Relat Res 218–225
55. Togari A, Arakawa S, Arai M et al (1993) Inhibition of in vitro mineralization in osteoblastic cells and in mouse tooth germ by phosphatidylinositol-specific phospholipase C. Biochem Pharmacol 46:1668–1670
56. Hsu HH, Morris DC, Davis L et al (1993) In vitro Ca deposition by rat matrix vesicles: is the membrane association of alkaline phosphatase essential for matrix vesicle-mediated calcium deposition? Int J Biochem 25:1737–1742
57. Sugawara Y, Suzuki K, Koshikawa M et al (2002) Necessity of enzymatic activity of alkaline phosphatase for mineralization of osteoblastic cells. Jpn J Pharmacol 88:262–269
58. Whyte MP, Valdes R, Ryan LM et al (1982) Infantile hypophosphatasia: enzyme replacement therapy by intravenous infusion of alkaline phosphatase-rich plasma from patients with Paget bone disease. J Pediatr 101:379–386
59. Whyte MP, McAlister WH, Patton LS et al (1984) Enzyme replacement therapy for infantile hypophosphatasia attempted by intravenous infusions of alkaline phosphatase-rich Paget plasma: results in three additional patients. J Pediatr 105:926–933
60. Weninger M, Stinson RA, Plenk H et al (1989) Biochemical and morphological effects of human hepatic alkaline phosphatase in a neonate with hypophosphatasia. Acta Paediatr Scand Suppl 360:154–160
61. Millán JL, Narisawa S, Lemire I et al (2008) Enzyme replacement therapy for murine hypophosphatasia. J Bone Miner Res 23:777–787
62. Whyte MP, Greenberg CR, Salman NJ et al (2012) Enzyme-replacement therapy in life-threatening hypophosphatasia. N Engl J Med 366:904–913
63. Krawitz PM, Schweiger MR, Rödelsperger C et al (2010) Identity-by-descent filtering of exome sequence data identifies PIGV mutations in hyperphosphatasia mental retardation syndrome. Nat Genet 42:827–829
64. Weiss MJ, Cole DE, Ray K et al (1988) A missense mutation in the human liver/bone/kidney alkaline phosphatase gene causing a lethal form of hypophosphatasia. Proc Natl Acad Sci U S A 85:7666–7669
65. Whyte MP (2010) Physiological role of alkaline phosphatase explored in hypophosphatasia. Ann N Y Acad Sci 1192:190–200
66. Fallon MD, Teitelbaum SL, Weinstein RS et al (1984) Hypophosphatasia: clinicopathologic comparison of the infantile, childhood, and adult forms. Medicine (Baltimore) 63:12–24
67. Anderson HC, Hsu HH, Morris DC et al (1997) Matrix vesicles in osteomalacic hypophosphatasia bone contain apatite-like mineral crystals. Am J Pathol 151:1555–1561
68. Yadav MC, Simão AM, Narisawa S et al (2011) Loss of skeletal mineralization by the

simultaneous ablation of PHOSPHO1 and alkaline phosphatase function: a unified model of the mechanisms of initiation of skeletal calcification. J Bone Miner Res 26:286–297
69. Millán JL (2012) The role of phosphatases in the initiation of skeletal mineralization. Calcif Tissue Int (10.1007/s00223-012-9672-8)
70. Fedde KN, Blair L, Silverstein J et al (1999) Alkaline phosphatase knock-out mice recapitulate the metabolic and skeletal defects of infantile hypophosphatasia. J Bone Miner Res 14:2015–2026
71. Zalatan JG, Fenn TD, Brunger AT et al (2006) Structural and functional comparisons of nucleotide pyrophosphatase/phosphodiesterase and alkaline phosphatase: implications for mechanism and evolution. Biochemistry 45:9788–9803
72. Ciancaglini P, Yadav MC, Simão AM et al (2010) Kinetic analysis of substrate utilization by native and TNAP-, NPP1-, or PHOSPHO1-deficient matrix vesicles. J Bone Miner Res 25:716–723
73. Murshed M, Harmey D, Millán JL et al (2005) Unique coexpression in osteoblasts of broadly expressed genes accounts for the spatial restriction of ECM mineralization to bone. Genes Dev 19:1093–1104
74. Fernley HN, Walker PG (1967) Studies on alkaline phosphatase. Inhibition by phosphate derivatives and the substrate specificity. Biochem J 104:1011–1018
75. Hessle L, Johnson KA, Anderson HC et al (2002) Tissue-nonspecific alkaline phosphatase and plasma cell membrane glycoprotein-1 are central antagonistic regulators of bone mineralization. Proc Natl Acad Sci U S A 99:9445–9449
76. Harmey D, Hessle L, Narisawa S et al (2004) Concerted regulation of inorganic pyrophosphate and osteopontin by akp2, enpp1, and ank: an integrated model of the pathogenesis of mineralization disorders. Am J Pathol 164:1199–1209
77. Thouverey C, Bechkoff G, Pikula S et al (2009) Inorganic pyrophosphate as a regulator of hydroxyapatite or calcium pyrophosphate dihydrate mineral deposition by matrix vesicles. Osteoarthritis Cartilage 17:64–72
78. Foster B, Nagatomo K, Tso H et al (2013) Tooth root dentin mineralization defects in a mouse model of hypophosphatasia. J Bone Miner Res 28:271–282
79. Foster BL, Nagatomo KJ, Nociti FH et al (2012) Central role of pyrophosphate in acellular cementum formation. PLoS One 7:e38393
80. McKee MD, Nakano Y, Masica DL et al (2011) Enzyme replacement therapy prevents dental defects in a model of hypophosphatasia. J Dent Res 90:470–476
81. Yadav MC, de Oliveira RC, Foster BL et al (2012) Enzyme replacement prevents enamel defects in hypophosphatasia mice. J Bone Miner Res 27:1722–1734
82. Tesch W, Vandenbos T, Roschgr P et al (2003) Orientation of mineral crystallites and mineral density during skeletal development in mice deficient in tissue nonspecific alkaline phosphatase. J Bone Miner Res 18:117–125
83. Thompson MD, Killoran A, Percy ME et al (2006) Hyperphosphatasia with neurologic deficit: a pyridoxine-responsive seizure disorder? Pediatr Neurol 34:303–307
84. Gospe SM (2006) Pyridoxine-dependent seizures: new genetic and biochemical clues to help with diagnosis and treatment. Curr Opin Neurol 19:148–153
85. Narisawa S, Fröhlander N, Millán JL (1997) Inactivation of two mouse alkaline phosphatase genes and establishment of a model of infantile hypophosphatasia. Dev Dyn 208:432–446
86. Narisawa S, Wennberg C, Millán JL (2001) Abnormal vitamin B6 metabolism in alkaline phosphatase knock-out mice causes multiple abnormalities, but not the impaired bone mineralization. J Pathol 193:125–133
87. Baumgartner-Sigl S, Haberlandt E, Mumm S et al (2007) Pyridoxine-responsive seizures as the first symptom of infantile hypophosphatasia caused by two novel missense mutations (c.677T>C, p.M226T; c.1112C>T, p.T371I) of the tissue-nonspecific alkaline phosphatase gene. Bone 40:1655–1661
88. Mornet E (2007) Hypophosphatasia. Orphanet J Rare Dis 2:40
89. Hanics J, Barna J, Xiao J et al (2012) Ablation of TNAP function compromises myelination and synaptogenesis in the mouse brain. Cell Tissue Res 349(2):459–471
90. Kermer V, Ritter M, Albuquerque B et al (2010) Knockdown of tissue nonspecific alkaline phosphatase impairs neural stem cell proliferation and differentiation. Neurosci Lett 485:208–211
91. Díez-Zaera M, Díaz-Hernández JI, Hernández-Álvarez E et al (2011) Tissue-nonspecific alkaline phosphatase promotes axonal growth of hippocampal neurons. Mol Biol Cell 22:1014–1024
92. Ermonval M, Baudry A, Baychelier F et al (2009) The cellular prion protein interacts with the tissue non-specific alkaline phosphatase in membrane microdomains of bioaminergic neuronal cells. PLoS One 4:e6497
93. Hatoff DE, Hardison WG (1981) Bile acids modify alkaline phosphatase induction and

bile secretion pressure after bile duct obstruction in the rat. Gastroenterology 80: 666–672
94. Hatoff DE, Hardison WG (1982) Bile acid-dependent secretion of alkaline phosphatase in rat bile. Hepatology 2:433–439
95. Chida K, Taguchi M (2005) Localization of alkaline phosphatase and proteins related to intercellular junctions in primary cultures of fetal rat hepatocytes. Anat Embryol (Berl) 210:75–80
96. Chida K, Taguchi M (2011) Localization of alkaline phosphatase and cathepsin D during cell restoration after colchicine treatment in primary cultures of fetal rat hepatocytes. Acta Histochem Cytochem 44:155–158
97. Ogawa H, Mink J, Hardison WG et al (1990) Alkaline phosphatase activity in hepatic tissue and serum correlates with amount and type of bile acid load. Lab Invest 62:87–95
98. Halling Linder C, Englund UH, Narisawa S et al (2013) Isozyme profile and tissue-origin of alkaline phosphatases in mouse serum. Bone 53(2):399–408
99. Kaplan MM, Righetti A (1970) Induction of rat liver alkaline phosphatase: the mechanism of the serum elevation in bile duct obstruction. J Clin Invest 49:508–516
100. Poelstra K, Bakker WW, Klok PA et al (1997) A physiologic function for alkaline phosphatase: endotoxin detoxification. Lab Invest 76:319–327
101. Alvaro D, Benedetti A, Marucci L et al (2000) The function of alkaline phosphatase in the liver: regulation of intrahepatic biliary epithelium secretory activities in the rat. Hepatology 32:174–184
102. Becq F, Jensen TJ, Chang XB et al (1994) Phosphatase inhibitors activate normal and defective CFTR chloride channels. Proc Natl Acad Sci U S A 91:9160–9164
103. Becq F, Fanjul M, Merten M et al (1993) Possible regulation of CFTR-chloride channels by membrane-bound phosphatases in pancreatic duct cells. FEBS Lett 327:337–342
104. Hori Y, Takeyama Y, Ueda T et al (1998) Impaired transport of lipopolysaccharide across the hepatocytes in rats with cerulein-induced experimental pancreatitis. Pancreas 16:148–153
105. Kanistanon D, Powell DA, Hajjar AM et al (2012) Role of Francisella lipid A phosphate modification in virulence and long-term protective immune responses. Infect Immun 80:943–951
106. Rietschel ET, Kirikae T, Schade FU et al (1994) Bacterial endotoxin: molecular relationships of structure to activity and function. FASEB J 8:217–225
107. Bentala H, Verweij WR, Huizinga-Van der Vlag A et al (2002) Removal of phosphate from lipid A as a strategy to detoxify lipopolysaccharide. Shock 18:561–566
108. Tuin A, Huizinga-Van der Vlag A, van Loenen-Weemaes AM et al (2006) On the role and fate of LPS-dephosphorylating activity in the rat liver. Am J Physiol Gastrointest Liver Physiol 290:G377–G385
109. Poelstra K, Bakker WW, Klok PA et al (1997) Dephosphorylation of endotoxin by alkaline phosphatase in vivo. Am J Pathol 151:1163–1169
110. Nouwen EJ, De Broe ME (1994) Human intestinal versus tissue-nonspecific alkaline phosphatase as complementary urinary markers for the proximal tubule. Kidney Int Suppl 47:S43–S51
111. March JG, Simonet BM, Grases F (2001) Determination of pyrophosphate in renal calculi and urine by means of an enzymatic method. Clin Chim Acta 314:187–194
112. Baumann JM, Bisaz S, Felix R et al (1977) The role of inhibitors and other factors in the pathogenesis of recurrent calcium-containing renal stones. Clin Sci Mol Med 53:141–148
113. Laminski NA, Meyers AM, Sonnekus MI et al (1990) Prevalence of hypocitraturia and hypopyrophosphaturia in recurrent calcium stone formers: as isolated defects or associated with other metabolic abnormalities. Nephron 56:379–386
114. Goldberg RF, Austen WG, Zhang X et al (2008) Intestinal alkaline phosphatase is a gut mucosal defense factor maintained by enteral nutrition. Proc Natl Acad Sci U S A 105:3551–3556
115. Russell RG, Bisaz S, Fleisch H (1976) The influence of orthophosphate on the renal handling of inorganic pyrophosphate in man and dog. Clin Sci Mol Med 51:435–443
116. Rachow JW, Ryan LM (1988) Inorganic pyrophosphate metabolism in arthritis. Rheum Dis Clin North Am 14:289–302
117. Moochhala SH, Sayer JA, Carr G et al (2008) Renal calcium stones: insights from the control of bone mineralization. Exp Physiol 93:43–49
118. Ho AM, Johnson MD, Kingsley DM (2000) Role of the mouse ank gene in control of tissue calcification and arthritis. Science 289:265–270
119. Carr G, Sayer JA, Simmons NL (2007) Expression and localisation of the pyrophosphate transporter, ANK, in murine kidney cells. Cell Physiol Biochem 20:507–516
120. Kapojos JJ, Poelstra K, Borghuis T et al (2003) Induction of glomerular alkaline phosphatase after challenge with lipopolysaccharide. Int J Exp Pathol 84:135–144

121. Lencel P, Magne D (2011) Inflammaging: the driving force in osteoporosis? Med Hypotheses 76:317–321
122. Lencel P, Hardouin P, Magne D (2010) Do cytokines induce vascular calcification by the mere stimulation of TNAP activity? Med Hypotheses 75:517–521
123. Doherty TM, Fitzpatrick LA, Inoue D et al (2004) Molecular, endocrine, and genetic mechanisms of arterial calcification. Endocr Rev 25:629–672
124. Mintz GS, Popma JJ, Pichard AD et al (1995) Patterns of calcification in coronary artery disease. A statistical analysis of intravascular ultrasound and coronary angiography in 1155 lesions. Circulation 91:1959–1965
125. Vengrenyuk Y, Carlier S, Xanthos S et al (2006) A hypothesis for vulnerable plaque rupture due to stress-induced debonding around cellular microcalcifications in thin fibrous caps. Proc Natl Acad Sci U S A 103:14678–14683
126. Tyson KL, Reynolds JL, McNair R et al (2003) Osteo/chondrocytic transcription factors and their target genes exhibit distinct patterns of expression in human arterial calcification. Arterioscler Thromb Vasc Biol 23:489–494
127. Ishimura E, Okuno S, Kitatani K et al (2002) Different risk factors for peripheral vascular calcification between diabetic and non-diabetic haemodialysis patients–importance of glycaemic control. Diabetologia 45:1446–1448
128. Shanahan CM, Cary NR, Salisbury JR et al (1999) Medial localization of mineralization-regulating proteins in association with Mönckeberg's sclerosis: evidence for smooth muscle cell-mediated vascular calcification. Circulation 100:2168–2176
129. Lomashvili KA, Garg P, Narisawa S et al (2008) Upregulation of alkaline phosphatase and pyrophosphate hydrolysis: potential mechanism for uremic vascular calcification. Kidney Int 73:1024–1030
130. Narisawa S, Harmey D, Yadav MC et al (2007) Novel inhibitors of alkaline phosphatase suppress vascular smooth muscle cell calcification. J Bone Miner Res 22:1700–1710
131. Markello TC, Pak LK, St HC et al (2011) Vascular pathology of medial arterial calcifications in NT5E deficiency: implications for the role of adenosine in pseudoxanthoma elasticum. Mol Genet Metab 103:44–50
132. Munroe PB, Olgunturk RO, Fryns JP et al (1999) Mutations in the gene encoding the human matrix Gla protein cause Keutel syndrome. Nat Genet 21:142–144
133. Lomashvili KA, Wang X, Wallin R et al (2011) Matrix Gla protein metabolism in vascular smooth muscle and role in uremic vascular calcification. J Biol Chem 286:28715–28722
134. Leroux-Berger M, Queguiner I, Maciel TT et al (2011) Pathologic calcification of adult vascular smooth muscle cells differs on their crest or mesodermal embryonic origin. J Bone Miner Res 26:1543–1553
135. Li L, Chang L, Pellet-Rostaing S et al (2009) Synthesis and evaluation of benzo[b]thiophene derivatives as inhibitors of alkaline phosphatases. Bioorg Med Chem 17:7290–7300
136. Dahl R, Sergienko EA, Su Y et al (2009) Discovery and validation of a series of aryl sulfonamides as selective inhibitors of tissue-nonspecific alkaline phosphatase (TNAP). J Med Chem 52:6919–6925
137. Li JJ, Zhu CG, Yu B et al (2007) The role of inflammation in coronary artery calcification. Ageing Res Rev 6:263–270
138. Stompór T, Pasowicz M, Sułlowicz W et al (2003) An association between coronary artery calcification score, lipid profile, and selected markers of chronic inflammation in ESRD patients treated with peritoneal dialysis. Am J Kidney Dis 41:203–211
139. Aikawa E, Nahrendorf M, Figueiredo JL et al (2007) Osteogenesis associates with inflammation in early-stage atherosclerosis evaluated by molecular imaging in vivo. Circulation 116:2841–2850
140. Al-Aly Z, Shao JS, Lai CF et al (2007) Aortic Msx2-Wnt calcification cascade is regulated by TNF-alpha-dependent signals in diabetic Ldlr-/- mice. Arterioscler Thromb Vasc Biol 27:2589–2596
141. Tintut Y, Patel J, Parhami F et al (2000) Tumor necrosis factor-alpha promotes in vitro calcification of vascular cells via the cAMP pathway. Circulation 102:2636–2642
142. Tintut Y, Patel J, Territo M et al (2002) Monocyte/macrophage regulation of vascular calcification in vitro. Circulation 105:650–655
143. Doherty TM, Asotra K, Fitzpatrick LA et al (2003) Calcification in atherosclerosis: bone biology and chronic inflammation at the arterial crossroads. Proc Natl Acad Sci U S A 100:11201–11206
144. Iqbal K, Wang X, Blanchard J et al (2010) Alzheimer's disease neurofibrillary degeneration: pivotal and multifactorial. Biochem Soc Trans 38:962–966
145. Díaz-Hernández M, Gómez-Ramos A, Rubio A et al (2010) Tissue-nonspecific alkaline phosphatase promotes the neurotoxicity effect of extracellular tau. J Biol Chem 285:32539–32548
146. Vardy ER, Kellett KA, Cocklin SL et al (2012) Alkaline phosphatase is increased in both

brain and plasma in Alzheimer's disease. Neurodegener Dis 9:31–37
147. Glickman RM, Alpers DH, Drummey GD et al (1970) Increased lymph alkaline phosphatase after fat feeding: effects of medium chain triglycerides and inhibition of protein synthesis. Biochim Biophys Acta 201:226–235
148. Hodin RA, Graham JR, Meng S et al (1994) Temporal pattern of rat small intestinal gene expression with refeeding. Am J Physiol 266:G83–G89
149. Narisawa S, Huang L, Iwasaki A et al (2003) Accelerated fat absorption in intestinal alkaline phosphatase knockout mice. Mol Cell Biol 23:7525–7530
150. Nakano T, Inoue I, Koyama I et al (2007) Disruption of the murine intestinal alkaline phosphatase gene Akp3 impairs lipid transcytosis and induces visceral fat accumulation and hepatic steatosis. Am J Physiol Gastrointest Liver Physiol 292:G1439–G1449
151. Pool C, Nutting DF, Simmonds WJ et al (1991) Effect of Pluronic L81, a hydrophobic surfactant, on intestinal mucosal cholesterol homeostasis. Am J Physiol 261:G256–G262
152. Mahmood A, Yamagishi F, Eliakim R et al (1994) A possible role for rat intestinal surfactant-like particles in transepithelial triacylglycerol transport. J Clin Invest 93:70–80
153. Lynes M, Narisawa S, Millán JL et al (2011) Interactions between CD36 and global intestinal alkaline phosphatase in mouse small intestine and effects of high-fat diet. Am J Physiol Regul Integr Comp Physiol 301:R1738–R1747
154. Malo MS, Alam SN, Mostafa G et al (2010) Intestinal alkaline phosphatase preserves the normal homeostasis of gut microbiota. Gut 59:1476–1484
155. Ramasamy S, Nguyen DD, Eston MA et al (2011) Intestinal alkaline phosphatase has beneficial effects in mouse models of chronic colitis. Inflamm Bowel Dis 17:532–542

Chapter 4

Robotic Implementation of Assays: Tissue-Nonspecific Alkaline Phosphatase (TNAP) Case Study

Thomas D.Y. Chung

Abstract

Laboratory automation and robotics have "industrialized" the execution and completion of large-scale, enabling high-capacity and high-throughput (100 K–1 MM/day) screening (HTS) campaigns of large "libraries" of compounds (>200 K–2 MM) to complete in a few days or weeks. Critical to the success these HTS campaigns is the ability of a competent assay development team to convert a validated research-grade laboratory "benchtop" assay suitable for manual or semi-automated operations on a few hundreds of compounds into a robust miniaturized (384- or 1,536-well format), well-engineered, scalable, industrialized assay that can be seamlessly implemented on a fully automated, fully integrated robotic screening platform for cost-effective screening of hundreds of thousands of compounds. Here, we provide a review of the theoretical guiding principles and practical considerations necessary to reduce often complex research biology into a "lean manufacturing" engineering endeavor comprising adaption, automation, and implementation of HTS. Furthermore we provide a detailed example specifically for a cell-free in vitro biochemical, enzymatic phosphatase assay for tissue-nonspecific alkaline phosphatase that illustrates these principles and considerations.

Key words High-throughput screens, HTS, Chemical libraries, Laboratory automation, Automated assays, Robotics, Phosphatase assays, Assay development, Implementation, Enzymes, Alkaline phosphatases, Tissue-nonspecific alkaline phosphatase (TNAP)

1 Introduction

Historical perspective. High-throughput screening (HTS) is a relatively young science [1] whose rapid evolution of associated multidisciplinary approach, technologies, best practices, and suppliers in just the past decade have been reviewed [2–6]. The initial emphasis was on the excitement of and investments in the underlying technologies in assay miniaturization [7–9], automation, data capture, data analysis, new detection technologies, and bioassay formats thus enabled [1, 8, 10–15]. This was followed by appreciation of the organizational [16] and infrastructural best practices [2, 17–21] as developed in pharmaceutical companies. And now has matured to

José Luis Millán (ed.), *Phosphatase Modulators*, Methods in Molecular Biology, vol. 1053,
DOI 10.1007/978-1-62703-562-0_4, © Springer Science+Business Media, LLC 2013

where its impact and value to the overall success of drug discovery have been scrutinized in light of diminishing NCE pipeline productivity [1, 22–25]. Finally, with the transformation of the NIH Molecular Libraries Program (http://mli.nih.gov/mli/mlpcn/) into the National Center for Advancing Translational Sciences (http://www.ncats.nih.gov/) and the growing embrace of open innovation strategies [5] by pharmaceutical companies, this well-trod approach of HTS has reemerged as a key tool that integrates novel assay formats, assay miniaturization, and automation to enable cost-effective and successful chemical genomics and translational programs through the identification of chemical probes and/or starting point for drug discovery in the academic sector [5, 26–29] and more recently with the Internet publication of the NIH-sponsored "Assay Guidance Manual (AGM) [Internet]" [30], which will be a "living" resource for the principles and practical aspects of screening with a multitude of case examples of specific assays that are relevant to compound screening-centered approaches. The AGM also has a comprehensive glossary of quantitative biology terms common to HTS and drug discovery.

Key HTS success factors. Pharmaceutical companies' best practices for successful HTS operations and campaigns, comprising proper target (e.g., disease relevance, chemical tractability) and assay format selection, assay development and validation, proper automation, chemical library selection, compound management, data management, cheminformatics [4, 28], and leverage of commercial vendors of HTS reagents, instrumentation, and services [29], have been recapitulated in academic screening centers [27, 30]. However, all HTS campaigns have as their genesis conversion of a "benchtop" assay suitable for basic research on a few tens of compounds into a robust, scalable, and automated assays implemented on robotic screening platforms with the throughput (10–100 K wells/day) and capacity (100 K–2 MM) to screen large chemical collections. Despite the challenge of an incredible diversity of assay formats (biochemical, cell-based, or high-content/phenotypic), this chapter attempts to summarize the overall process and provide guiding principles that can be applied to the conversion of most "benchtop" assays into the most appropriate assay formats and types suitable for automation; the practical considerations of scalability (especially reagent procurement, cost, and time) that are intimately connected to the relative merits and limitations of any assay format, miniaturization, signal generation and detection technologies, available instrumentation, and particular automation and robotic platforms chosen; and the quality and performance metrics that validate an HTS assay's performance during development, and that need to be monitored when implemented under the actual automated robotic screening protocol and process for the duration of a full HTS campaign against a large chemical collection. Finally, these higher level and general principles and guidances will be illustrated with the specific example of a homogenous, luminescent, 384-well format, biochemical enzyme assay for tissue-nonspecific alkaline phosphatase (TNAP, EC 3.1.3.1).

2 Materials

Prepare all solutions using ultrapure water (reverse-osmosis MilliQ™ deionized water to a resistance of at least 18 MΩ cm at 25 °C), the same lot of ultrapure analytical grade chemical reagents for buffers and solvents for stock solutions of reference compounds, and the same lot (insofar as possible) of biological and tissue culture media for the entire assay development and HTS campaigns. In cases where there are insufficient lots of commercially available reagents, chemicals, and biological for full HTS and follow-up, demonstration of equivalent assay performance with the lots in question removes this concern. Alternatively lots may be mixed to ensure a consistent lot for the entire HTS campaign and follow-ups. Prepare and store chemical reagents at room temperature (unless indicated otherwise). Biological reagents should be kept at −80 °C for long-term storage, −20 °C for storage of concentrated working stocks for intermediate storage, and refrigerated (~5 °C) for daily working solutions.

2.1 Automated System(s) and Functional Components

1. The particular automation systems whether independent workstations, partially integrated systems, or fully integrated system with a robotic arm are too numerous to specify, and the merits of full robotic integration [31] versus semi-automated workstations with stackers [32] have been debated. However, any system chosen must have been fully validated and passed factory and site acceptance tests and all components of the system and the overall system should be on preventative maintenance contracts and schedules as per vendor/integrator. The system must have safety interlocks, safety barriers, detection systems to ensure an instantaneous cessation of all movement in case of crash/collision of the robotic arm or microplate gripper, a hard system shutdown "panic button," and an industrial power surge protector and reserve power battery to allow managed system "power down" in case of a power failure. All critical liquid-handling components should be calibrated periodically according to the manufacturer's specifications or internal QA/QC procedures. It should be pointed out that as a HTS laboratory design and operational principle, every "online" system component should have an equivalent "off-line" unit on the research bench to ensure that any unit operation for liquid dispensing, aspiration/washing, and signal detection developed "off-line" will implement smoothly to the "online" robotic environment.

2. Figure 1a shows a schematic of the single robotic arm (1-POD) radially configured fully integrated system from HighRes Biosolutions (HRB, Woburn, MA) that was utilized for the automation and execution of the TNAP HTS campaign by our La Jolla automation team in San Diego, CA. Table 1 provides the specifications and the unit operation (or step) that each system component performs on a microplate during HTS. The 6-axis anthropomorphic Stäubli arm outfitted with a custom microplate gripper is the master controller that transports each microplate amongst all the individual system components. Figure 1b provides a schematic of our larger 3-POD system at our sister Lake Nona facility in Orlando, FL, which has a higher throughput through redundant components. A common operational scheduling software (Cellario™) capable of interleaving and multithreading between the two systems and analogous functional components allow automation protocols to be seamlessly transferred between both systems. Both systems were designed, integrated, and installed by HRB (http://www.highresbio.com) of Woburn, MA, and share the patented modularity feature achieved by mounting all unit components on movable rolling Microcarts™ that can be "plug and played" into fixed Microdocks™ that provide all services, power, gasses, and Internet IP connectivity that allows the main operating system to be "aware" of all current locations of all devices and make the appropriate radial coordinate transformations to allow continued operations without "reteaching" each position. This allows for rapid reconfiguration or repositioning and switching out of components as needed to optimize operations for each HTS campaign.

2.2 Key Custom, Unique or Costly Reagents, and/or Cells and Cell Lines

Every assay has a "key" critical reagent(s) such as purified protein target(s) and substrate(s), hopefully in their biologically "relevant" form(s) for biochemical assays, or engineered cells or cell lines that reflect the unique biology of the system. For most cutting-edge targets or targeted pathways, these reagents are often "unique" by definition, not available from commercial suppliers, and require custom expression and purification or molecular engineering, cloning, and selection of cell engineered to express the target pathway, hopefully in the relevant species and cell type. From the perspective of implementation of robotic HTS, the key guiding principle is to have sufficient quantities of the key reagent(s) in hand to allow assay development and validation, robotic HTS implementation, HTS, and all follow-up including SAR support to be conducted on the *same lot of material* throughout. These considerations are integrally and recursively connected to considerations of the assay format and level of microplate miniaturization achieved. Table 2 provides a summary of the specifications for commercially available 96-, 384-, and 1,536-well microplates (in SBS format and footprint

a

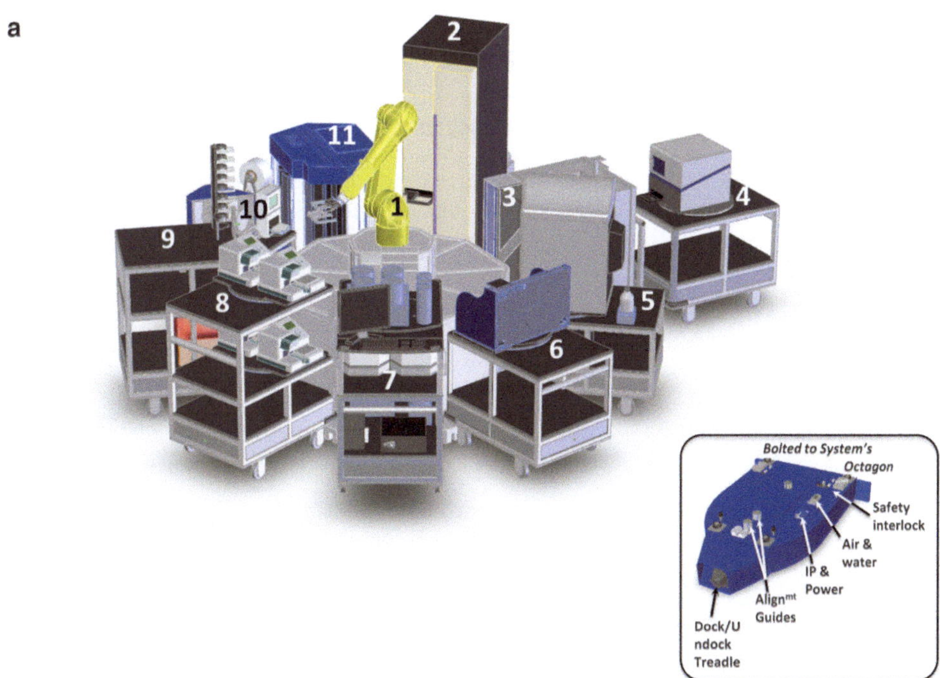

b

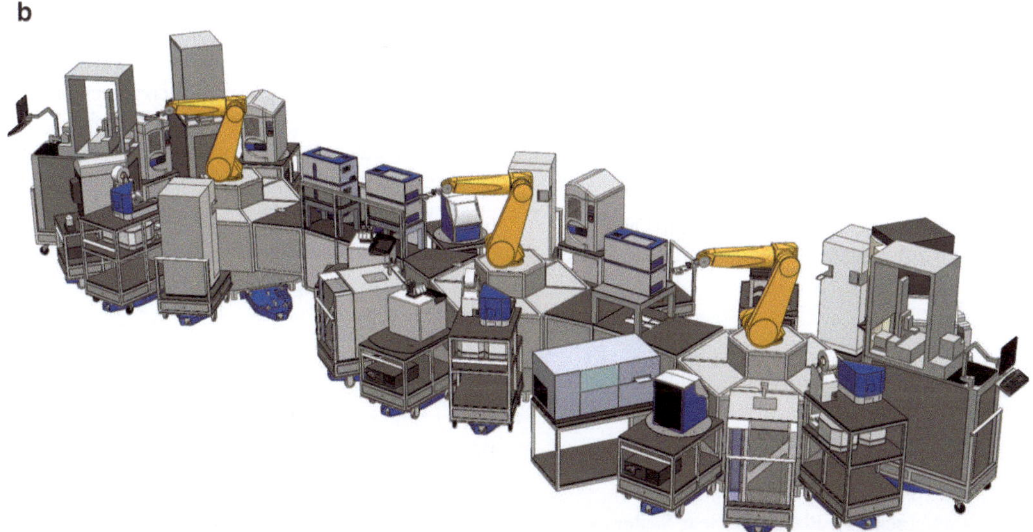

Fig. 1 Solidworks™ 3D rendition of (**a**) the 1-Arm/POD (Stäubli TX90XL) fully integrated robotic screening system utilized for automation of the example TNAP inhibitor biochemical assay in our San Diego Facility. Each MicroCart™ or major components are numbered and their vendor and functional specifications are detailed in Table 1. Note that Component No. is shown "undocked" from the main system. 4. *Inset*: a MicroDock™ showing MicroCart™ alignment guides and covered ports for gasses, electrical power, and IP connections. This system has a maximum throughput of ~320,000 wells per day; (**b**) 3-arm/POD (with larger Stäubli RX-160L with a CS8 controller) fully integrated robotic screening systems in our Orlando facility is shown for illustrative purposes of modular scaling. This system has a theoretical maximal throughput of ~2,200,000 wells per 24-h day (Fig. 1a. *Inset*: Details of a MicroDock™ to which a MicroCart™ with unit operation device is mounted)

Table 1
Integrated 1-arm robotic system component vendor and functional specifications

No.	Component (vendor and name)	Function
1	Stäubli TX90XL robotic arm	Controls each microplate transport between component devices
2	PerkinElmer ViewLux uHTS Imager	Multimode CCD image detection (UV/Vis, TRF, TR-FRET luminescence, fluorescence) simultaneously from all wells of a microplate
3	Liconic Incubator (210 plate capacity)	Non-ambient temperature incubation (25–42 °C) with humidity and CO_2 control
4	BMG LabTech GmbH PHERAstar FS	High-sensitivity PMT detection of FPA, TRF, TR-FRET, AlphaScreen, Luminescence
5	Labcyte Echo™ 555 w/pump	Noncontact Acoustic Drop Ejection of 2.5–500 nL compound DMSO solutions from source to destination plates for "cherry-picking" or dose–response series
6	V&P Scientific MicroPin w/3 Cole-Palmer MasterFlex L/S Peristaltic pumps	Simultaneous, contact transfer of fixed volumes (10, 25, 50 nL) of compound DMSO solutions with slotted stainless steel pins in 384- or 1,536-well formats
7	Top: BioTek EL406 Aspirator/dispenser Mid: 2 Velocity11 VSpin Centrifuges w/ Access 2	Top: 384, 1,536 washer/aspirator for gentle, tangential washing of poorly adherent cells on tissue culture plates; mid: rapidly spins plate (15 s) to collect all μL volumes into the bottom of wells and avoid hang-ups of microdrops
8	4× Thermo Electron Co. MultiDrop Combi 836 w/consumable 8-channel peristaltic cassettes	"Bulk" (1–20 μL) peristaltic dispensing of common reagent/buffer to all wells (15–30 s/ plate) in 384- or 1,536-well formats, with 8-parallel dispense heads
9	High Res Biosolutions Microplate Lid Station; EXAIR Ionizing Bar and Barcode reader	Suction cups hold lids during "de-lidding" operations; ionizing bar removes static to prevent microdrop "pop-off"; barcode reader positively identifies plate
10	Nexus Biosystems XPeel XPA Velocity 11 PlateLoc Heat Sealer	Continuous roll of aluminized polymer tape for automated heat sealing of microplates (seal and cut); continuous "sticky tape" allows "peeling off" of heat seals; system specified for 150–200 seal–peel cycles w/o deforming microplates
11	HighRes Biosolutions AmbiStore D (480 plate)	High-capacity and access speed RT storage of microplates (compound and assay)

All peripheral unit functional devices are mounted on HighRes Biosolution's rolling MicroCart™ which "plug and play" onto the *blue* MicroDocks™ (*see* figure inset) that are rigidly attached to the central octagonal robotic arm hub. Each MicroCart has two to three tiers of shelves to accommodate devices on multilevels or side by side on the same shelf

Table 2
Specifications of commercially available microplate SBS format and impact on scalability of assay automation (*adapted and updated from* [3])

Microplate density	96-well	384-well	1,536-well	Notes
Array format (rows × cols)	8 × 12	16 × 36	32 × 48	*4×96 into 1×384 (A1, A2, B1, B2)*
Capacity increase over 96-well	–	4-fold	16-fold	*But note impact of No. of control wells*
Center-to-center spacing (pitch) Well diameter Assay volume (useful range)	9 mm 7 mm 50–300 μL	4.5 mm 4.5 mm 20–100 μL	2.25 mm 1.5 mm 0.5–10 μL	*For cylindrical wells only; 384* and *1,536 also in square prismatic* and *half-height well shapes w/different vols*
Surface/contained volume ratio (cm^{-1})	18 at 50 μL 6 at 300 μL	22 (19) at 20 μL 15 (11) at 100 μL	240 (72) at 0.5 μL 16 (29) at 10 μL	*Calculated assuming no meniscus for cylinder* and *(sq. prism) at high* and *low volume extremes*
Availability of different plate types[a]	Numerous	Numerous	Increasing	*Only three suppliers of 1,536 plates in 1997*
No. of sample wells No. of control wells	88 (80) 8 (16)	352 (320) 32 (64)	1,408 (1,280) 128 (256)	*(No. of Cols 1* and *12 are controls in original pre-compressed 96-well)*
No. of plates for 365 K cmpd HTS	4,148	1,037	260	*Assuming 88 cmpd/96-well; +1 for partial plates*
Total assay volume for 365 K HTS[b]	19.9 L at 50 μL 53.8 L at 300 μL	7.96 L at 20 μL 3.58 L at 100 μL	0.200 L at 0.5 μL 2.99 L at 10 μL	*See note "b" below. Cellular assay requires culture media* and *labor to expand, propagate,* and *seed cells*
Potential reagent savings	–	4-fold	At least tenfold	*Fold savings dependent on final [critical reagents]*
Average cost/plate (total)	<$1	$2–$4	$15–$25	*Bulk discounts possible; ~same $/well basis*
Project plate costs for 365 K HTS	$4,000–$5,000	$2,074–$4,148	$3,900–$6,500	*5–10 % overage for plate "failures"* and *waste*

[a]Multiple suppliers; materials (COC, polystyrene, polypropylene, glass, low autofluorescent); coatings and surface treatments (tissue culture, poly-lysine, BIND, low protein binding, streptavidin); masking (white opaque, clear, black w/clear bottoms, white w/clear bottoms, and specialty microplates (extra-thin clear bottom tissue culture-treated polystyrene with black wall or glass) for imaging applications

[b]Calculated as No. of plates (per format) required for HTS (includes control wells) × nominal assay volumes/well × 125 % (to account for losses due to priming and dead-volume waste in liquid-handling devices for each batch of plates run/day). To include reagents for use of the HTS for subsequent "hit confirmation" of a typical ~1 % hit rate (3,650) resupplied "cherry-picked" stock compound solutions at the original screening concentration in triplicate, materials must be in hand for an additional 10,950 (i.e., 3,650 × 3) samples and controls. For a typical confirmation rate of 50 %, up to 1,825 compounds may be advanced into full triplicate 12-point dose–response analysis for an estimated 65,700 additional sample wells or reagents, just to complete hit validation. If this HTS is envisioned to support SAR rounds, then an equivalent number of addition wells may be expended in this endeavor alone. Therefore our "rule of thumb" is for sufficient material to be in hand for 150 % of the calculated nominal number of compounds to be screened, with 200 % being preferred. Clearly this burden is significantly impacted by the final assay miniaturization and microplate format achieved

[33]) and provides calculated cumulative total volumes of assay reagent in each plate format to complete an example 365,000 compound HTS at the typical ranges of assay volumes in each format that informs on the impact on scalability of critical reagent production and procurement on implementation of an automated robotic HTS campaign (*see* **Note 1**), where at the extremes, up to ~54 L of total assay volume would be required for a 300 μL assay in a 96-well microplate format, while only ~3 L are required for a 10 μL in a 1,536-well format (an 18-fold reduction).

1. For biochemical assays with protein target reagents, one needs to only multiply the final concentrations of key reagents to estimate the total amount sufficient for the HTS project (*see* **Note 1**). It is assumed that the infrastructure for protein expression, purification, and QC are within the HTS lab or accessible either through other core services or on an outsourced basis.
2. For adherent cell-based assays the critical parameter to calculate the total number of cells required to complete an HTS project is the optimized number of cells seeded per assay well necessary to yield the desired level of confluency during the assay multiplied by the total number of wells to be run. However, as cells live, grow, and expand, practically it is the scalability to maintain the total cells necessary to support the number of assay wells (plates) for each day's runs seeded at days 1, 2, and 3 for assay runs on days 3, 4, and 5 for example that are critical, rather than the cumulative total number of cells for the entire HTS project (*see* **Note 2**). While our laboratories only perform continuous cell culture during cell-based HTS, a number of pharmaceutical companies have had success in thawing prefrozen cells in media and seeding cells to attach for just a few hours prior to their use in HTS assays. This then allows massive cell expansion, scale-up, freeze down, and cryogenic storage of cell bank aliquots commensurate with the batch size of HTS runs (*see* **Note 2**). It is assumed that the HTS labs have access to proximal tissue culture hoods and facilities and the requisite incubators for cell expansion, cell propagation, and sterile assay plate seeding. Instrumentation for cell banking and cryogenic freezing and storage of cell stocks will also be desired.

2.3 Microplates and Disposable Labware

Table 2 provides an appreciation of the large number (4,148, 1,037, and 260) of microplates necessary just to complete primary HTS of a 365 K compound library in 96-, 384-, and 1,536-well formats, respectively (*see* **Note 3**). These must all be on hand in the particular material and "masking" format (clear bottom, black masked bottoms, black walls, white walls, etc.) necessary for the particular assay optimized before the start of production HTS (also of the same lot if possible) and typically an average of 25–50 % is recommended for assay development and HTS follow-up. Many robotic systems have multiple (2–4) bulk dispensers that utilize custom cassettes (8 or 16 channel) of peristaltic tubing ($400–$800 each) and are typically good for just 1 or 2 HTS campaigns; so they must be regarded as laboratory consumables (*see* also **Note 3**, about use and cost of commercial ready-made kits).

2.4 Chemical Library Collections: Stock Solutions, Powders, and Compound Management

Numerous reviews cover the selection, construction, and curation of chemical libraries, debates about their size and composition, operational costs, infrastructure, and best practices of compound management of stock solutions and dry powders [2, 3, 20]. For the purposes of implementation of automated robotic assays, the practical aspects are that the entire chemical library exists as working stocks (typically 2 mM in 100 % DMSO) of individual compounds stored in heat-sealed (*see* also **Note 4**), "daughter" barcoded microplates of well density format matching the final screening format (96-, 384-, or 1,536-well) in a plate material consistent w/chemical inertness (cycloelefin copolymer (COC) or polypropylene) or optimized for acoustic dispensing (Echo-certified COC or "diamond" plates).

2.5 Chemical and Biological Database, LIMS, Data Analysis, and Visualization Software

In addition to the robotic arm controller, scheduling software, and run logs necessary to operate the actual robotic screening, the HTS laboratory must also have sophisticated laboratory information management systems that can automatically accumulate in real time or transfer in large batches the resultant data files (in appropriate formats comprising plate ID, well location, and resultant signal) into a relational database that allows mapping of assay plates to compound plates to establish compound identity versus biological activity data. Ideally, these applications allow processing and calculations on HTS data to allow data normalization, correction for edge effects, statistical analysis, and visualization of data on a plate-by-plate and overall HTS basis for evaluation of assay robustness, and screen performance. Furthermore, these applications and databases must be able to communicate with each other or at minimum generate outputs that can be understood amongst them. Our HTS group uses a customized Web-based Chemical and Biological Information System (CBIS) comprising a Web-based portal, applications for structure rendering, data normalization, and curve fitting developed by ChemInnovation (San Diego, CA). We review plate "heatmaps" of the primary screening data, correct for plate "edge effects" with GeneData Screener® (Basel, Switzerland) software, and visualize overall assay performance with TIBCO Spotfire® (Sommerville, MA).

3 Methods

The overall process and critical activities for robotic implementation of automated HTS from intake of the researcher's "benchtop" assay through to the identification of a validated, optimized lead series of compounds around a bona fide chemical scaffold are shown in Fig. 2. The downstream steps of hit confirmation/validation and structure–activity relationship (SAR) elucidation and hit to lead (HTL) have been included to emphasize that the same robotic assay used for production HTS can and should be used by the HTS team to perform high-volume hit confirmation and first-line automated dose–response analysis of confirmed hits to obtain quantitative potency (IC_{50}) rankings to elucidate the nascent SAR of emergent hit series

3.1 Review of the "Benchtop" Assay

A thorough review and discussion with the researchers about their "benchtop" assay, written protocol, and list of available reagents will help the HTS assay development team understand and assess the chosen target selection, biological rationale, proof of concept, competitive landscape, and prior art. If these are sound, then the assay development's appropriate role is to assess the technical feasibility of developing an automated robotic HTS assay. Figure 3 illustrates a simple 4-tier (2×2) "decision matrix" for prioritizing proposed HTS campaigns against the strategic factors of high impact, need and rationale *versus* the likelihood of technical success. Tier-1 assays are clearly ready to advance into HTS assay development and screening. Tier-2 projects are worth investing the expertise of the team to improve the technical feasibility, since

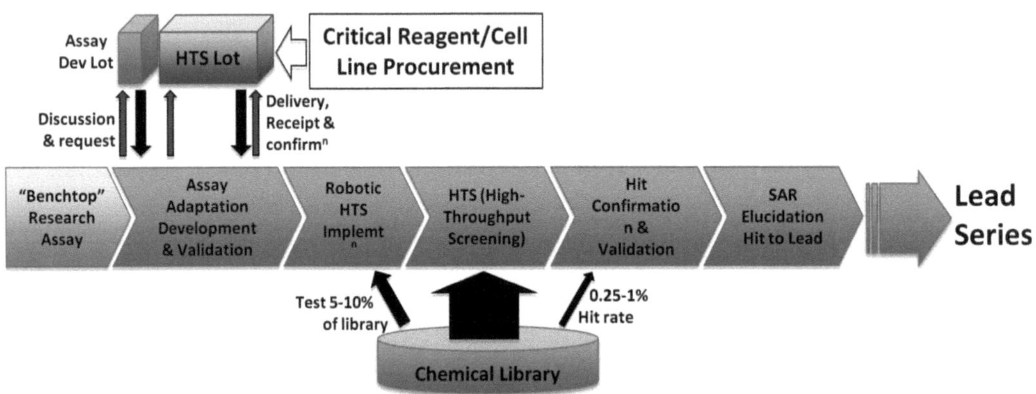

Fig. 2 Schematic pipeline diagram with major steps from intake of a "benchtop assay" through to generation of a lead series. This chapter reviews the processes up to HTS execution. The sourcing of critical reagent in a test development lot and production lot for HTS is indicated above the assay adaptation, development, and validation chevron by *blue* and *light red bricks*. The chemical library is represented by the cylindrical tank below the HTS chevron

HTS Prioritization Matrix

- Should we do it *vs.* can it be done (Important? *versus* doable?)

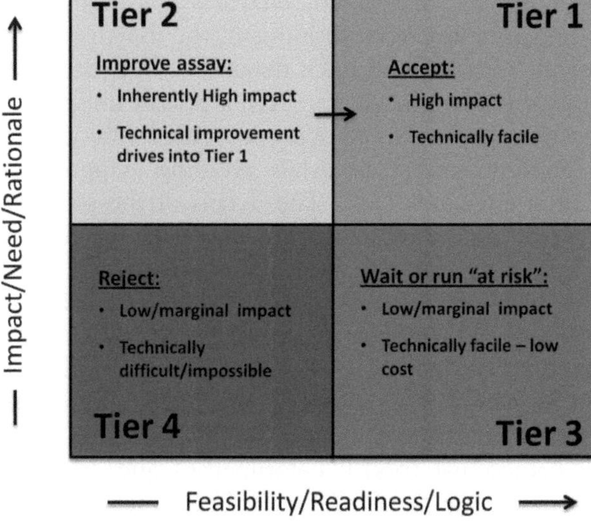

Tier 1 (Accept - execute)

- Inherently high impact/unmet need/lack of prior art/clear R_x & scientific rationale/innovation of approach/target and/or assay
- Meet Tech reqs /assay(s)readiness

Tier 2 (Worth Improving assay – *invest effort***)**

- Still inherently high impact/unmet need/lack of prior art/clear R_x/scientific rationale
- Improving feasibility drives to Tier 1

Tier 3 (Wait - Improve impact or *if very facile assay, might screen at risk***)**

- Seek out or monitor for improvement of impact/need/rationale that could drive to Tier 1
- Verify if facile/cheap merits screen for borderline impact or potential tool compound to provide proof-of-concept

Tier 4 (Reject – *Send back to PI for improvement***)**

- Low impact/unclear need/prior art/weak R_x & scientific rationale
- Very hard to do as well, so no clear reason to invest any time currently
- PI should invest time to bolster impact 1st

Fig. 3 A 2 × 2 prioritization matrix summarizing the quadrants to which proposed HTS projects can be assigned for prioritizing whether to execute or delay

the impact is inherently high. Tier-3 projects are judgment calls where the team/organization may elect to delay while monitoring the literature that improves project impact or if the assay is very facile and of low cost may elect to implement and run "at risk," with the idea that it might provide a compound that provides a "proof of concept."

It is important to point out that while most of the burden and expertise for impact are borne by the proposing researcher, it is often the chemists on the overall drug discovery team that have the appropriate domain expertise to evaluate the competitive landscape. It is also important for the assay development team to review and verify the assay development and optimization data of the benchtop assay. Sometimes the fundamental assumptions made are faulty, for example that the steady-state conditions are not met for an enzyme assay (i.e., nonlinear progress curves with too high a level of substrate consumption), that endpoint measurements were taken when a progress curve had already plateaued (this is a surprisingly common error, whose impact has been analyzed) [*see Fig. 3 in* [4] *for simulation, discussion of reasons for nonlinearity with respect to substrate depletion, and workable off-linearity for HTS*], or that substrate concentration-to-K_m ratio is set inappropriately and so will not allow detection of the desired types of inhibitors [*see Fig. 1 and associated text in* [4] *which provides an excellent discussion and kinetic equations for various inhibitor-favored conditions*].

Specific example of automating a phosphatase assay—TNAP. Alkaline phosphatases (EC 3.1.3.1) are dimeric enzymes that catalyze the hydrolysis of phosphomonoesters. In humans, the fourth isozyme is tissue-nonspecific (TNAP) and has a major role in hydrolyzing extracellular inorganic pyrophosphate (PP_i), a potent inhibitor of bone mineralization, thus maintaining homeostatic levels of PP_i to ensure proper bone mineralization while avoiding inappropriate deposition of bone elsewhere [34]. The details of the solid and high biological rationale (target selection), enzymatic mechanism, parent clinical assay, standard colorimetric assay, and specific details of a 1,000-fold more sensitive assay with a 100-fold increased dynamic range have been elegantly described previously [35] and this assay was used successfully to screen ~200 K compounds in an HTS (Pubchem Bioassay AID: 518, 1012, and 1135) and focused SAR study (Summary Pubchem Bioassay AID: 1548 and associated AIDs for orthogonal (pNPP colorimetric) and selectivity assays against intestinal (IAP) and placental (PLAP) alkaline phosphatases) to identify novel TNAP inhibitors with diverse modes of action [36]. This project had a high unmet medical need for the relatively small affected population (orphan disease), but solid biological rationale, and in addition had a solid technical underpinning in a series of robust precedented colorimetric assays for both the primary and secondary targets (SAR selectivity) that were improved by a fairly rapid conversion into a luminescent primary assay, and the HTS assay was fully implemented for HTS and downstream hit confirmation, validation, and SAR support.

3.2 Critical Reagent/ Cell Line Procurement

As Fig. 2 illustrates, procurement of critical reagents is often a two-step process wherein after initial discussion with the researcher and investigation of sources for procurement, a small test batch of reagents should be procured from commercial vendors or an expression facility, used to verify the feasibility and performance of the initial assay format, detection technology, form of the critical reagent (e.g., holoenzyme or subunit form of enzyme, protein tagging technology), and quality of the critical reagents chosen. Some initial optimization of assay parameters that allows preliminary estimate of the total amount of reagents necessary to complete assay development, implementation, HTS, and follow-up should be completed. However, very soon after studies are completed in 2–6 weeks, these data should be shared with the protein expression or tissue culture facilities to allow more definitive cost and delivery time estimates for, ideally, a single large batch of reagents upon which all of the steps from the final assay development, optimization, and validation through robotic implementation and full deck screening will be performed (*see* **Note 1**). The goal is to have reagents in relatively pure and appropriate form, in sufficient amounts for the entire assay development, implementation, HTS, and follow-up that are stable and not contaminated with any activities that would invalidate assay data.

3.3 Assay Adaptation, Assay Development, and Validation (for Robotic Automation)

1. *Adaptation*. It is assumed for the purposes of this chapter on implementing robotic assays that the "benchtop" assay provided to the implementation team meets sound basic research assay principles alluded to in the previous paragraphs, such as the following: (1) proper choice of the form of the target (e.g., purity, holoenzyme vs. catalytic domain, with accessory proteins, or full-length receptor clones in proper cell and coupling pathway background, with matched sham vector controls); (2) proper choice and concentrations of substrate, inducer, and cofactor; (3) proper choice of assay formats from both fundamental assay configuration (e.g., proper E/S ratio to allow detection of inhibitor types desired or ensuring that coupling enzymes and substrates are not limiting and only the target enzyme is rate determining) and practical (e.g., fluorescent tags used are red-shifted to minimize library compound optical interferences) considerations; (4) optimized for time, temperature, order of addition of components, buffer/media pH, and growth conditions; (5) assessment of the assay tolerance to solvents (especially DMSO), carrier proteins, and detergents; and (6) general stability of critical assay components.

   However, even a sound, properly optimized and validated "benchtop" assay is usually unsuitable or underdeveloped for an HTS [26] robotic implementation. Indeed one of the first adaptations necessary will be to match the HTS assay method to the target type, as often the non-HTS researcher may not be familiar or have access to these technologies. This critical choice and selection will have the most profound effect of the technical feasibility of the assay. Table 3 provides some guidance on HTS-compatible assay formats (biochemical, cell-based, or phenotypic and high-content imaging) for various target classes. This list is no means exhaustive or definitive as no "label-free" or specialized "chip-based" biosensors have been included. For any given target the experimentalist is advised to research the ever-growing HTS, screening, and assay development literature to find specific examples of their particular or similar target(s) as implemented and used for HTS-driven projects. Additionally, specific application notes and assay kits in many of these formats are readily available from commercial reagent suppliers and can be found through online search engines (as noted previously). These searches will also find a number of useful online HTS community bulletin boards and listserves, where technical questions can be posted.

Table 3
HTS-compatible assay technologies for general assay formats to target classes or types (*adapted and updated from* [4])

Target class or type	Assay formats	Cell based [46]	
	Biochemical/biophysical	***Non-imaging "plate reader"***	***High-content imaging***
GPCRs	FP, FIDA, thermal melt	FLIPR®-like, reporter gene, aequorin, TR-FRET, AlphaScreen®, EFC	Transfluor® β-arrestin redistribution
Ion channels	FP	FLIPR®-like, FRET, AAS, automated patch clamp	–
Nuclear hormone receptors	FP, TR-FRET, AlphaScreen®	Reporter gene—luciferase, colorimetric, fluorescent, luminescent	Receptor translocation with labeled receptor
Kinases	FP, TR-FRET, IMAP, thermal melt	Lysed—FRET, AlphaScreen®, Alpha-LISA®; Intact—PathHunter® kinase [47]	Specific α-P-protein mAb staining fixed and permeabilized cells
Proteases	FP, FRET, TR-FRET, thermal melt	Reporter gene—luciferase, colorimetric, fluorescent, luminescent	Labeled protein substrates
Phosphatase	Fluorescence, luminescence, absorbance, thermal melt	–	–
Other enzymes	FP, FRET, TR-FRET, fluorescence, luminescence, absorbance		
Protein–protein or protein–DNA/RNA	FP, FRET, TR-FRET, BRET, ECL, AlphaScreen®	BRET, reporter gene	BRET, localization, specific α-protein mAb staining on fixed-permeabilized cells
Whole organisms or isolated organelles		Reporter gene, biosensor, viability (ATP)	True phenotypic or physiologic responses, morphology, and motility

SPA = scintillation proximity assay is possible for all biochemical assays on isolated targets; however, the need for radioactivity has been supplanted by homogeneous, nonradioactive formats. Abbreviations: *FP* fluorescence polarization, *FIDA* fluorescence intensity distribution, *(TR)-FRET* (time-resolved)-fluorescence resonance energy transfer, *IMAP* immobilized metal affinity for phosphochemicals, *BRET* bioluminescence resonance energy transfer, *EFC* enzyme fragment complementation, ***Molecular** DevicesTransfluor*® and *FLIP*® fluorescent imaging plate reader, *PerkinElmer*—AlphaScreen® and AlphaLISA, *DiscoveRx*—PathHunter®

Even if the benchtop assay appears a suitable starting point for direct automation-compatible HTS assay development, it behooves the assay development team to obtain starting material from the assay provider and to replicate, at least manually, that the assay performs according to the assay provider's protocol. Table 4 provides a comparison of some of the mechanical and technical parameters that need to be adjusted in the process of adapting a "benchtop" assay into an HTS (automatable robotic) format. The general guidelines and practical principles are as follows:

(a) Reduce the number of steps, combine certain steps, eliminate washing steps, and convert assay into homogenous nonradioactive "mix-and-measure" formats. This simplifies robotic liquid handling steps, speeds overall screening rates, reduces timing issues, and reduces error propagation with resulting improvements in assay variability. Homogenous assays are easier to automate and miniaturize, require fewer steps, avoid difficulty in automating separation (e.g., centrifugation) or washing steps that generate large waste stream, and are fast, robust screening assays. While they may be less sensitive requiring higher amounts of enzyme and substrates (biochemical assays) or may be more susceptible to compound optical interferences, overall their advantages outweigh their disadvantages [37].

(b) Miniaturize the assay into the highest density microplate format feasible (384- or 1,536-well), subject to the signal intensity and signal/background achievable in these formats. As shown in Table 2, this assay miniaturization reduces assay volumes, reagent procurement needs, and overall costs in terms of purchase and more importantly the cost of producing custom critical reagents; however, one must take care that assays are not subject to surface-dependent effects (*see* **Note 6**).

(c) Convert all incubations to room temperature where possible. For live cell-based assays automated tissue culture incubators are available; however, care must be taken to ensure that each microplate spends most of its time in incubator relative to the time it is at ambient temperature while being processed. This is often the case for cell reporter assays, where expression/induction takes 5 h to overnight versus the 5–10 min at ambient temperature required for robotic manipulations (*see* **Note 5**).

(d) The modified protocols should be developed and tested on the replicate off-line unit devices (liquid handlers and plate readers) that replicate those on the online system to

Table 4
Comparison of parameters between laboratory "benchtop" and HTS assays

Parameter	Benchtop	HTS
Protocol	Numerous often complex steps, aspirations, washes	Few (5–10) simple steps, "add only" homogenous preferred
Assay volumes	0.1–1 mL	<1–100 μL
Reagents	Quantity may be limited, batch variation acceptable, may be unstable	Sufficient quantity for single batch, stable to storage and over prolonged periods at working concentrations
Plate, reagent, liquid handling	Manual—un/capping, pipetting, mixing, shaking, inverting, rocking, pouring	Workstations with manual plate transport or stackers between stations or fully integrated robotic system
Variables	Many often not be fully controlled or logged during experiment or between experiments	Preset, monitored, and logged; compound, compound concentration
Assay container	Varied—tube, slide, microtiter plates (6-, 12-, 24-, 48-, and 96-well), Petri dishes, cuvette, animals	Microtiter plates—96, 384, and 1,536-well
Assay duration/ time and type of measurement	Seconds to months duration; milliseconds to days measurements; endpoint, multiple time points, or continuous "real-time" measurements	Minutes to days duration; seconds to minutes measurement; typically endpoints preferred with incubation times long enough for automation (30–60 min); pre-reads possible
Format/signal generation technologies	Heterogenous—wash, e.g., ELISA, SEAP reporters; colorimetric, absorbance; homogeneous—luminescence, fluorescence, FRET	Homogeneous—luminescence, fluorescence, TR-FRET, FPA, AlphaScreen™
Incubation temperature	Multiple temperatures and methods	Ambient—room temperature preferred Available automated tissue culture incubators, if necessary for long incubations; cell grown off-line
Output formats	Cuvette or plate reader, microscope images, videos, radioactivity (scintillation or Geiger counting)	Plate reader—lists formats suitable for automated deposition into databases for processing and visualization. HCS—image files and calculated parameters and metadata
Reporting formats	Manually tabulated spreadsheets, calculated results, graphs, figures on "representative" data; statistical analysis of manually curated datasets	Automated reports for assay performance, HTS performance, QA plate-by-plate statistics and overall run statistics, visualizations on all data over campaign

ensure microplate automation capability and limitations of those steps, and to accurately estimate the achievable total signals and signal/backgrounds for the formats chosen.

2. *Assay development* and *optimization*. With these guidelines in mind, further assay development and optimization of assay parameters should be conducted in several replicate wells ($n \geq 5$ with >20 preferred) of high signal and background signal controls, in the actual microtiter plates (material, well density, masking) and assay formats selected, using the liquid-handling devices and plate readers equivalent to those on the robotic screening system while performing the critical incubation times and steps manually between these plate-based devices with a mind towards the eventual robotic implementation (that robots take ~1–2 min per unit operation). Assuming that temperature, buffer, pH, ionic strength, osmolarity, concentration of monovalent ions (e.g., K^+, Na^+, Cl^-) and divalent cations (e.g., Mg^{2+}, Ca^{2+}, Mn^{2+}, Zn^{2+}), and general parameters of the baseline assay have been established, some of the additional critical parameters to be optimized [4, 26] are: (1) establishment of the linear portion of a progress curve; (2) selection of the proper incubation time within the linear portion; (3) selection of the optimal concentration of the target protein (e.g., enzyme or binding protein) or number of cells seeded into a well, expressing the target of interest (e.g., GPCR receptor in U2OS cell line); (4) optimization of additional buffer components such as carrier protein, nonionic, non-denaturing detergents to reduce nonspecific compound aggregation or general protein "stickiness" (*see* **Note** 7); and (5) selection of the optimal concentration of substrate, inducer, or reference compound that will provide the best balance amongst signal-to-background (S/B), signal-to-noise (S/N), variation (%CV) in signal or background, and the Z-factor [38]. Table 5 summarizes the formulas used to calculate these values, an interpretation of their meaning, and typical values for assays with "acceptable" performance. The *Preliminary Studies* sections of current (PAR-12-058 and -059) and past (PAR-09-129) Funding Opportunity Announcements (FOA) concerning HTS assays cite the technical prerequisites of HTS-ready assays as one with a Z-factor of at least 0.5, which can be translated to an S/B of at least four combined with a %CV of 10 %.

3. *Cell-based assay considerations.* Macarrón and Hertzberg [4] provide a list of important consideration for cell-based assays and note that they generally have more variables than biochemical assays as they are conducted in cell media of complex formulations containing variable biological preparations (e.g., fetal calf serum, conditioned media), so it is imperative to try to reserve the same lot of these biological media additives from

Table 5
Assay development and HTS statistical parameter used to evaluate HTS assay quality (*adapted from* [4])

Parameter	Formula[a]	Comments and references
Signal-to-background	$S/B = M_{signal}/M_{background}$ *Acceptable assays usually have* $S/B \geq 5$; *though* $S/B \sim 2$ *still acceptable if* $Z' > 0.5$ *(very small SDs)*	Shows separation between positive and negative controls. But is a poor indicator of assay quality, as variability in S or B is not considered [38]
Coefficient of variation (*calculate separately for S or B*)	$CV\ (\%) = 100 \times SD/M$ *Acceptable assays usually have* $CV(\%) \leq 10\ \%$	Relative measure of variability due to assay/reagent stability, precision, and reproducibility of liquid-handling and signal detection instruments
Signal-to-noise	$S/N = (M_{signal} - M_{background})/\sqrt{(SD_{signal})^2 + (SD_{background})^2}$	Measure of signal separation relative to variability in both signal and background signals. Not classic S/N [38]
Signal window		
Z-factor	$Z' = 1 - 3 \times (SD_{signal} + SD_{background})/\lvert M_{signal} - M_{background}\rvert$ $Z' < 0$; *S and B populations overlap significantly—no assay* $Z' = 0$; *S and B populations touch—"coin toss" or "yes/no" assay* $Z' \geq 0.5$; *S and B populations well separated—acceptable assay* $Z' = 1$; *infinite separation between S and B populations*	Widely accepted [4, 48] dimensionless parameter that is an elegant combination of S/B and variability that indicates the separation of the signal and background population (assuming normal distributions and random error) [38]. A minimally acceptable assay has a Z' of at least 0.4

[a] *Where* M *is the average and* SD *is the standard deviation of any group of n values of any measurement*

reputable suppliers, and use extreme diligence and maintain documentation of cell culture stocks (*see* **Note 2**) (viability and cell plating efficiency post thaw, verification of mycoplasma-free status), cell banking protocols (passage and generation numbers), and protocols for culturing, maintaining, expanding, and seeding cells into final assay plate while maintaining sterility through careful tissue culture procedure or including antimycotics and antibiotics in the culture media (verifying these do not have deleterious effect on assay signal). The functional phenotype and assay performance of cell monolayers are highly dependent upon initial seeding density, state of confluency, and passage number, so the effect of these on all the parameters above must be verified during assay development [30, 39]. In general stable recombinant cell lines that reflect the physiology of the relevant primary cell type engineered to express moderate levels of the desired target, that have proven to maintain stable functional responses through several passage, are preferred to transiently transfected cells, and early passages are desired. As mentioned in previous sections, cell-based assays are particularly challenging in 1,536-well when long culture times post cell seeding and long incubations for compound treatment and signal development are proposed. Even despite humidified tissue culture incubators, evaporation, and uneven gas exchange lead to pronounced "edge-effects" that are exacerbated as the total incubation time increase beyond 1–2 days (*see* **Note 5**).

4. *Importance of a control or a reference compound.* We note finally that it is desirable to have a control reference compound (e.g., a reference inhibitor) available that can be used to define and experimental assay background (e.g., at 100× IC_{50}) that accounts for all liquid-handling and indeterminate sources of experimental error, rather than an artificial one by leaving out a critical reagent, such as leaving out the measured label, substrate, or enzyme. Also, an ability of the assay to replicate the IC_{50}, Hill coefficient ($n \sim 1$), and maximal efficacy ($\%I_{max}$ at plateau) of this compound from a full-dose response analysis relative to literature values under the miniaturized, HTS conditions provides strong validation of the assay. For promoter gene reporter assays, this is particularly useful, as often the reference standard cited for assay performance is co-transfection of a stimulating transcription factor into the reporter system and it is often unclear if a small molecule can ever achieve the efficacy (E_{max}) even at the highest concentrations of small molecules. Therefore setting the apparent activation threshold (% Activation) too high based on maximal efficacy of a transcription factor may lead to a bias against high-potency (low EC_{50}) but low-efficacy compounds (20–60 % of E_{max}), which may prove to be useful starting point to optimize and improve efficacy or may also provide potential for design of inverse agonists.

5. *Assay tolerance to DMSO.* Particularly for HTS assays, as most compound libraries exist as concentration stock solutions with DMSO as the preferred solvent, the tolerance of the assay to the final concentration of DMSO is important to evaluate, including the effect of both the positive and negative controls, and if possible with or without reference controls (whether inhibitor or inducer) dispensed from stocks under the eventual protocol to be robotized. Typically, the assay performance (S, B, S/B, S/N, %CV, and Z-factors) is evaluated over a range of concentrations of DMSO (0–10 % (v/v) in 0.25–0.5 % increments) and DMSO concentration(s) that yields <10–20 %

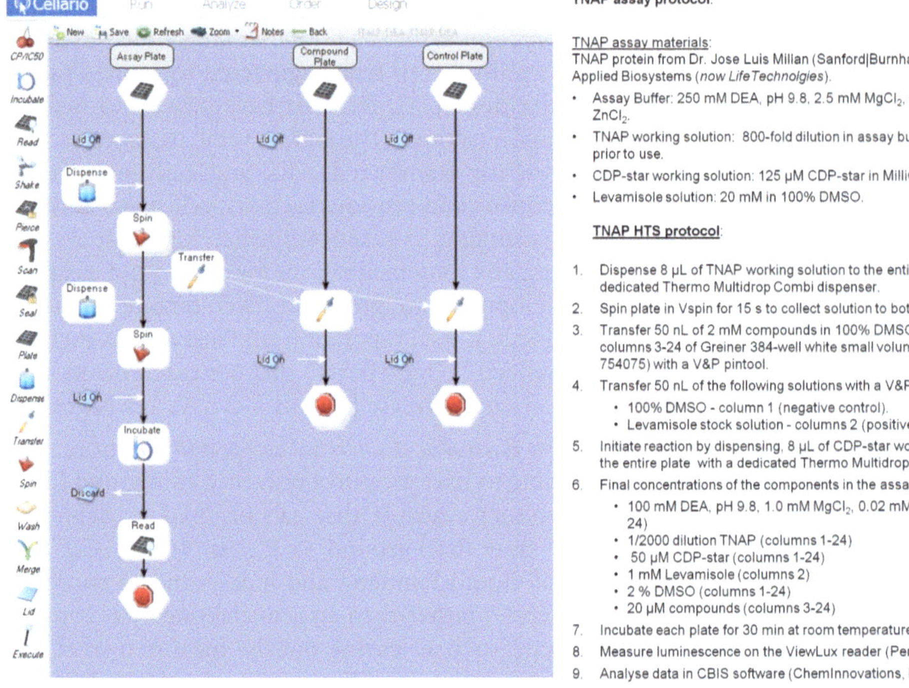

Fig. 4 Screen capture of the HighRes Biosolutions Cellario™ graphical user interface (GUI) showing the icon-driven sequence of steps and process paths for the assay, compound, and control plates. The written protocol corresponding to the GUI flowchart is shown on the *right*

change in signal or S/B relative to no DMSO controls is usually acceptable. Generally enzyme assays are tolerant up to 5–10 % DMSO, while many cell-based assays only tolerate 0.25–0.5 % DMSO. A high tolerance to DMSO expands the range of usable stock solution concentrations, avoids ultralow volume transfer of stock compound solutions, and potentially improves compound solubility in the final assay (*see* **Notes 4** and **8**). This consideration of DMSO tolerance also affects the use of a control compound for establishing the assay performance, insofar as %DMSO affects compound solubility and therefore the highest concentration achievable.

3.4 Robotic HTS Implementation

1. *Programming and factor for robotic implementation.* While scheduling and programming of a robotic screening system are highly specific to each system, the programming interfaces broadly fall into editable line-by-line executable commands or object-oriented programming with a graphical user interface (GUI) presenting icons representative of devices and operations that connect them. Cellario™ is an example of a GUI (Fig. 4) with icons representing objects, devices, and operations, with arrows indicating the direction and sequence position where the operation is performed.

In this example the sequence of operations performed upon each of the three types of plates (assay, compound, and control plates) laid across the top of the screen is represented in a series of vertically descending black arrows connecting key operation and device station (spin, incubate, read), with white horizontal arrows indicating the additional operations (relatively rapid) done to that plate ("lidding"/"unlidding," dispense, transfer). The two arrows connecting one *Transfer* step to *Assay Plate* to the two different *Transfer* steps on the *Compound Plate* and *Control Plate* indicate a "pooling" of sequential transfers of control compounds the appropriate controls well on the *Assay Plate* from a constant master plate containing control compounds, followed by transfer of test compounds from one of the stack of compound source plates. In general, programs are written from a microplate centric perspective of (1) where a plate is currently on the system (at what device) and where it is being moved to, (2) what operation(s) is being performed on a plate, and (3) when is this operation going to be performed and what the duration of this operation is. In the HRB system (Fig. 1a) the central robotic arm is issued a sequence of "pick-up" and "move" commands to place a microplate onto the plate loading positions in any of the ten peripheral devices so that they can perform their unit operation. The flowchart of steps represented in Fig. 4 does not not show the duration of steps, cumulative duration of a complete assay for one plate, nor the timing of an entire batch of plates. This is the function of the scheduling software that controls the timing between these steps and also maximizes the utilization of the robotic arm, by allowing it to perform operations on another plate while the previous plate's operation is being completed. For example, when the microplate is placed into an incubator for a relatively long incubation, the arm does not simply wait until incubation for that one plate is complete, but rather, it will continue to process plates according to the sequence of steps until the first plate needs to be removed from the incubator. In Cellario™, the scheduler also allows a simulation of an entire run of a batch of plates (Fig. 5a), to check for potential "resource conflicts" where the robot would be asked to perform two different operations in two different plates at the same time. The programmer can then adjust the incubation times to accommodate more plates before the first assay plate creates conflict with subsequent plates. For example, with a 30-min incubation (Fig. 5a) only nine control plate accessions can be completed before the first plate needs to be removed from the incubator, whereas by increasing the incubation time by 10 min, accession of the control plate three more times (Fig. 5b) is possible before the first plate has to be moved out of the incubator. The programmer can also insert

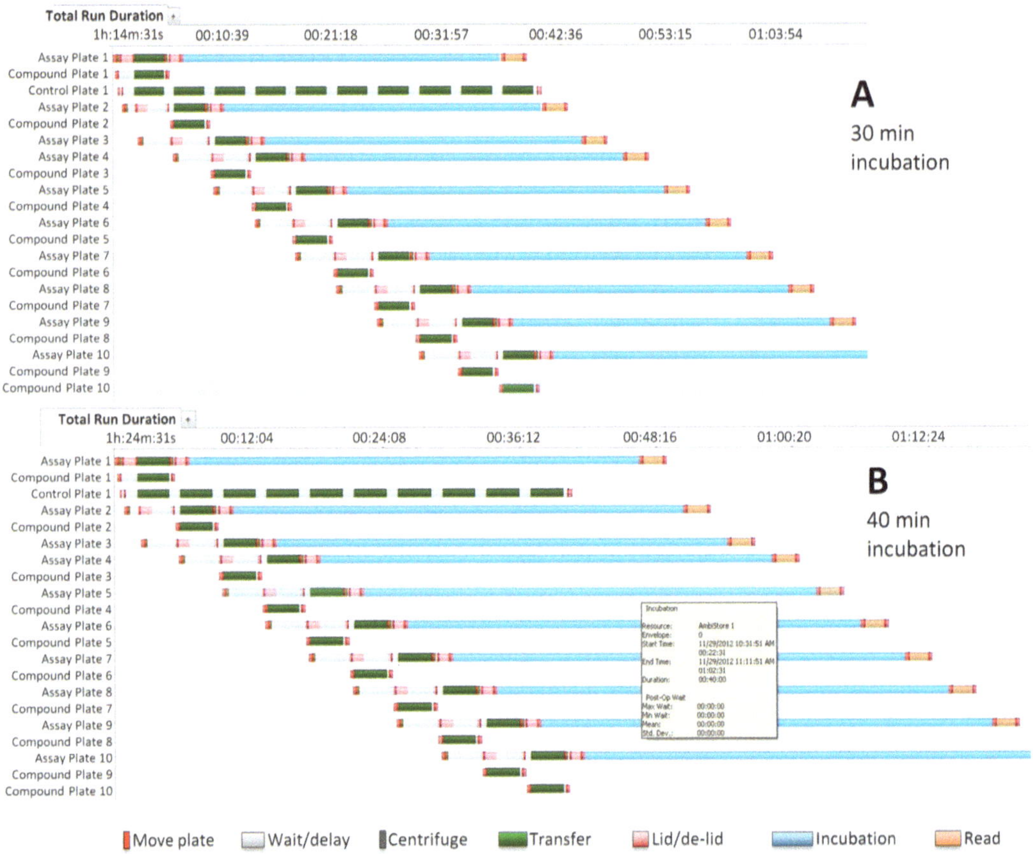

Fig. 5 Screen capture of a 10-plate run simulation for the TNAP protocol of Fig. 4 for a (**a**) 30-min incubation and a (**b**) 40-min incubation for each plate. The various color-coded bars correspond to unit operations that are performed on each plate. Their duration is indicated by the length of the bars and the color code key is indicated at the bottom of the figures for the seven major operations

additional "wait" or "delay" periods to allow the robot arm to resolve these conflicts. Such cycles of simulation adjustment allow the programmer to maximize the overall throughput and batch capacity while avoiding potential system crashes due to resource conflicts for the entire run on a batch of plates. Comparison of Fig. 5a and b shows that the total elapsed time for a 10-plate run differs only by about 10 min, the difference in the incubation time, rather than 100 min (10 min × 10 plates) because the sophisticated programming allows "nesting" of operations on multiple plates, rather than strict linear sequence of steps for each step until completion of an entire plate. The sequence of robotic steps and timings of unit operations are very reproducible and tightly controlled and thus follow six sigma standards for assembly line manufacturing processes (i.e., "industrialized") so that each microtiter plate experiences the same set of mechanical operations, whether one plate or 100 plates are processed during a run. However, this continuous infinitely scalable manufacturing analogy fails because often the key biological reagents (enzymes, cells, antibodies) are not indefinitely stable, often degrading in a matter of 3–5 h once brought to the working dilution in the reservoirs on the robotic system, even with refrigeration of the reservoir (*see* **Note 9**).

2. *Pilot robotic screening on a sampling of the chemical library collection.* Once programming of the robot for a plate batch has been completed, it should be used to actually run several compound plates representing 1–5 % of the total chemical collection at several potential screening concentrations (5, 10, or 20 μM) along with several close to full control plates, with many replicates of high and low signal controls over most of the assay plate, and a full dilution series of any available reference compound. Ideally for control plates, careful analysis and calculation of assay performance parameters on a per plate, per column, per row, or per plate quadrant basis and "heatmap" visualization of an entire plate will help to identify any systematic liquid handling errors. Common errors are "striping" due to a clogged bulk dispenser or a low first row signal due to insufficient priming of supply lines to ensure fresh reagents. Additionally, a tabulation of the Z′, S/B, CV%, and S/N on a plate-to-plate basis, as well as derived IC_{50}, Hill coefficients, and I_{max} for the reference compounds from different plate replicates will indicate the stability of the assay from a plate-to-plate basis. The assay performance parameters calculated for the control wells of the actual compound plates at different concentrations will further corroborate the performance of the baseline assay. A finding that the robust Z-factor (Z), where the mean and standard deviations of the signal obtained in the presence of compounds are used in place of the no-compound control in the Z-factor equation, is within 20 % the Z-factor (Z′) of the controls or remains above 0.5 indicates that the baseline assay is robust and does not break down in the presence of the random screening collection at the concentrations tested. Finally, upon setting the hit threshold (50 % of control activity or three standard deviations from the mean signal in the presence of compounds), the number of hits and hit rate at each test concentration is obtained to estimate the total number of hits one might expect for the entire collection to be tested. For a typically acceptable hit rate of 0.2–2 %, a 365,000 compound collection would generate 730–7,300 initial hits that could be robotically cherry picked from stock solutions.

3.5 Robotic HTS of Large Chemical Collections

1. *Scaling and monitoring HTS batch runs.* Once the implemented robotic HTS has been validated and pilot library screens meet specifications for acceptability, and a screening

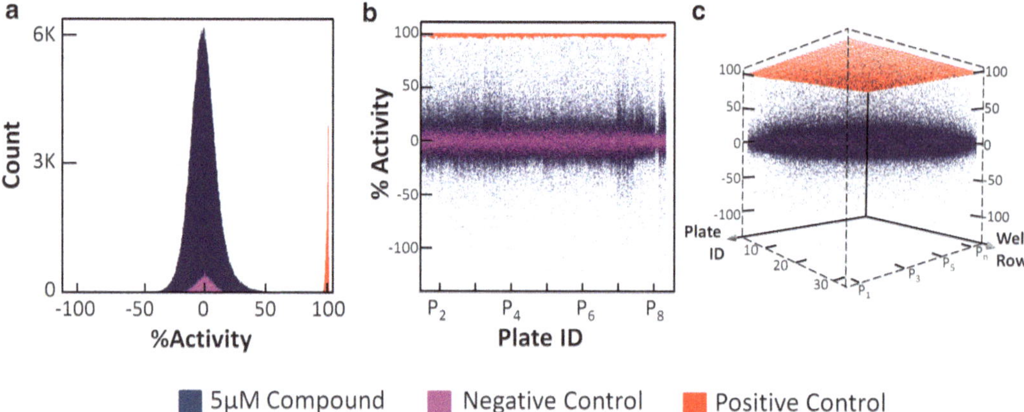

Fig. 6 Example of useful visualizations of an HTS campaign on 365,000 compounds of the NIH molecular libraries small molecule repository in an example enzyme assay. The Z-factor for this HTS campaign over all the normalized controls was Z′ ~ 0.82 and the robust Z-factor for the screened compounds was degraded only slightly to Z ~ 0.8, and the signal-to-background of ~24.2 and signal-to-noise of ~143.3 indicated an excellent assay and HTS campaign. The (**a**) frequency distribution of the activities, the (**b**) 2-D scattergram of activities as a function of plate order, and (**c**) a 3-D view of the plates and wells against their activities, all indicate a robust close to normally distributed screen, with no serious plate-based artifacts

concentration has been selected based on hit rate, the final programming is adjusted to accommodate the desired batch size of a given run (*see* **Note 9**). Typically the full chemical collection is run in three to five batches with continuous monitoring of data generated (raw reader data in a plate map view or in a system-generated "heatmap" view of activity) and a day between batches to calculate and critically review the assay parameters, performance, and distribution of activity values of compounds in the HTS (*see* **Note 10**). It is very useful to have a running tabulation of the key assay parameters: Z′, Z, S/B, S/N, and CV% on a plate-by-plate, run-by-run, and day-by-day basis to allow any adjustments in protocols or schedule, or to troubleshoot and diagnose any mechanical system failures. When the entire chemical library screen is completed, the assay parameters for the entire run should be calculated and verified to be within acceptable assay specifications. Also several visualizations of the entire run are useful to verify an acceptable HTS campaign. A frequency distribution plot (Fig. 6a) of the activity of the chemical collection overlaid on the population distribution of the positive and negative controls is particularly useful to verify the expected true normal symmetrical Gaussian distribution of controls and the near-normal distribution of the centroid of compound activities near baseline activity, however, with a desirable slight "tailing" or "smearing" of this distribution towards higher activities, which indicate that the library has a significant number of compounds spanning a range of activities. Indeed a true "normal" distribution of a chemical collection would indicate that this entire population of compounds behaves as inactive baseline "noise" with *no separable hits* under the screening conditions. Scattergrams (Fig. 6b) plotting %Activity (inhibition or enhancement) as a function of wells tested by plate or by compound ID allow visualization of the assay window and emphasize the individually active wells that "pop up" over the main manifold of background activities. Finally, 3D visualization (Fig. 6c) of all plates across all runs allows confidence that all plates are performed within specifications and no particular run is corrupted by large run artifact.

4 Notes

1. *Key reagent procurement.* Furthermore, the calculations for total volume of assays in Table 2 are for *homogenous* assays, where assay components are simply added and no additional washing steps that would generate 5–10× more waste volumes are included. Indeed reagent procurement is often the major bottleneck in the HTS process (also in the author's personal experience of over 20 years of managing HTS projects and organizations). If initial materials cannot be obtained or prove inadequate, it can delay assay development. If production methods do not scale-up properly and yields of protein reagents or cells are insufficient of inappropriate quality, implementation and HTS will also be significantly delayed. In certain cases, only partial batches will be able to be produced and a decision will have to be made as to whether the reagent production processes yield consistent batches to allow ongoing screening or whether the screen should be delayed until all batches can be pooled before the start of production screening. It is also assumed that the batches are subdivided sensibly and stored (−80 °C) in sufficient aliquots that match the needs of assay development, production HTS screening run size, and HTS follow-up steps. Stability of reagents to freeze–thaw cycles, working dilution and storage on ice during HTS runs, and long-term storage also need to be established. Also, while stabilizing agents may be added for storage of concentrated reagents (e.g., 50 % glycerol for proteins), their concentrations may need to be minimized in the working dilutions to ensure proper dispensing characteristics (e.g., low viscosity and hang-up droplets).
2. *Cell line procurement.* This assumes that the functional phenotype of the engineered cells or cell lines is stable during expansion and passage, remains mycoplasma free during passage,

and grows and tolerates a relatively high split ratio during passage to allow expansion from seed cell stocks, and maintain a relatively high plating efficiency. For the use of frozen cell, the long-term cryogenic stability as measured by assay response and expression phenotype of the banked engineered cell lines, a consistent thawing and seeding protocol, must be established prior to and monitored during the entire HTS campaign.

3. *Microplates, labware, and assay kits.* This represents an inescapable and significant cost associated w/HTS projects, though production HTS groups are recommended to develop as many assays into common plate types to allow consolidation and projection of plate needs that allow core labs to negotiate favorable bulk discounts and coordinate production runs from suppliers to ensure single microplate lots, and even set up standing orders, or off-siting of inventories for just-in-time delivery to minimize storage requirements. In Table 2, disposable plastic tips for multichannel pipette tips were not included though some workstations still make use of them. These tips are typically priced at \$0.03–\$0.06 each except for specialized tips (w/filter plugs, conductive volume sensing, or ultralow volume or long reach) which may be as high as \$0.25 per tip, and represent more a marginal cost during very early assay development. Furthermore, most production HTS labs now have practically eliminated the need for them by using dispense-only bulk reagent dispensers (1–10 μL) for common assay reagents/buffer/controls and pin tools (20–100 nL) or noncontact acoustic drop ejection dispensers (2.5–50 nL) for critical volume transfers of DMSO stock solutions of compounds. In certain cases, readymade commercial kits that prove critical or offer significant performance, speed, and ease of implementation may be used. These kits often represent a largest cost of an assay, though again bulk discounts and just-in-time delivery may be negotiated with suppliers.
4. *Chemical libraries.* Keep compound source plates heat sealed between use and avoid freeze–thaw to keep compounds in solution. We also note that depending upon the choice of number of control wells in the parent source plate formats (which often originates as first empty column of a 96-well plate) from which higher density format plate are compressed (in a 4:1 A1, A2, B1, B2 compression grid), the number of empty control wells can be a significant number of wells that detract from the screening capacity of a plate, and represent a significant "overage" of reagents not being used to test compounds. This issue can be removed if an organization has limited the number of control wells, by directly creating source plates in 1,536- or 384-well formats with fewer control wells. Management of compound DMSO solutions also is subject of

much review, especially with regard to letting compound plates reach room temperature before unsealing, to avoid "cold-trap" condensation of ambient moisture into the DMSO stock. Similarly, minimizing the time a stock solution well is unsealed reduces the amount of water absorption into the deliquescent DMSO. Much of the anecdotal "instability" of chemical library compounds was determined to be due to increasing insolubility of compounds and subsequent precipitation and loss from solution, when %DMSO levels drop due to increasingly absorbed water in the stock solutions that are badly maintained or poorly used. It is also important to avoid intermediate dilution of compounds into aqueous stocks with attendant reduction in %DMSO or to store them for any periods of time before their use in screening as precipitation of compounds increases with decreasing %DMSO. Direct pre-dispensing into dry microtiter plates, with short term (<1 day) storage is preferred to any intermediate dilution, as long as mixing is not a problem [40]. It is also important that any serial dilutions are made into neat DMSO; each stock solution in the series is added to the assay at a constant volume to achieve the desired final constant DMSO concentration for the assay. It has sometimes been the case that the potency of a reference compound is consistently and reproducibly lower than literature values and upon reviewing the details of referenced literature, it is found that the serial dilution protocols intermediate aqueous solutions of compounds at lower DMSO/solvent, or serial dilution into aqueous solutions with increasingly lower concentration of DMSO, which has the effect of reducing apparent potency, due to precipitation of compound in the assay.

5. *Incubation temperature and duration*. At the other extremes of cell-based assays, often research benchtop assays are proposed that required 3–10-day differentiations, followed by 1-day compound pretreatment, and 3–5 days of signal development for an overall assay duration of 6–15 days. This leads to logistical issue for cell line seeding and maintenance and scheduling of tissue culture support. Also practically, it is very difficult to prevent significant "edge effects" due to high evaporation rates at edge wells, especially in 1,536-well plates even in well-humidified tissue culture incubators. At these volumes, cells are seeded in about half-volume of a full microwell (2–4 μL); the additional media with assay reagents are simply added (homogeneous) instead of washing or exchanging off media. Gas exchange in 1,536-well plates is less efficient versus 96- and 384-well tissue culture plates since the former present relatively narrow columns of liquid with a small meniscus area. Therefore, practically our group has achieved acceptable results at most for 1–3-day growth post seeding w/1–2-day treatment

and incubation before signal reads for 2–5-day overall cell-based assays in 1,536-well assays, while in 384-well assay we are able to achieve 1–2 days of extra duration. At the other extreme in live-cell kinetic imaging assays for GPCR Ca^{2+} flux measurements (e.g., FLIPR or FDSS) in cases where the cell lines do not function at ambient temperature, the entire assembly of plates and reagents must be allowed to equilibrate to 37 °C inside the thermo-controlled cabinets of these instruments prior to addition of agonists to start signaling reactions. This may require as much as 5–10 min per plate as *plastic is a very poor thermal conductor of heat*. This limits the overall timing and throughput of these instruments. In the case of high-content imaging assay, post biology, cells are fixed and stained (washes required), so actual imaging is asynchronous to assay biology.

6. *Surface-dependent effects*. It is an underappreciated fact that 384- and 1,536-well formats have a proportionately higher contacted surface area versus contained assay volume (*see* Table 2), so any surface-dependent phenomena is amplified with both positive (e.g., plate coating or ELISA assays) and deleterious (e.g., limiting enzyme depletion by absorption) effects.
7. *Nonspecific compound aggregation and protein stickiness*. Williams and Scott [26] point out that proteins are inherently "sticky," so assays with very low protein concentrations or lacking solubilizing detergents may tend to "find" more nonselective compounds [41, 42]. Many library compounds are known to form aggregates in solution, which nonspecifically inhibits enzymes, and low concentrations of additives such as bovine serum albumin (0.1 % BSA), casein, Tween-80, and Triton X-100 [43–45] have been found to mitigate against this. Additionally, these additives can mitigate against the stickiness of both test compounds and target proteins towards the plastic well surfaces of microtiter plates (especially untreated polystyrene) and as noted in Table 2 surface absorption is exacerbated in miniaturized assays as the surface area-to-contained volume ratio increases from 3- to 15-fold in going from 96- to 1,536-well formats. In the case of a glycosidase assay [26], the apparent instability of an enzyme was determined from preincubation studies and amelioration with 0.01 % TX-100 to be due to such surface plastic binding.
8. *Assay tolerance to DMSO*: This is one key determining factor in what screening concentration can be achieved and whether a high concentration (e.g., 10 mM) "mother" stock solution or more plentiful diluted "daughter" stock solution (e.g., 2 mM) can be utilized as a compound source during the HTS run. For example, an adherent cell-based assay with only 0.1 %

DMSO tolerance requires a 1,000-fold dilution of DMSO, so while a 10 mM stock achieves a final desired 10 μM screening concentration, a 2 mM stock could not. Furthermore, this affects the level of practical miniaturization of the assays achievable from a liquid transfer and mixing perspective. In the present example, since the volume of stock solution to be transferred must be 1/1,000th of the final desired assay volume, miniaturization to a 5 μL 1,536-well assay is robust achievable only with an acoustic dispense of 5 nL, whereas the 50 nL required for a 50 μL 384-well assay could be dispensed with either an acoustic dispenser or a pintool. With pintools, all wells simultaneously see compounds, and the pins can be used to facilitate mixing of well contents to minimize compound and DMSO concentration gradients through rapid repeated cycles of insertion–withdrawal of the pins. On the other hand, acoustic dispensers while very accurate and precise are rather slow (7–10 min/1,536-well plate), so the first well will be in contact with compound and DMSO for a longer time than the last well dispensed, so the response time of the cells to compounds must be slower than the cumulative dispense time or the incubation time for generation of signal must be several lifetimes longer than the compound transfer time. Additionally, the acoustic dispenser requires inversion of seeded cell plates, and each well receives the ejected compound stock DMSO solution onto the assay meniscus which is away from the cell monolayer, thus preventing an undesirable "bolus" of the denser DMSO settling on the cell monolayer. However, active mixing is not possible and relies on diffusion to achieve solution uniformity (>5 min), which may be incomplete. If the cells are not adherent (suspension), 5 nL of compounds could be pre-dispensed to "dry" plates and dispensing of 2–3 μL of cell suspension should be adequate for mixing in the well.

9. *Implementation of robotics.* Therefore often the key determinant of implementing a robotic HTS assay is the working stability of the key reagents that affects the stability of the overall assay parameters to stay within acceptable specifications (e.g., ±10 % of initial time values). This time window of stability affects the total number of microplates or "batch size" that can be practically processed in a single HTS "run" to yield acceptable data, before the "expired" critical reagents must be flushed from the system and replenished. The cumulative "dead volume" in the inaccessible volume of reagent reservoirs, supply lines, connection, and valves and required "overages" of reagents involved in an automated system can be significant. A good robotic engineer will minimize these by using narrow and short lengths of biocompatible tubing (e.g., microbore HDPE tubing) consistent with acceptable backpressure and

distance of the reservoir to the device. Furthermore, this "dead volume" often stands static and reaches ambient temperatures where stability of the biological reagent may be poorer before the start of the run and between additions to plates; therefore the systems may have to be "flushed" or "primed" before the first addition and perhaps even before each plate addition (if the data confirm this), which will add to the overage of reagents when calculating reagent needs.

10. *Production robotic HTS*. It may be naïvely assumed, at this point, that the full chemical library can be robotically screened simply as a matter of "pushing the button" on the robotic system, and letting the system run unattended 24/7 until the screen is completed. Indeed in a 1,536-well format (containing 1,408 compounds/plate) an entire library of 365,000 compounds would comprise (Table 2) just 260 plates and with a typical robot specified for 1.5–2 min per unit operation, these 260 plates could each be started at these intervals to the entire run and could theoretically be completed in 7.2–9.3 h (260 plates × (1.5–2 min/plate) + 30-min incubation + 10 min for five other operations for the last plate). However, as a practical matter this theoretical maximum may never be attempted nor achieved due to several key reasons. The foremost reason is the very high cost of run or assay failure, followed by the limitations of the stability of key reagents at the working dilutions in the reservoirs on the system or the necessity to seed batches of cells into plates for cell-based assays, and finally the human and organizational factors involved in a research environment where staffs usually do not perform shift work. Catastrophic run failures with the robots are infrequent, an usually arise from a dropped or jammed plate, that can lead to a system shut down and a need to manually "recover" the system. This results in a loss of some plates, and a delay in the overall screen in which the later plates may be compromised due to reagent instability at the unanticipated longer times. More insidious run errors are where no major mechanical operation jams occur, but where a particular valve or dispenser has a clogged tip and jeopardizes all data, or where an unanticipated contaminant or reagent instability presents itself only during the full production extended run. When these failures occur, the often very valuable and difficult-to-replace critical reagents (or cells) are already diluted and will expire whether the system can be recovered or not, or if the data generated is suspect, then the run represents a loss of this valuable reagent with the attendance of actual costs and high-time-sensitive opportunity cost. Therefore, more typically, the library run is divided into three to five batch runs with a day between batches to analyze the performance parameters to ensure assay and run stability. Real-time inspection of data generation from readers in a plate-grid format allows ready notice of any clogged tips from "striping"

or pronounced edge effects, though it is important to note that even with this level of monitoring, all plates that have been processed before the first plate is read may still be corrupted, before the run can be interrupted, and problem diagnosed and repaired before continuing a run.

Acknowledgments

This work was supported by the NIH Roadmap Molecular Libraries Program Grant U54 HG005033 and the *Conrad Prebys* Center for Chemical Genomics at the Sanford-Burnham Medical Research Institute. The referenced example TNAP assay was originally supported by the NIH Roadmap Initiative U54 HG003916, NIH Grant RC1 HL101899, and NIH Grant R03 MH077602-01.

The author wishes to thank Dr. Fu-Yue Zeng and Mr. Carlton Gasior for providing the examples of the programming interface, example schedule, and 10-plate run simulations on the robotic screening system from the GUI of the HRB, Cellario™ system software. The author also thanks Dr. Eduard Sergienko for his thorough reading and useful comments and for his guidance on the assay and HTS history of the TNAP project.

References

1. Macarron R, Banks MN, Bojanic D et al (2011) Impact of high-throughput screening in biomedical research. Nat Rev Drug Discov 10:188–195
2. Houston JG, Banks MN, Binnie A et al (2008) Case study: impact of technology investment on lead discovery at Bristol-Myers Squibb, 1998-2006. Drug Discov Today 13:44–51
3. Houston JG, Banks M (1997) The chemical-biological interface: developments in automated and miniaturised screening technology. Curr Opin Biotechnol 8:734–740
4. Macarrón R, Hertzberg R (2011) Design and implementation of high throughput screening assays. Mol Biotechnol 47:270–285
5. Roy A, McDonald PR, Sittampalam S et al (2010) Open access high throughput drug discovery in the public domain: a Mount Everest in the making. Curr Pharm Biotechnol 11: 764–778
6. Silverman L, Campbell R, Broach JR (1998) New assay technologies for high-throughput screening. Curr Opin Chem Biol 2:397–403
7. Banks MN, Cacace AM, O'Connell J , Houston JG (2005) High-throughput screening: evolution of technology and methods. In: Gad WC (ed) Drug discovery handbook. John Wiley & Sons 559–602
8. Burbaum JJ, Sigal NH (1997) New technologies for high-throughput screening. Curr Opin Chem Biol 1:72–78
9. Comley JCW, Binnie A, Bonk C et al (1997) A 384-HTS for human factor VIIa: comparison with 96- and 864-well formats. J Biomol Screen 2:171–178
10. Rogers MV (1997) Light on high-throughput screening: fluorescence-based assay technologies. Drug Discov Today 2:156–160
11. Sittampalam GS, Kahl SD, Janzen WP (1997) High-throughput screening: advances in assay technologies. Curr Opin Chem Biol 1:384–391
12. Hertzberg RP, Pope AJ (2000) High-throughput screening: new technology for the 21st century. Curr Opin Chem Biol 4:445–451
13. Entzeroth M, Flotow H, Condron P (2009) Overview of high-throughput screening. Curr Protoc Pharmacol Chapter 9, Unit 9 4
14. Eglen RM, Singh R (2003) Beta galactosidase enzyme fragment complementation as a novel technology for high throughput screening. Comb Chem High Throughput Screen 6:381–387
15. Fan F, Wood KV (2007) Bioluminescent assays for high-throughput screening. Assay Drug Dev Technol 5:127–136
16. Ormand J, Bruner J, Birkemo L et al (2000) A centralized global automation group in a

decentralized organization. J Autom Methods Manag Chem 22:195–198
17. Fox S (2006) High-throughput screening: update on practices and success. J Biomol Screen 11:864–869
18. Hamilton SD (1991) Managing an automation development group. J Autom Chem 13:23–27
19. Janzen WP, Popa-Burke IG (2009) Review: advances in improving the quality and flexibility of compound management. J Biomol Screen 14:444–451
20. Matson SL, Chatterjee M, Stock DA et al (2009) Best practices in compound management for preserving compound integrity and accurately providing samples for assays. J Biomol Screen 14:476–484
21. Radu C, Adrar HS, Alamir A et al (2012) Designs and concept reliance of a fully automated high-content screening platform. J Lab Autom 17:359–369
22. Cuatrecasas P (2006) Drug discovery in jeopardy. J Clin Invest 116:2837–2842
23. Frantz S (2007) Pharma faces major challenges after a year of failures and heated battles. Nat Rev Drug Discov 6:5–7
24. Gribbon P, Andreas S (2005) High-throughput drug discovery: what can we expect from HTS? Drug Discov Today 10:17–22
25. Macarron R (2006) Critical review of the role of HTS in drug discovery. Drug Discov Today 11:277–279
26. Williams K, Scott JE (2009) Enzyme assay design for high-throughput screening. In: Janzen WP, Bernasconi P (eds) High throughput screening. Methods Mol Biol 565: 107–126
27. Inglese J, Johnson RL, Simeonov A et al (2007) High-throughput screening assays for the identification of chemical probes. Nat Chem Biol 3:466–479
28. Frearson JA, Collie IT (2009) HTS and hit finding in academia—from chemical genomics to drug discovery. Drug Discov Today 14:1150–1158
29. Baker M (2010) Academic screening goes high-throughput. Nat Methods 7:787–792
30. Devanarayan V, Sawyer BD, Montrose C et al (2012) Glossary of quantitative biology terms. http://www.ncbi.nlm.nih.gov/books/NBK92002/ In: Sittampalam GS, Gal-Edd N, Arkin M et al (eds) Assay guidance manual. Bethesda (MD): Eli Lilly & Company and the National Center for Advancing Translational Sciences. http://www.ncbi.nlm.nih.gov/books/NBK53196/
31. Houston JG (1999) HTS: productivity or oblivion? J Biomol Screen 4:229–230
32. Sills MA (1997) Integrated robotics vs task-oriented automation. J Biomol Screen 2:137–138
33. SLAS (2004) http://www.thefreelibrary.com/ANSI Approves Microplate Standards Drafted by SBS Committee; New...-a0112221338
34. Millán J (2006) Alkaline phosphatases. Purinergic Signal 2:335–341
35. Sergienko EA, Millan JL (2010) High-throughput screening of tissue-nonspecific alkaline phosphatase for identification of effectors with diverse modes of action. Nat Protoc 5:1431–1439
36. Sergienko E, Su Y, Chan X et al (2009) Identification and characterization of novel tissue-nonspecific alkaline phosphatase inhibitors with diverse modes of action. J Biomol Screen 14:824–837
37. Wu J (2002) Comparison of SPA, FRET, and FP for kinase assays. In: Janzen W (ed) High throughput screening. Humana, pp 65–85
38. Zhang JH, Chung TD, Oldenburg KR (1999) A simple statistical parameter for use in evaluation and validation of high throughput screening assays. J Biomol Screen 4:67–73
39. Gupta S, Indelicato SR, Jethwa V et al (2007) Recommendations for the design, optimization, and qualification of cell-based assays used for the detection of neutralizing antibody responses elicited to biological therapeutics. J Immunol Methods 321:1–18
40. Smith T, Ho P-i, Yue K , Itkin Z, MacDougall D, Paolucci M, Hill A, Auld DS (2013) Comparison of compound administration methods in biochemical assays: effects on apparent compound potency using either assay-ready compound plates or pin tool-delivered compounds. J Biomol Screen 18:14–25
41. Ryan AJ, Gray NM, Lowe PN et al (2003) Effect of detergent on "promiscuous" inhibitors. J Med Chem 46:3448–3451
42. Knowles J, Gromo G (2003) Target selection in drug discovery. Nat Rev Drug Discov 2: 63–69
43. McGovern SL, Helfand BT, Feng B et al (2003) A specific mechanism of nonspecific inhibition. J Med Chem 46:4265–4272
44. Feng BY, Shelat A, Doman TN et al (2005) High-throughput assays for promiscuous inhibitors. Nat Chem Biol 1:146–148
45. Feng BY (2007) A high-throughput screen for aggregation-based inhibition in a large compound library. J Med Chem 50:2385–2390
46. Sundberg SA (2000) High-throughput and ultra-high-throughput screening: solution- and cell-based approaches. Curr Opin Biotechnol 11:47–53
47. Smith C (2011) Cell-based kinase assays: many routes to the same information. Biocompare http://www.biocompare.com/Editorial-Articles/41838-Cell-based-Kinase-Assays-Many-Routes-to-the-Same-Information/
48. Iversen PW, Eastwood BJ, Sittampalam GS et al (2006) A comparison of assay performance measures in screening assays: signal window, Z′ factor, and assay variability ratio. J Biomol Screen 11:247–252

Chapter 5

Inhibitors of Tissue-Nonspecific Alkaline Phosphatase (TNAP): From Hits to Leads

Peter Teriete, Anthony B. Pinkerton, and Nicholas D.P. Cosford

Abstract

The optimization of active hits, commonly derived from high-throughput screening campaigns (*see* Chapters 2 and 4), into promising small-molecule lead compounds is one of the fundamental steps in early drug discovery. Directions taken during this stage can have important consequences reaching through lead optimization into preclinical development and beyond. Considering the ever-increasing costs of preclinical as well as clinical development phases (DiMasi et al., J Health Econ 22:151–185, 2003) the choices made at the early stages of drug discovery can have a real impact on the likelihood of the best lead becoming a viable candidate (Bleicher et al., Nat Rev Drug Discov 2:369–378, 2003). Thus it is important to utilize proven and robust methodologies to turn promising hits into suitable lead series with propitious characteristics. Here, we describe such an approach using the example of a tissue-nonspecific alkaline phosphatase (*see* Chapter 3) inhibitor developed in our group (Sidique et al., Bioorg Med Chem Lett 19:222–225, 2009).

Key words Absorption, distribution, metabolism, excretion, and toxicity (ADME/T), High-throughput screening (HTS), Hit-to-lead, Inhibitor design, Medicinal chemistry, Pharmacokinetics (PK), Pharmacophore, Rational design, Structure–activity relationship (SAR)

1 Introduction

The development of potent and selective small-molecule inhibitors of tissue-nonspecific alkaline phosphatase (TNAP) from a range of initial screening hits was achieved by following a number of steps common to many hit-to-lead projects. Our aim in this discourse is to demonstrate the process by presenting a real example that resulted in the development of a highly potent and selective inhibitor of TNAP and thus is intended as an illustration representing more general approaches. Where appropriate we will make reference to alternative approaches; however, our primary aim is to focus on the path taken in this project for the development of phosphatase inhibitors, thus keeping in spirit with the subject of this treatise. Essential steps performed during the process are listed

José Luis Millán (ed.), *Phosphatase Modulators*, Methods in Molecular Biology, vol. 1053,
DOI 10.1007/978-1-62703-562-0_5,

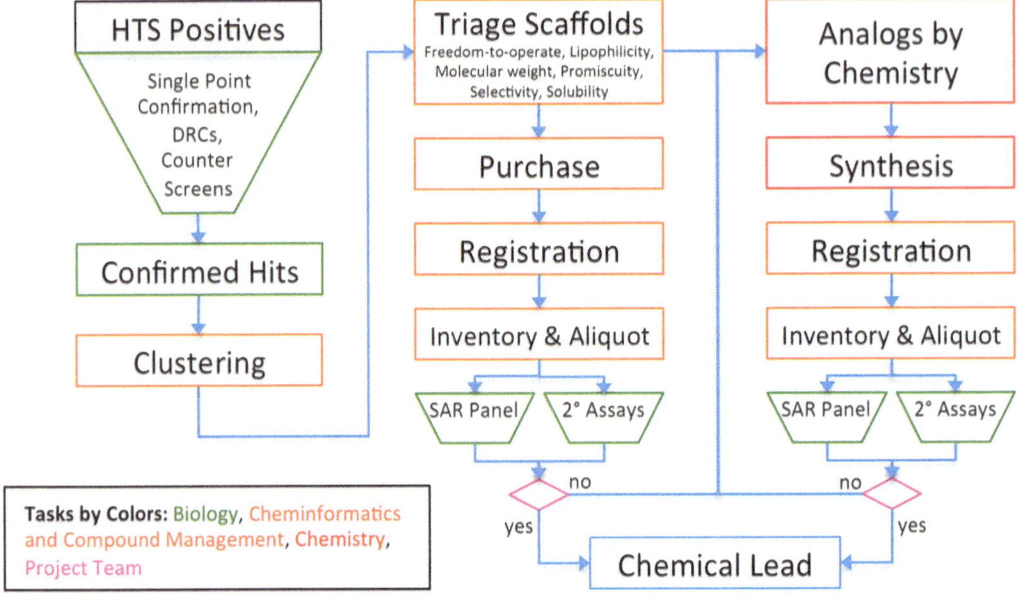

Fig. 1 Work-flow schematic of the hit-to-lead process commonly followed. Tasks are colored according to the expertise predominantly required for them

below. Each step is elaborated in the subsequent sections and further illustrated using our own experience. In practice, the steps depicted are not necessarily taken in the consecutive order described herein. Results from one experiment invariably influence the design of another, while the desire for efficiency often dictates the need to run things in parallel. The general schematic at the end shows how a workflow can look in comparable projects (Fig. 1):

- Hit selection
- First-round SAR studies
- Pharmacophore development
- 2nd to *n*th round of SAR
- Pharmacokinetics and pharmacodynamics
- Mechanistic studies

2 Hit Selection

Recent advances in robotics, cheminformatics, and assay formats, including miniaturization, have made the generation of small-molecule hits by high-throughput screening (HTS) campaigns a more or less routine procedure in the pharmaceutical and increasingly the academic sectors (*see* also Chapters 1 and 2). Active compounds identified from HTS are then validated in confirmatory assays and their respective activities characterized in more detail (i.e., dose–response curves, orthogonal assays, and selectivity assays). An alternative method that has gained increasing traction in recent years is the utilization of in silico approaches which allow the ultra high-throughput virtual screening of ligand libraries. This is largely due to considerable advances in computational resources available for even small-scale projects with limited resources as well as improved algorithms and superior physicochemical descriptors of ligand–receptor systems. We will not describe this approach further since it was not employed in our studies and there are many suitable publications and review articles covering the latest developments and best use scenarios [4, 5]. Subsequent to initial characterization, hit compounds are ranked according to a number of readily available properties. Most important are affinity, molecular weight, and lipophilicity (usually measured as cLogP), which can be linked in single parameters such as ligand efficiency and lipophilic efficiency to assess drug likeness. Furthermore, it is common practice to triage the initial hits by analyzing them against the pan assay interference compounds (PAINS) database [6, 7], as well as publicly available databases such as Pubchem and commercial tools (e.g., SciFinder) to identify and exclude highly promiscuous scaffolds. Furthermore, depending on the circumstances it might be advisable to engage in preliminary patent searches on the best compounds to determine potential intellectual property (IP) issues and to perform a "freedom to operate" (FTO) analysis. The results of the latter can have a substantial influence on the selection criteria for the top ranked hits. What follows is a detailed evaluation of the remaining scaffolds based on their respective properties for chemical tractability and thus their suitability in first-round structure–activity relationship (SAR) studies. To reiterate, properties evaluated at this stage are:

- Activity (single-point inhibition)
- Potency (IC_{50})
- Molecular weight
- Solubility
- Lipophilicity
- Selectivity
- Promiscuity
- Known adverse properties
- FTO

Using the TNAP inhibitors developed in our lab as a case study what follows is a more detailed account of how these properties were assessed and utilized in a hit-to-lead optimization procedure.

An HTS campaign of 66,000 compounds using a luminescence-based assay was performed under the MLSCN

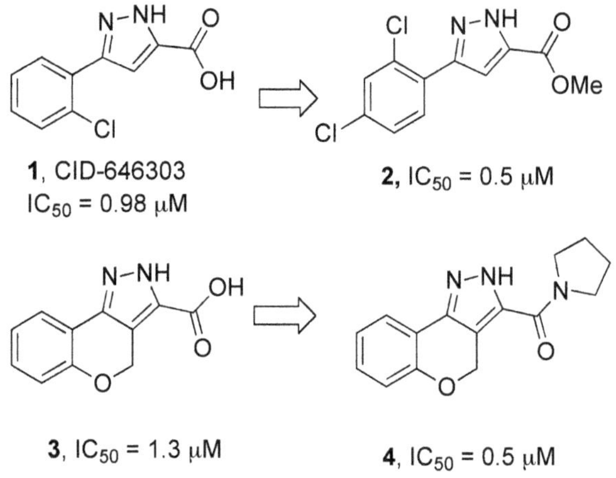

Fig. 2 Initial hits from screening and commercial analogues

initiative [8, 9]: PubChem link to AID 1056 (http://pubchem.ncbi.nlm.nih.gov/assay/assay.cgi?aid=1056). Initial HTS was performed at a concentration of 20 μM and compounds with greater than 50 % inhibition of TNAP were defined as *actives* from the primary screening. Hit confirmation was performed using both luminescent and orthogonal colorimetric assays to verify inhibitory activity against TNAP in dose–response mode performed in duplicate using a 10-point, twofold serial dilution of the hit compounds in DMSO. Compounds that showed good *solubility*, i.e., >10 μg/mL [10], and demonstrated reproducible IC_{50} values in the range of analyzed concentrations remained classified as "active." This resulted in an effective cutoff at 15.8 μM for the IC_{50} values of hits that proceeded to the next stage of evaluation. This screening effort led to the identification of several compound classes (54 compounds in ~33 distinct subsets, with a Tanimoto coefficient of 0.75 or greater) of low- and submicromolar inhibitors of TNAP [8]. *Selectivity* was assessed by testing compounds against the isozymes placental and intestinal alkaline phosphatase (PLAP and IAP, respectively) in luminescence-based assays. In addition, the remaining compounds were ranked against their activity in unrelated assays (PubChem) to determine their level of *promiscuity*. Finally, compounds with low promiscuity that were active in dose–response mode against TNAP, soluble in the range relevant to their potency, and inactive against PLAP and IAP were prioritized for synthetic chemistry follow-up. Applying the selection criteria described above to rank order the hits we identified the pyrazole derivative CID-646303 (Fig. 2) as one of the most promising hits. Our selection criteria also took into consideration the availability of commercial analogues (Fig. 2) and the prediction of synthetic accessibility of the scaffold (including availability of starting materials) by experienced medicinal chemists.

3 First-Round SAR Studies

SAR studies are at the heart of hit-to-lead development projects. The identification of key functional groups essential for improved activity is often achieved through a mix of experience, thoughtful planning of synthetically accessible derivatives, careful analysis of the results, and serendipity. The first and last factors are not easily implementable at will, however sound planning and thorough analysis are, and are never optional in any aspect of science.

Exploitation of commercially available analogues of the selected hits is often an early step in the first round of SAR studies. This is usually followed by the use of basic medicinal chemistry to modify these analogues to sample a reasonable fraction of the overall chemical space. In parallel it is essential to spend time researching synthetic routes to generate the basic scaffold and selected synthons with the objective to incorporate more, extended, or diverse functional groups and thus expand the sampled chemical space considerably.

The process of first-round SAR studies can be itemized by the following steps:

- Commercially available analogues
- Derivatization or conversion of readily accessible functional groups
- Identification of a key synthon
- Identification of synthetic route(s)
- Synthesis of a focused library of analogues

How this was performed using the example of TNAP inhibitors is illustrated in the following section.

Preliminary hit follow-up was accomplished by performing similarity searches on databases of *commercially available analogues* (Fig. 2). In this initial phase, 50 commercial analogues were identified, purchased, and tested for their ability to inhibit TNAP. This allowed us to define some important features of the SAR. For example, the potency in this series was improved from $IC_{50} = 0.98$ μM for the lead pyrazole **1** to $IC_{50} = 0.50$ μM for the 2,4-dichlorophenyl ester derivative **2** (Fig. 2). Furthermore, *conversion* of the tricyclic *derivative* **3**, with an IC_{50} value of 1.33 μM, to the pyrrolidine amide analogue **4** led to a threefold improvement in potency ($IC_{50} = 0.50$ μM). Encouraged by these results we developed a *synthetic route* (Scheme 1) and synthesized two focused libraries of substituted pyrazole amide analogues. In order to optimize the potency of the hit structure the pyrazole acid scaffold **8** was selected as the key *synthon* for the preparation of amide analogues. Reaction of acetophenone derivative **5** with sodium methoxide and dimethyl oxalate yielded the 1,3-diketone derivative

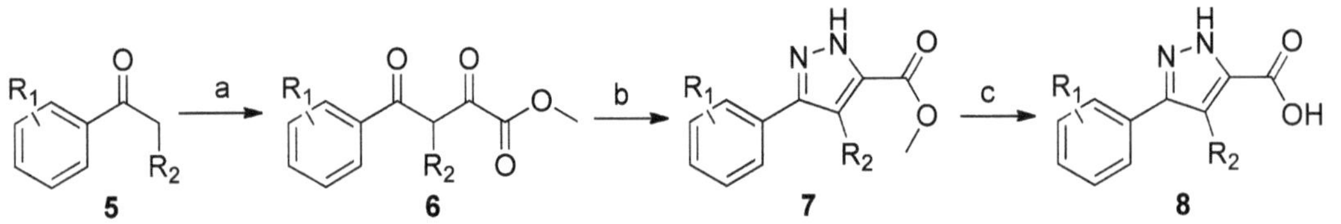

Scheme 1 Reagents and conditions: (**a**) i. NaOMe, Et_2O, dimethyl oxylate, 25 °C, 4–12 h, ii: HOAc (75–90 %); (**b**) i. N_2H_4, HOAc, 100 °C, 12 h (50–85 %) ii. LiOH, THF, MeOH, reflux, iii. HCl (90–95 %)

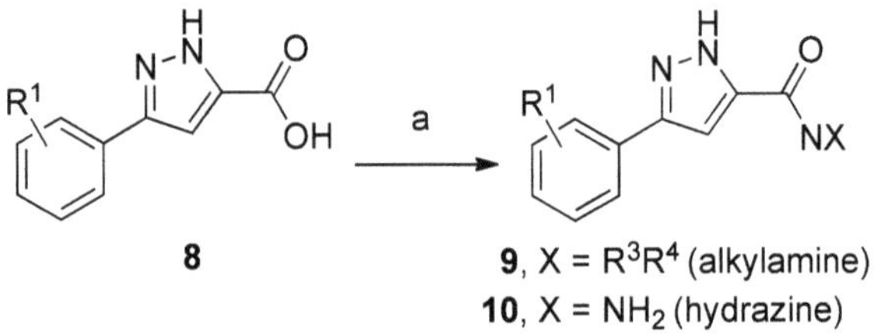

Scheme 2 Reagents and conditions: (**a**) EDC, HOBT, DMF, DIEA, NH_2X (85–95 %)

6 in excellent yields (75–90 %). Compound **6** was then reacted with hydrazine to give the corresponding pyrazole ester **7**. Saponification of the methyl ester provided access to the pyrazole acid derivative **8**.

The synthetic chemistry used for hit optimization by analogue production and testing is shown in Scheme 2. The pyrazole acid **8** was treated with HOBT, EDC, and DIEA to produce the amide **9** or the hydrazide derivative **10**. In light of the preliminary data generated from the HTS hits and commercial analogues our goal was to determine the key components of the SAR required for potency. For the *focused library* synthesis we selected a 2,4-dichloro and 2,4-dichloro-5-fluoro substitution pattern for the phenyl ring based on the initial SAR data. In the first library, 26 compounds were synthesized and tested in the in vitro assay. This led to the identification of four analogues with potency values of 0.1 μM or better (Table 1). The incorporation of a hydroxyl group on the amide alkyl substituent generally increased potency (**9a** and **9j**). In all cases the 2,4-dichloro analogues were more potent than the corresponding 2,4-dichloro-5-fluoro analogues (Table 1).

4 Pharmacophore Development

Development of an appropriate and useful pharmacophore model can substantially facilitate the rational design of advanced analogues (*see* Subheading 5). Pharmacophores are abstract descriptors of molecular features crucial for the interactions between ligands and receptors. There are two fundamental ways to generate pharmacophore models. The first takes the interaction between the

Table 1
Summary of in vitro data from first focused library

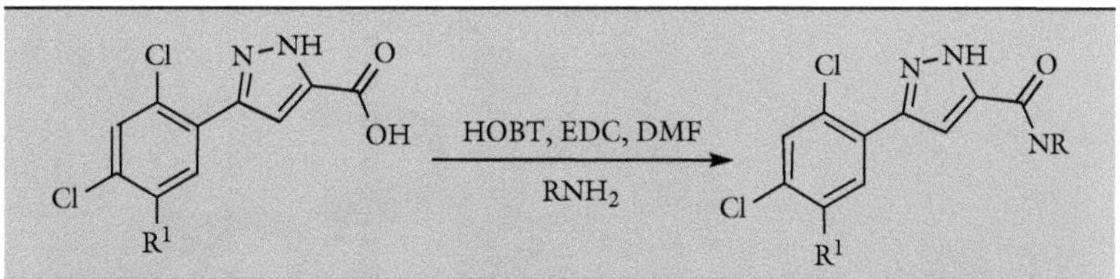

HOBT, EDC, DMF

RNH_2

Compound	R^1	RNH_2	IC_{50} (μM)
9a	H	$NH_2CH_2CH_2OH$	0.044
9b	H	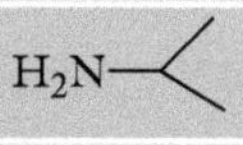 H_2N–	0.051
9c	H	 HN	0.119
9d	H	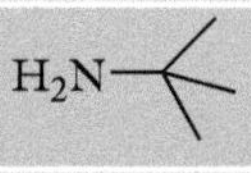 H_2N–	0.398
9e	H	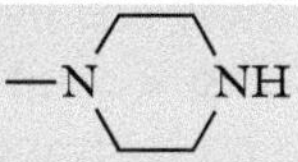 –N NH	0.828
9f	H	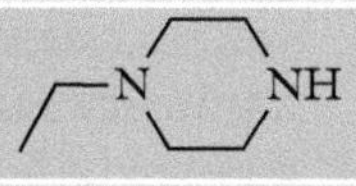 N NH	0.488
9g	H	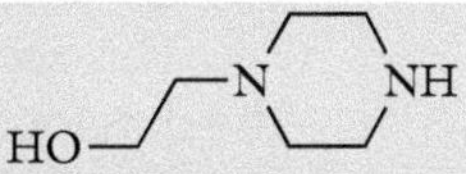 HO N NH	0.375
9h	H	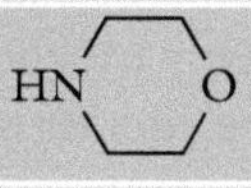 HN O	0.574
9i	H	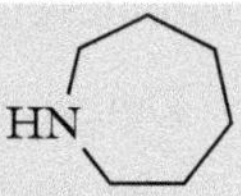 HN	1.450
9j	F	$NH_2CH_2CH_2OH$	0.100
9k	F	 HN	0.324
9l	F	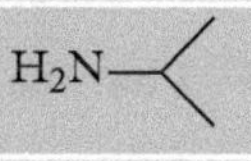 H_2N–	4.850
10	H	NH_2NH_2	0.044

ligand and the receptor into account as well as activity data resulting from assays performed on the focused analogue library (*see* Subheading 3). It is usually based on the availability of a high-resolution complex structure of the receptor and a ligand closely related to the hit series, but docking models and even the use of homology-modeled structures of the receptor can be useful. However, it is important to judge results more carefully the further

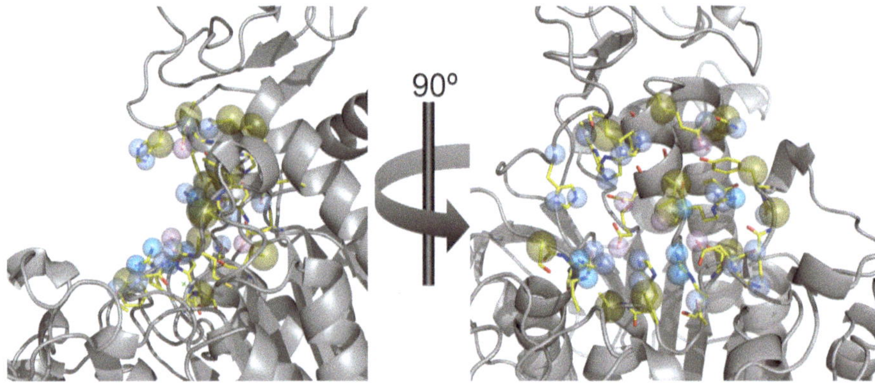

Fig. 3 Side view and full view of the pharmacophoric features of residues (*yellow*) found in the active site of TNAP (homology model based on 1EW2; human placental alkaline phosphatase). Hydrogen bond donor and acceptors are highlighted in *blue* and *red spheres*, respectively, while features for potential hydrophobic interactions are shown as *green spheres*

away one moves from original structures of the system investigated. The second way is based on the availability of a set of analogues and corresponding bioactivity data. Particularly in the latter case it will be essential to have analogues with activities spanning as large a potency range as possible. In an ideal case one would like to see a range of 4 magnitudes (commonly the nM to μM range).

This section is not intended to provide an in-depth guide to the development of pharmacophore models, but we hope that the benefit of including this type of analysis is self-evident. There are many excellent guides and publications on this subject available in the literature [11–13]. The basic steps performed for pharmacophore development are as follows:

- Determination of suitable receptor structures
- List of active compounds with bioactivity data (IC_{50}, EC_{50}, K_m, etc.)
- Conformational analysis
- Generating pharmacophore models
- Pharmacophore model validation

There is, as of yet, no high-resolution X-ray or NMR structure of TNAP elucidated that is suitable for the development of a protein–ligand-based pharmacophore. However, using the Swiss-Model project [14] we created a homology model of TNAP based on the high-resolution X-ray structure of human PLAP (PDB code 1EW2). Although the sequence identity between these two proteins is only 56 %, an actual homology of 74 % made it a promising approach. Residues involved in the active site of TNAP have been determined previously [15] and we used this information to establish a pharmacophoric profile of potential interactions in and adjacent to these residues. Figure 3 shows a frontal and side view

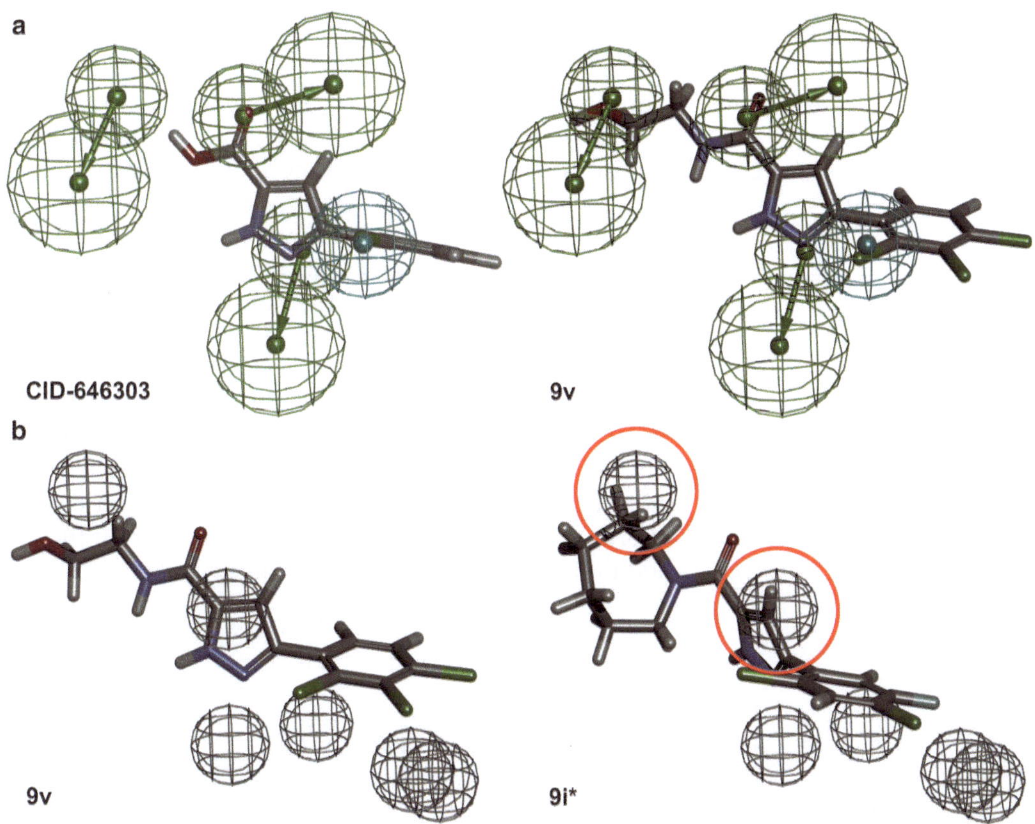

Fig. 4 (**a**) CID-646303, the original HTS hit, and **9v** overlaid on the HypoGen pharmacophore model with the best overall score. Favorable interactions providing hydrogen bond capabilities are shown by *green mesh spheres*. (**b**) Occupation of exclusion volumes (*grey mesh spheres*) by atoms of pyrazole analogues are highlighted in *red*. **9v**, one of the most active analogues, avoids all exclusion volumes while fulfilling favorable hydrogen bond potentials. On the other hand **9i*** (compound **9i** with an additional fluorine on the R1 position) clashes in two of the exclusion volumes, although it fulfills most of the hydrogen bond potentials

of this profile with the type of interactions possible depicted by colored spheres. However, the study delineating the residues involved [15] did not provide sufficient information to clearly identify the most crucial interactions utilized by the pyrazole compounds and thus we developed a ligand-only-based pharmacophore model. Forty analogues from the pyrazole series were chosen and divided up into a training set (30 compounds) and test set (10 compounds). Care was taken to select compounds with activities reaching across 4 orders of magnitude for both sets. Published protocols were closely followed to generate suitable pharmacophore models [11]. Figure 4 illustrates the best HypoGen pharmacophore model determined using Discovery Studio [16, 17]. Results from the pharmacophore development were utilized to guide the design of analogues during subsequent rounds of SAR.

5 Second to *n*th Round of SAR

Subsequent rounds of SAR are used to further drive the potency in the primary assay and to begin to address other issues that are crucial for the development of advanced leads. These factors include selectivity against closely related targets (in this case other alkaline phosphatases such as intestinal and placental alkaline phosphatase) as well as broader off-target activity (defined below). Also, at this point the preliminary in vitro absorption, distribution, metabolism, excretion, and toxicity (ADME/T) profile of one or more of the most potent analogues is determined, and potential issues defined. Further iterative rounds of SAR will then attempt to address these liabilities. Some of the important parameters are defined in the pharmacokinetics (PK) and pharmacodynamics (PD) section below. The design approach to second and ensuing rounds of SAR studies is similar to that employed during the first round. Additional steps often undertaken are:

- Exploration of available chemical space
- Advanced medicinal chemistry approaches
- Broader selectivity studies
- Detailed pharmacophore analysis
- Results from in vitro ADME/T studies
- PK and PD characteristics

For the development of optimized TNAP inhibitors we pursued the following strategy for subsequent rounds of SAR:

A second-generation set of pyrazoles consisting of a library of 28 compounds were synthesized next (Table 2). In this series we found that branching of the amide substituents generally decreased potency in the in vitro assay, especially when the chain length was greater than three carbon atoms. We also observed that amides with chain lengths of three carbons or less were the most active (Tables 1 and 2).

In the second series, when the hydroxyethyl chain was increased by one or two additional carbon atoms (**9m**, **9n**), or a second hydroxyethyl side chain was added (**9p**), these changes did not affect the potency of the initial hit **9a** (Tables 1 and 2). The most potent compound in this series was the 2,3,4-trichlorophenyl analogue **9v** (Table 3). This compound showed exceptional activity with an in vitro IC_{50} of 5 nM (Table 3). Furthermore, compound **9v** was inactive ($IC_{50} > 10$ μM) against both the related PLAP isozyme and the housekeeping enzyme glyceraldehyde-3-phosphate dehydrogenase (GAPDH), indicating a selectivity for TNAP of at least 2,000-fold.

Results from PK and ADME/T studies performed in parallel to the iterative SAR studies are detailed below.

Table 2
Summary of in vitro data from second focused library

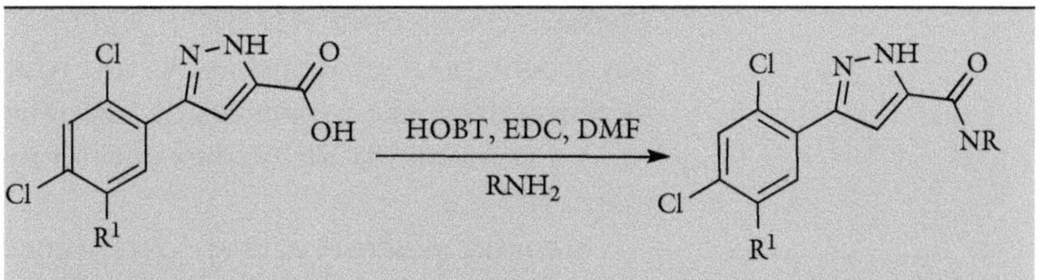

Compound	R^1	RNH_2	IC_{50} (μM)
9m	H	$NH_2CH_2CH_2CH_2OH$	0.031
9n	H	$NH_2CH_2CH_2CH_2CH_2OH$	0.035
9o	H	$NH_2CH_2CH_2OMe$	0.046
9p	H	$NH_2(CH_2CH_2OH)_2$	0.047
9q	H	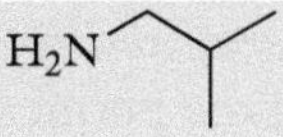 H_2N	0.082
9r	H	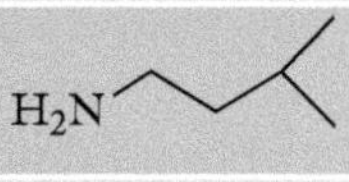 H_2N	0.459
9s	H	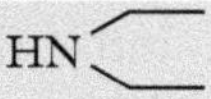 HN	0.285
9t	F	$NH_2CH_2CH_2CH_2OH$	0.056
9u	F	$NH(CH_2CH_2OH)_2$	0.127

Table 3
Summary of in vitro data for trichloro- and trifluorophenyl derivatives

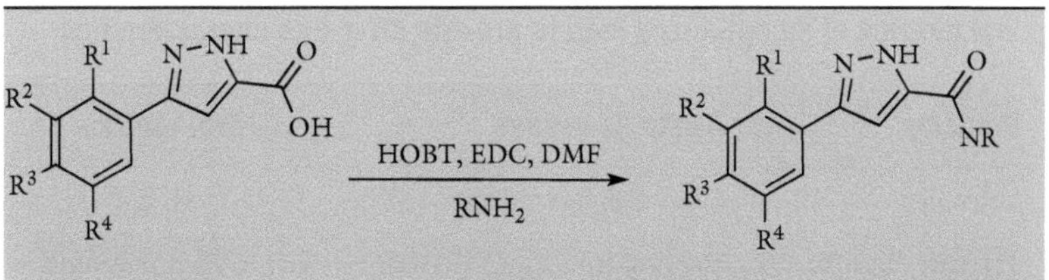

Compound	R^1	R^2	R^3	R^4	RNH_2	IC_{50} (μM)
9v	Cl	Cl	Cl	H	*$NH_2CH_2CH_2OH$*	0.005
9w	F	H	F	F	$NH_2CH_2CH_2OH$	0.134
9x	F	H	F	H	$NH_2CH_2CH_2OH$	0.035

6 ADME/T and PK Studies

Preliminary in vitro ADME/T profiling: In vitro ADME/T and physicochemical profiling assays are routinely employed to optimize the drug-like properties of analogues and to aid in the selection of compounds for further development (*see* Table 4 for ideal metrics).

Aqueous solubility data are determined at pH 5.0, 6.2, and 7.4 with UV detection. Compounds with the most favorable solubility profiles (>10 μg/mL) will be advanced. *Free drug concentrations in plasma* are determined using rapid equilibrium dialysis, which is the most quantitative method for determining the levels of plasma protein binding [18]. *Metabolic stability* in human, mouse, and rat liver microsomes is determined by incubating compounds in the presence of microsomes (1 mg/mL); the metabolites are quantitated using LC/MS methods. *Membrane permeability* data are determined using a parallel artificial membrane permeability assay (PAMPA). This in vitro method assesses the passive diffusion of compounds across a layer of specialized mixtures of phospholipids that mimic (a) the gut epithelium, and (b) brain capillary endothelial cells, the primary barrier to absorption into the brain [19]. Optimal permeability (P_{app}) is $>3\times10^{-6}$ cm/s. *Cytochrome P450* (*CYP450*) isoform (CYP1A2, 2C9, 2D6, and 3A4) inhibition is determined in human liver microsomes. Inhibition of product formation for probe substrates is detected by luminescence using the isoform-specific P450-glo assay (Promega; Madison, WI) over ten concentrations of inhibi-

Table 4
Ideal metrics of an optimized lead to provide drug-like characteristics

Property	Available assays	Ideal metrics	References
Solubility	Chemilumiescence	$10\times IC_{50}$	[22]
Protein binding	Rapid equilibrium dialysis	< 98.5 % bound	[23]
Metabolic stability	Liver microsomes or S9 fraction (various species)	$t_{1/2} > 1$ h	[24]
Permeability	CNS PAMPA	$\mathrm{Log}P \cong 1.0–4.5$	[19]
CYP450 profiling	Inhibition reaction profiling P450glo-assay	$IC_{50} < 10$ μM or $100\times IC_{50}$	[25]
Toxicity	MTT Cell-titer glo multi-tox assay	≥50 μM IC_{50}	[26]
In vivo exposure	Comprehensive pharmacokinetic (PK) assessment	AUC ~ $10\times IC_{50}$ or $>IC_{90}$; %F ≥ 20	[27, 28]

tor and assessed at one time point (previously determined to be in the linear range for time and protein concentration). The IC_{50} is analyzed by a four-parameter logistic fit. Appropriate positive control inhibitors are used for each enzyme. Mechanism-based inhibition will be investigated where warranted using the established method. *Covalent modification* is studied on selected compounds for the unacceptable propensity to form reactive metabolites in vitro. An assay that detects glutathione-trapped reactive intermediates after incubation with a liver homogenate S9 fraction is employed, and glutathione adducts are detected by LC/MS/MS using a method routinely used in our lab [20]. *Drug transporter screening* is used to identify substrates of MDR-1 (P-glycoprotein) which have reduced CNS concentrations because of efflux at the blood–brain barrier. Compounds are tested using MDCK cells transfected with and expressing MDR-1 (Panlabs; Concord, OH). Optimal P_{app} (apical to basal) is $>5 \times 10^{-6}$ cm/s. *Cross-reactivity* (off-target activity) profiling is performed by a number of external services, including Panlabs and Life Technologies (Carlsbad, CA). Selected compounds are tested against panels of cloned human or rat receptors, ion channels, and transporters. In general, the desired profile is ≤ 50 % inhibition or competitive binding at 10 μM for any target other than TNAP. Compounds exceeding this threshold are tested in eight-point concentration–response mode to determine the EC_{50} or binding affinity (K_i). The results of this profiling dictate the need for additional in vitro selectivity assays.

In vivo PK: Lead compounds with favorable properties are chosen for comprehensive PK analysis to select those suitable for evaluation in in vivo efficacy models [21]. Specific rodent strains are used for PK experiments to match the background/strain used in subsequent efficacy studies. For example, a cohort of Wistar rats are divided into two groups: the test drug is administered to one group orally and a second group intravenously to measure key parameters. Standard formulations are evaluated, including hydroxypropyl methylcellulose, carboxymethylcellulose, and polyethylene glycol. For intravenous studies, compounds are administered via indwelling catheters in the jugular vein and samples are collected from the carotid artery. Blood samples are obtained over a 24-h period and plasma analyzed by predetermined analytical methods. These experiments provide basic PK parameters including peak plasma concentration (C_{max}), bioavailability (%F), exposure (AUC), half-life ($t_{1/2}$), clearance (CL), and volume of distribution (V_d).

7 Mechanistic Studies

Mode of action or mechanism of action (MOA) studies are performed to characterize the interaction of an active compound with its target in detail in order to understand how the compound interacts with the target and in turn how natural substrates at physiologic concentrations will modulate this activity. Knowing a compound is competitive is often very useful during the iterative rounds of SAR, since further chemical synthesis can be directed to mimic the natural substrate. However, competitive compounds with promising structures and potent biochemical activity might compete with substrate present in the tissue of relevance at high concentration which can lead to an apparent loss of in vivo activity. Alternatively, a greater increase of in vivo activity than is biochemically predicted from in vitro IC_{50} data might be related to unusual kinetic behaviors such as slow on and/or slow off rates (tight binding). As there is no single unique answer, biochemical MOA studies help in the interpretation of cell-based and in vivo activities, and provide further support for molecules with desirable characteristics to move forward in the hit-to-lead process. A range of biochemical and biophysical methodologies are available to establish the desired information. In this chapter we focus on the methods employed during the optimization of the pyrazole-based TNAP inhibitors.

Important characteristics one would like to resolve through MOA studies are:

- Type of modulation (competitive, uncompetitive, or noncompetitive)
- On and/or off rates
- Presence of single or multiple binding sites
- Cooperativity

We performed a series of experiments to elucidate the MOA of the novel TNAP inhibitors. The catalytic mechanism by which TNAP degrades PP_i consists of rapid phosphorylation of the active site in the presence of the phospho–donor substrate and a rate-limiting dephosphorylation by the phospho–acceptor substrate, either water or amino-containing alcohols (Fig. 5). All the TNAP inhibitors reported to date are uncompetitive with respect to phospho–donors and are likely to be non- or uncompetitive with diethanolamine (DEA). The latter conclusion is based on the fact that the majority of the alkaline phosphatase assays are performed in the presence of saturating concentrations of DEA or other phosphor–acceptors.

To characterize the MOA of the novel TNAP inhibitor series we selected compound **9v** for additional studies. By performing detailed kinetic studies, we demonstrated that **9v** is competitive with respect to both substrates, the water-soluble 1,2-dioxetane reagent disodium 2-chloro-5-(5′-chloro-4-methoxyspiro[1,2-dioxetane-3,20-tricyclo[3.3.1.13,7]decan]-4-yl)-phenol-1-(dihydrogen phosphate) (CDP-star) and DEA (Fig. 6). This is the first time that a competitive MOA has been established for an inhibitor of TNAP.

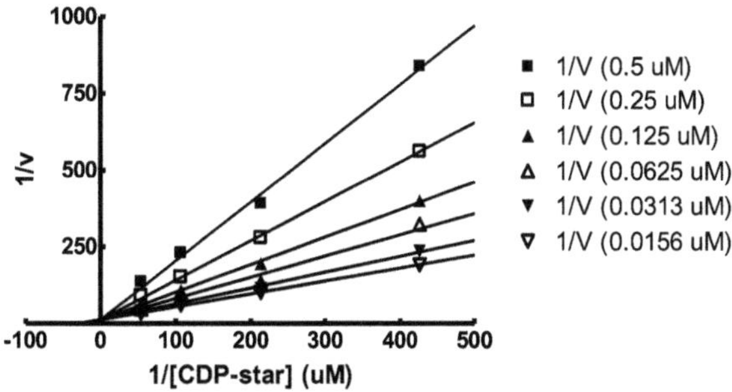

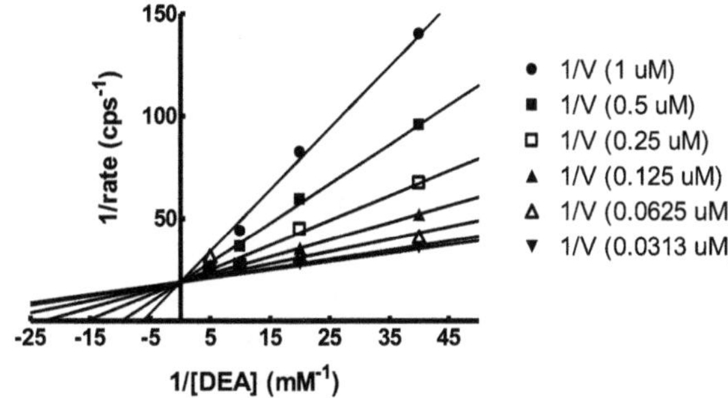

Fig. 5 Mechanism of action studies for compound **9v**: Lineweaver–Burk plots. TNAP kinetic data were obtained in the presence of (**a**) 6.25 mM DEA or (**b**) 18.8 μM CDP-star. The concentration of the second substrate was varied in the presence of different concentrations of **9v**. *Closed circle*, 1,000 nM; *closed square*, 500 nM; *open square*, 250 nM; *closed triangle*, 125 nM; *open triangle*, 62.5 nM; *closed upside-down triangle*, 31 nM; *open upside-down triangle*, 15.6 nM

$$E + DO\text{-}P_i \rightleftharpoons E\cdot DO\text{-}P_i \rightleftharpoons E\text{-}P_i\cdot DOH \rightleftharpoons E\text{-}P_i \rightleftharpoons E\cdot P_i \rightleftharpoons E + P_i$$

DOH (released from $E\text{-}P_i\cdot DOH$); AOH (added to $E\text{-}P_i$)

$$E\text{-}P_i\cdot AOH \rightleftharpoons E\cdot AO\text{-}P_i \rightleftharpoons E + AO\text{-}P_i$$

Fig. 6 The catalytic mechanism of the alkaline phosphatase reaction. The initial alkaline phosphatase (E)-catalyzed reaction consists of a substrate (DO–Pi) binding step, phosphate-moiety transfer to Ser-93 (in the TNAP sequence of its active site), and product alcohol (DOH) release. In the second step of the reaction, phosphate is released through hydrolysis of the covalent intermediate (E–Pi) and the non-covalent complex (E·Pi) of the inorganic phosphate in the active site. In the presence of nitrogen-containing alcohol molecules (AOH), such as the buffer diethanolamine (DEA), phosphate is also released via a transphosphorylation reaction

All the MOA data suggested that a single binding site is found on TNAP, and thus a further investigation into MOAs concerning cooperativity was not needed. However, the literature provides plenty of excellent examples of how to determine the exact properties of protein–ligand interactions not following single-mode binding events.

8 Conclusion

By following the described steps in this chapter we were able to take a promising hit from an HTS screen and turn it into a lead compound with high potency and excellent selectivity for the intended target. The resulting optimized TNAP inhibitor provided us with a useful small-molecule probe to further investigate the involvement of TNAP in a range of disease models as well as aid the development of a drug candidate, the logical next step.

In conclusion, it is clear that successful hit-to-lead campaigns rely on a sound strategy. It is imperative that the characteristics one needs in an effective lead compound are clearly delineated in advance and that the various selection stages one encounters during the process are guided by the appropriate parameters. An essential component is the collaboration of experienced chemists, cheminformaticists, biologists, pharmacologists, and assay developers. Thus, assembling a strong team with the required expertise can be as important as identifying the ideal target.

References

1. DiMasi JA, Hansen RW, Grabowski HG (2003) The price of innovation: new estimates of drug development costs. J Health Econ 22:151–185
2. Bleicher KH, Bohm HJ, Muller K et al (2003) Hit and lead generation: beyond high-throughput screening. Nat Rev Drug Discov 2:369–378
3. Sidique S, Ardecky R, Su Y et al (2009) Design and synthesis of pyrazole derivatives as potent and selective inhibitors of tissue-nonspecific alkaline phosphatase (TNAP). Bioorg Med Chem Lett 19:222–225
4. Barh D, Ahmad S, Bhattacharjee A (2012) In silico and ultrahigh-throughput screenings (uHTS) in drug discovery: an overview. Pharmaceutical biotechnology. Wiley-VCH Verlag, Weinheim, Germany, pp 451–490
5. Zoete V, Grosdidier A, Michielin O (2009) Docking, virtual high throughput screening and in silico fragment-based drug design. J Cell Mol Med 13:238–248
6. Baell JB, Holloway GA (2010) New substructure filters for removal of pan assay interference compounds (PAINS) from screening libraries and for their exclusion in bioassays. J Med Chem 53:2719–2740
7. Pinkerton AB, Vernier J-M, Cube RV, Hutchinson JH (2006) Inventor Heterocyclic indanone potentiators of metabotropic glutamate receptors. USA WO Patent WO/2006/047237
8. Sergienko E, Su Y, Chan X et al (2009) Identification and characterization of novel tissue-nonspecific alkaline phosphatase inhibitors with diverse modes of action. J Biomol Screen 14:824–837
9. Sergienko EA, Millan JL (2010) High-throughput screening of tissue-nonspecific alkaline phosphatase for identification of effectors with diverse modes of action. Nat Protoc 5:1431–1439
10. Kerns EH, Di L, Carter GT (2008) In vitro solubility assays in drug discovery. Curr Drug Metab 9:879–885

11. John S, Thangapandian S, Arooj M et al (2011) Development, evaluation and application of 3D QSAR Pharmacophore model in the discovery of potential human renin inhibitors. BMC Bioinformatics 12 Suppl 14, S4
12. Binns M, de Visser SP, Theodoropoulos C (2012) Modeling flexible pharmacophores with distance geometry, scoring, and bound stretching. J Chem Inf Model 52:577–588
13. Tai W, Lu T, Yuan H et al (2011) Pharmacophore modeling and virtual screening studies to identify new c-Met inhibitors. J Mol Model 1–14
14. Arnold K, Bordoli L, Kopp J et al (2006) The SWISS-MODEL workspace: a web-based environment for protein structure homology modelling. Bioinformatics 22:195–201
15. Kozlenkov A, Le Du MH, Cuniasse P et al (2004) Residues determining the binding specificity of uncompetitive inhibitors to tissue-nonspecific alkaline phosphatase. J Bone Miner Res 19:1862–1872
16. Taha MO, Bustanji Y, Al-Bakri AG et al (2007) Discovery of new potent human protein tyrosine phosphatase inhibitors via pharmacophore and QSAR analysis followed by in silico screening. J Mol Graph Model 25:870–884
17. Zampieri D, Mamolo MG, Laurini E et al (2009) Synthesis, biological evaluation, and three-dimensional in silico pharmacophore model for sigma(1) receptor ligands based on a series of substituted benzo[d]oxazol-2(3H)-one derivatives. J Med Chem 52:5380–5393
18. Kariv I, Cao H, Oldenburg KR (2001) Development of a high throughput equilibrium dialysis method. J Pharm Sci 90:580–587
19. Kansy M, Senner F, Gubernator K (1998) Physicochemical high throughput screening: parallel artificial membrane permeation assay in the description of passive absorption processes. J Med Chem 41:1007–1010
20. Dahl R, Bravo Y, Sharma V et al (2011) Potent, selective, and orally available benzoisothiazolone phosphomannose isomerase inhibitors as probes for congenital disorder of glycosylation Ia. J Med Chem 54:3661–3668
21. Lin JH, Lu AY (1997) Role of pharmacokinetics and metabolism in drug discovery and development. Pharmacol Rev 49:403–449
22. Lipinski CA (2000) Drug-like properties and the causes of poor solubility and poor permeability. J Pharmacol Toxicol Methods 44:235–249
23. Banker MJ, Clark TH et al (2003) Development and validation of a 96-well equilibrium dialysis apparatus for measuring plasma protein binding. J Pharm Sci 92:967–974
24. Di L, Kerns EH et al (2008) Applications of high throughput microsomal stability assay in drug discovery. Comb Chem High Throughput Screen 11:469–476
25. Guengerich FP (1989) Oxidation of halogenated compounds by cytochrome P-450, peroxidases, and model metalloporphyrins. J Biol Chem 264:17098–17205
26. Mosmann T (1983) Rapid colorimetric assay for cellular growth and survival: application to proliferation and cytotoxicity assays. J Immunol Methods 65:55–63
27. Korfmacher WA, Cox KA et al (2001) Cassette-accelerated rapid rat screen: a systematic procedure for the dosing and liquid chromatography/atmospheric pressure ionization tandem mass spectrometric analysis of new chemical entities as part of new drug discovery. Rapid Commun Mass Spectrom 15:335–340
28. Mei H, Korfmacher WA et al. (2006) Rapid in vivo oral screening in rats: reliability, acceptance criteria, and filtering efficiency. AAPS J 8:E493–500

Chapter 6

A Method for Direct Assessment of Tissue-Nonspecific Alkaline Phosphatase (TNAP) Inhibitors in Blood Samples

Eduard A. Sergienko, Qing Sun, and Chen-Ting Ma

Abstract

Tissue nonspecific alkaline phosphatase (TNAP) is one of four human alkaline phosphatases (AP), a family of exocytic enzymes that catalyze hydrolysis of phospho-monoesters in bone, liver, kidney, and various other tissues. Overexpression of TNAP gives rise to excessive bone and soft tissue mineralization, including blood vessel calcification. Our prior screening campaigns have found several leads against this attractive therapeutic target using in vitro assay with a recombinant enzyme; these compounds were further optimized using medicinal chemistry approaches. To prioritize compounds for their use in animal models, we have designed and developed a biomarker assay for in situ detection of TNAP activity within human and mouse blood samples at physiological pH. This assay is suitable for screening compounds in 1,536-well plates using blood plasma from different mammalian species. The user may choose from two different substrates based on the need for greater assay simplicity or sensitivity.

Key words Alkaline phosphatase, Tissue-nonspecific alkaline phosphatase, Blood plasma, Biomarker, Colorimetric assay, High throughput screening

1 Introduction

Tissue-nonspecific alkaline phosphatase (TNAP) is one of four human alkaline phosphatases (AP), a family of exocytic enzymes that catalyze hydrolysis of phospho-monoesters. TNAP is ubiquitously expressed in various tissues, with higher levels observed in bone, liver, and kidney. Overexpression of TNAP gives rise to excessive bone and soft tissue mineralization, including blood vessel calcification [1, 2]. Thus, TNAP is an attractive therapeutic target whose inhibition is expected to help alleviate soft tissue calcification. Our successful high-throughput screening (HTS) campaigns using a biochemical assay with recombinant enzyme [3] led to identification of several TNAP inhibitor leads [4] that were further optimized using medicinal chemistry approaches [5–7].

To prioritize the compounds for studies in the animal models, it is important to be able to predict how these compounds would behave

José Luis Millán (ed.), *Phosphatase Modulators*, Methods in Molecular Biology, vol. 1053, DOI 10.1007/978-1-62703-562-0_6,

in a biologically relevant environment. For example, compound binding to plasma proteins results in apparent decrease of their potency in biological assays. In addition, it is known that due to certain specifics of TNAP enzymatic reaction leading to rate-limiting release of its product orthophosphate, reaction reversibility and relatively high concentration of phosphate in biological samples, majority of TNAP in these samples is present in a phosphate-bound state [8]. This change in the enzyme binding state is expected to change its catalytic properties and sensitivity to inhibition by certain classes of compounds [9]. We decided to take advantage of TNAP presence in mammalian plasma samples and establish an in situ assay measuring the enzyme activity and inhibition with compounds within plasma samples, preserving their pH and composition as close to its original state as possible.

Alkaline phosphatase activity is a well established and perhaps the most common clinical blood test parameter; the level of AP activity and its isozyme spectrum are predictive of a number of pathophysiological conditions [10]. However, all up-to-date alkaline phosphatase assays rely on highly diluted samples and use of high concentration strongly alkaline buffers [11, 12]. Here, we present a detailed description of a newly developed assay for measuring blood TNAP activity, the in situ biomarker assay for direct detection of TNAP activity within human and mouse blood samples. This assay is robust and is also applicable for screening compounds in 1536-well plates, as was successfully confirmed through our in-house screening. This versatile assay is applicable across different mammalian species, and the selection of *para*-nitrophenylphosphate (pNPP) or phenolphthalein monophosphate (PPMP) substrate reduces assay complexity or offers greater detection sensitivity for the end user, respectively.

2 Materials

Prepare all solutions using ultrapure water (prepared by purifying deionized water to attain resistivity greater than 18.0 MΩ-cm at 25 °C) and analytical grade reagents. Prepare and store all reagents at room temperature (unless indicated otherwise). Diligently follow all waste disposal regulations when disposing waste materials.

1. 1 M Tris-HCl pH 7.5. Add 150 ml water to glass beaker. Weigh 24.2 g Tris and add to beaker. Dissolve Tris powder and pH to 7.5 with concentrated HCl. Add water to bring volume up to 200 ml. Pass through 0.22 μm filtration unit. Store at 4 °C.
2. 1 M $MgCl_2$. Add 75 ml water to glass beaker. Weigh 20.3 g $MgCl_2$ hexahydrate and add to beaker. Dissolve $MgCl_2$ powder and add water to bring volume up to 100 ml. Pass through 0.22 μm filtration unit. Store at 4 °C.

3. 50 mM $ZnCl_2$. Add 9 ml water to 15 ml disposable conical tube. Weigh 68 mg $ZnCl_2$ and add to conical tube. Dissolve $ZnCl_2$ powder by vortexing, and add water to bring volume up to 10 ml. Pass the solution through a 0.22 μm syringe filter into a new 15 ml conical tube. Store at 4 °C.
4. Mouse plasma: C57BL6 mouse plasma, mixed sex, lithium heparin anticoagulant (Sanford-Burnham Animal Facilities, La Jolla, CA, USA) (*see* **Note 1**). Store aliquots at -20 °C. Store thawed aliquot on ice and flash-freeze the remaining blood sample at end of assay day.
5. Human Plasma: pooled human plasma, mixed sex, sodium citrate anticoagulant (Innovative Research, Novi, MI, USA) (*see* **Note 1**). Store aliquots at -20 °C. Store thawed aliquot on ice and flash-freeze the remaining blood sample at end of assay day.
6. 13.5 mM *para*-nitrophenylphosphate. Weigh 5 mg pNPP powder (Sigma-Aldrich) and add to 1.6 ml Eppendorf tube. Add 1 ml of water and mix by vortexing. Store at -20 °C in smaller aliquots. Store thawed aliquot on ice and discard at end of assay day.
7. 20 mM phenolphthalein monophosphate bis(cyclohexyl ammonium) salt. Weigh 11.93 mg PPMP powder (Sigma-Aldrich) and add to 1.6 ml Eppendorf tube. Add 1 ml of water and mix by vortexing (*see* **Note 2**). Store at -20 °C in smaller aliquots. Store thawed aliquot on ice and discard at end of assay day.
8. 5 M sodium hydroxide. Weigh 8 g sodium hydroxide pellet and add to 50 ml conical tube. Add 20 ml of water and mix by vortexing. Add water to bring volume up to 40 ml.
9. 1 M sodium carbonate.
10. 200 mM sodium orthovanadate (Sigma-Aldrich), activated. Weigh 1.47 g sodium orthovanadate and add to small beaker (≤100 ml). Add 20 ml of water and mix by vortexing. Adjust the pH to 10 using NaOH or HCl. Add water to bring volume up to 40 ml. The yellow solution should be boiled until it turns colorless (roughly 10 min). Cool to room temperature and readjust the pH to 10 and repeat the boiling/cooling/titration process until the solution remains colorless and the pH stabilizes at 10. Store at -20 °C in smaller aliquots. Store thawed aliquot on ice and discard at end of assay day.
11. Agilent Bravo Automated Liquid Handling Platform (Agilent Technologies, Santa Clara, CA, USA).
12. P30 pipette tips in 384-well racks (Velocity11, Santa Clara, CA, USA).

13. Multidrop Combi Reagent Dispenser (Thermo Scientific, Waltham, MA, USA).
14. Small tube metal tip dispensing cassettes for Combi (Thermo Scientific).
15. Viaflow electronic multichannel pipette (Integra Biosciences, Hudson, NH, USA).
16. PHERAstar plate reader (BMG Labtech, Cary, NC, USA).
17. Echo 555 liquid handler (Labcyte, Sunnyvale, CA, USA).
18. 384-well polypropylene small volume intermediate plate (catalog# 784201) (Greiner Bio-One, Monroe, NC, USA).
19. 1,536-well clear polystyrene, HiBase, flat-bottom, square well assay plate (Catalog# 782101) (Greiner Bio-One).
20. Aluminum adhesive seals.

3 Methods

Carry out all procedures at room temperature unless otherwise specified. Prime all Combi tubing with the appropriate assay reagent, empty the tubing back into the conical tube holding the reagent, and prime again to ensure adequate mixing. Decontaminate Bravo liquid handler and PHERAstar plate reader at the end of the assay day with 70 % ethanol. Dispose of tips, tubes, and plates in biohazardous waste bin.

3.1 Optimization Substrate Concentration and Assay Duration

1. Mix 500 μl of 1 M Tris solution, pH 7.5, 80 μl of 1 M $MgCl_2$ solution, 40 μl of 50 mM $ZnCl_2$ solution, and 9,380 μl water in a 15 ml disposable conical tube, followed by thorough mixing by vortexer. This is the Assay Buffer referenced in subsequent steps, and it should be prepared fresh every assay day (*see* **Note 3**).
2. Mix 89.6 μl of 5 M sodium hydroxide solution, 140 μl of 200 mM vanadate solution, and 2,570 μl sodium carbonate solution in a 15 ml disposable round-bottom Falcon tube, followed by thorough mixing by vortexer. This is the Color Developer referenced in subsequent steps, and it should be prepared fresh every assay day. The pH value should be between 12.1 and 12.5 (*see* **Note 5**).
3. Thaw animal plasma sample on ice, and spin down the blood at >2,400 × *g* for 10 min to remove insoluble materials. In a sterile tissue-culture hood, fill one column of a 384-well intermediate plate with 24 μl of the blood. Mix 0.5 ml animal plasma and 25 μl of 200 mM vanadate solution by pipette. Fill another column of the 384-well intermediate plate with this vanadate-control group to monitor substrate degradation in

the absence of active phosphatase. Spin down plate at 182 × *g* for 1 min, and keep on ice until the next step.

4. Mix 50 μl of 1 M Tris solution, pH 7.5, 8 μl of 1 M $MgCl_2$ solution, 4 μl of 50 mM $ZnCl_2$ solution, 891 μl of 13.5 mM pNPP solution, and 47 μl water in a 1.6 ml disposable Eppendorf tube, followed by thorough mixing by vortexer. Transfer 500 μl of the mixture to a fresh Eppendorf tube with 500 μl Assay Buffer for twofold serial dilutions to reach the following concentrations: 12, 6, 3, 1.5, 0.75, 0.19, 0.094, 0 mM (*see* **Note 4**). Mix 50 μl of 1 M Tris solution, pH 7.5, 8 μl of 1 M $MgCl_2$ solution, 4 μl of 50 mM $ZnCl_2$ solution, 600 μl of 20 mM PPMP solution, and 338 μl water in a 1.6 ml disposable Eppendorf tube and perform similar mixing and serial dilutions. With the Viaflow electronic pipette, transfer the content of each tube into eight replicate wells in a 1,536-well assay plate, 1.5 μl each. For PPMP, transfer the content of each tube into at least 40 replicate wells in the 1,536-well assay plate so that there would be quadruplicate data for each time point (*see* **Note 5**).
5. Use Bravo liquid handler and one unused boxes of P30 tips to aspirate 19.5 μl from the blood intermediate plate and dispense 4.5 μl into quadruplicate occupied wells in the 1,536 well assay plate. Spin down plate and read on the PHERAstar plate reader at fixed time points, using 380 nm for pNPP substrate and 555 nm for PPMP substrate (*see* **Note 6**). For PPMP substrate, use Multidrop Combi and small metal tip cassette to dispense 2 μl Color Developer solution and spin once more before reading.
6. To start the analysis of data, plot optical density (OD) of assay wells against time and fit the linear region to a straight line (Fig. 1) (*see* **Note 7**). Use the time point at the end of the linear region for the final screening assay.
7. Obtain the background-subtracted enzymatic rates for the various concentrations and plot against the substrate concentration. Fit the data to the Michaelis–Menten equation for K_m value (Fig. 2). If the substrate K_m value is below 500 μM, use 500 μM in the final screening assay and proceed to the next step. If the K_m value is above 500 μM, use K_m value for the final screening assay (*see* **Note 8**). Perform **steps 6** and 7 in Subheading 3.1 for both pNPP and PPMP substrates.

3.2 Screening Assay for Human Plasma Using pNPP as Substrate

1. Prepare the plasma similar to **step 3** in Subheading 3.1. Place the plasma in a 100 ml reservoir or tip box lid and store on ice when not in use.
2. Mix 2,000 μl of 1 M Tris solution, pH 7.5, 40 μl of 1 M $MgCl_2$ solution, 16 μl of 50 mM $ZnCl_2$ solution, 1,484 μl of 13.5 mM

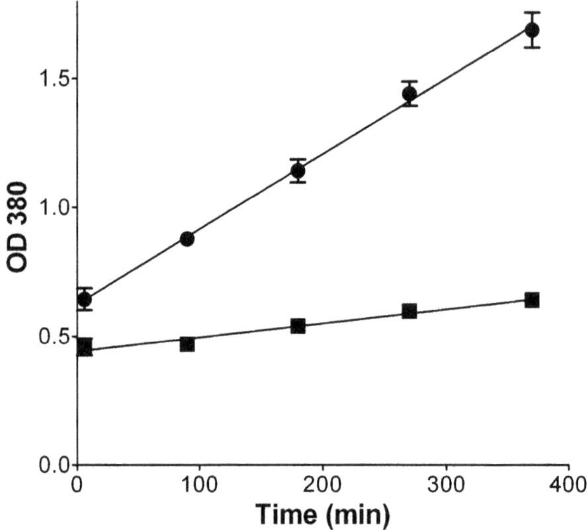

Fig. 1 Sample time course. Change in OD380 values are plotted against time points in 6 μl assay volume, and the data fitted linearly. Slopes are 0.0029 ± 0.00007 and 0.0005 ± 0.00006 for plasma sample (*filled circle*) and vanadate-spiked plasma control sample (*filled square*), respectively. 1 mM PNPP is included in all samples

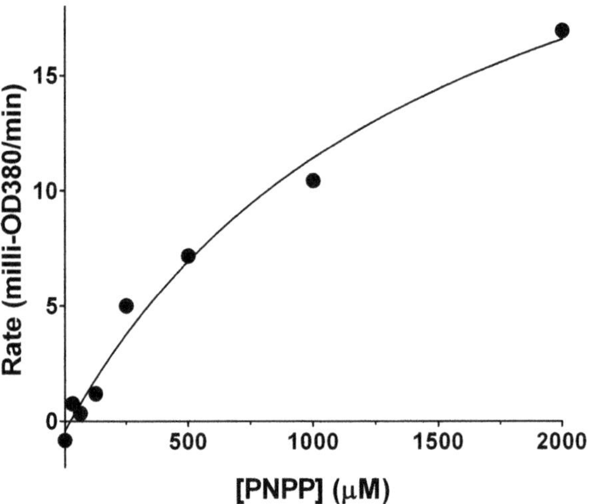

Fig. 2 Sample PNPP titration curve. Enzymatic rates are plotted against PNPP concentrations in 6 μl assay volume, and the data fitted to Michaelis–Menten equation. K_m value is 1,600 ± 550 μM for plasma sample

pNPP solution, and 6,460 μl water in a 15 ml round-bottom Falcon tube, followed by thorough mixing by vortexer. Transfer 3 ml into another round-bottom Falcon tube and add 150 μl of 200 mM vanadate solution and mix by vortexer. These two tubes contain the assay mix and control mix, respectively.

3. With the Echo liquid dispenser, dispense nanoliter amounts of compounds resuspended in DMSO into empty 1,536-well assay plate, followed by an equal amount of DMSO into columns reserved for no-enzyme positive control and vehicle negative controls.
4. Dispense with Multidrop Combi and small metal tip cassette 1.5 μl assay mix into columns 3–48 of the assay plate. Dispense with Combi 1.5 μl control mix into columns 1 and 2 of the assay plate.
5. Use the Bravo liquid handler to transfer 4.5 μl plasma to all wells of the assay plate, and spin down the plate and seal it with aluminum adhesive seal. After at least 17 h of incubation, read on PHERAstar plate reader at OD380 (*see* **Note 9**).

3.3 Screening Assay for Mouse Plasma Using pNPP as Substrate

1. Prepare the plasma similar to **step 3** in Subheading 3.1. Place the plasma in a 100 ml reservoir or tip box lid and store on ice when not in use.
2. Mix 2,000 μl of 1 M Tris solution, pH 7.5, 40 μl of 1 M $MgCl_2$ solution, 16 μl of 50 mM $ZnCl_2$ solution, 4,603 μl of 13.5 mM pNPP solution, and 3,341 μl water in a 15 ml round-bottom Falcon tube, followed by thorough mixing by vortexer. Transfer 3 ml into another round-bottom Falcon tube and add 150 μl of 200 mM vanadate solution and mix by vortexer. These two tubes contain the assay mix and control mix, respectively.
3. With the Echo liquid dispenser, dispense nanoliter amounts of compounds resuspended in DMSO into empty 1,536-well assay plate, followed by an equal amount of DMSO into columns reserved for no-enzyme positive control and vehicle negative controls.
4. Dispense with Multidrop Combi and small metal tip cassette 1.5 μl assay mix into columns 3–48 of the assay plate. Dispense with Combi 1.5 μl control mix into columns 1 and 2 of the assay plate.
5. Use the Bravo liquid handler to transfer 4.5 μl plasma to all wells of the assay plate, and spin down the plate and seal it with aluminum adhesive seal. After 5 h incubation, read on PHERAstar plate reader at OD380 (*see* **Note 9**).

3.4 Screening Assay for Human Plasma Using PPMP as Substrate

1. Prepare the plasma similar to **step 3** in Subheading 3.1. Place the plasma in a 100 ml reservoir or tip box lid and store on ice when not in use.
2. Mix 2,000 μl of 1 M Tris solution, pH 7.5, 40 μl of 1 M $MgCl_2$ solution, 16 μl of 50 mM $ZnCl_2$ solution, 1,260 μl of 20 mM PPMP solution, and 6,684 μl water in a 15 ml round-bottom Falcon tube, followed by thorough mixing by vortexer.

Transfer 3 ml into another round-bottom Falcon tube and add 150 μl vanadate solution and mix by vortexer. These two tubes contain the assay mix and control mix, respectively.

3. With the Echo liquid dispenser, dispense nanoliter amounts of compounds resuspended in DMSO into empty 1536-well assay plate, followed by an equal amount of DMSO into columns reserved for no-enzyme positive control and vehicle negative controls.
4. Dispense with Multidrop Combi and small metal tip cassette 1.5 μl assay mix into columns 3–48 of the assay plate. Dispense with Combi 1.5 μl control mix into columns 1 and 2 of the assay plate.
5. Use the Bravo liquid handler to transfer 4.5 μl plasma to all wells of the assay plate, and spin down the plate and seal it with aluminum adhesive seal. After at least 17 h of incubation, prepare the Color Developer similar to **step 2** in Subheading 3.1 and dispense with Combi 2 μl to all assay wells. After spinning the plate, read on PHERAstar plate reader at OD555 (*see* **Note 9**).

3.5 Screening Assay for Mouse Plasma Using PPMP as Substrate

1. Prepare the plasma similar to **step 3** in Subheading 3.1. Place the plasma in a 100 ml reservoir or tip box lid and store on ice when not in use.
2. Mix 2,000 μl of 1 M Tris solution, 40 μl of 1 M $MgCl_2$ solution, 16 μl of 50 mM $ZnCl_2$ solution, 1,000 μl of 20 mM PPMP solution, and 6,944 μl water in a 15 ml round-bottom Falcon tube, followed by thorough mixing by vortexer. Transfer 3 ml into another round-bottom Falcon tube and add 150 μl 200 mM vanadate solution and mix by vortexer. These two tubes contain the assay mix and control mix, respectively.
3. With the Echo liquid dispenser, dispense nanoliter amounts of compounds resuspended in DMSO into empty 1,536-well assay plate, followed by an equal amount of DMSO into columns reserved for no-enzyme positive control and vehicle negative controls.
4. Dispense with Multidrop Combi and small metal tip cassette 1.5 μl assay mix into columns 3–48 of the assay plate. Dispense with Combi 1.5 μl control mix into columns 1 and 2 of the assay plate.
6. Use the Bravo liquid handler to transfer 4.5 μl plasma to all wells of the assay plate, and spin down the plate and seal it with aluminum adhesive seal. After 5 h incubation, prepare the Color Developer similar to **step 2** in Subheading 3.1 and dispense with Combi 2 μl to all assay wells. After spinning the plate, read on PHERAstar plate reader at OD555 (*see* **Note 9**).

4 Notes

1. In the case of animal plasma preparation, an anticoagulant is necessary to maintain solubility of blood components, and sodium citrate or lithium heparin would be preferable over any alternatives that contain EDTA. EDTA will chelate magnesium and zinc ions and require much higher concentrations of the two ions in the assay buffer for acceptable final ion concentrations, possibly to the point of precipitation.
2. PPMP powder does not dissolve easily. Vortex every 5 min for short durations and eventually >95 % of the powder should be resuspended in solution.
3. The addition of Tris ensures physiological pH in the assay solution, and Tris serves as a phosphoacceptor that accelerates the turnover of the TNAP enzyme in blood [13]. An activator such as Tris is necessary for blood samples that contain very little TNAP or higher than normal phosphate and salt concentrations, both of which inhibit TNAP starting at physiological concentrations.
4. The 4× substrate solution is diluted by the addition of plasma sample, so the final concentration of substrate and buffer components in the assay well is ¼ the concentration in the substrate mix described in Subheading 3.
5. Whereas reaction progress with pNPP substrate can be monitored without stopping the reaction, PPMP reaction progress, determined from released phenolphthalein concentration, requires the Color Developer to change the color of phenolphthalein to pink. However, once pH 10 or higher is reached, the assay is no longer under physiological pH, and no data from the next time point can be obtained from the same assay wells. Thus, for each time point in the progress curve for PPMP, a new set of wells with different amounts of PPMP are needed. Sodium orthovandate is added into the Color Developer solution to prevent further TNAP reaction.
6. Due to the background absorbance of bilirubin in blood, 405 nm is no longer the optimal OD for detection of *para-nitrophenol* (PNP), and 380 nm offers the best signal/background ratio (data not shown). Phenolphthalein has highest extinction coefficient at around 555 nm, and none of the blood components contribute significant background absorbance at this wavelength. In summary, pNPP substrate offers one less dispensing step (no Color Developer is added) and thus simpler assay, while PPMP substrate offers greater assay sensitivity with lower background OD.

7. Steady-state kinetic conditions presumes catalytic amount of the enzyme and a large supply of the substrate, and the enzymatic rate does not change leading to constant increase in absorbance. The slope of the linear fit is the rate of catalysis and can be used directly in kinetic calculations and enzyme characterization. For each substrate concentration, subtract the vanadate-control rate from the original rate to remove background degradation from the calculations.
8. Due to high background absorbance at OD380 for pNPP substrate, no less than 500 μM pNPP is needed such that a consumption of 20 % substrate would yield an OD change of 0.45 OD, sufficiently high above the background OD of the plasma sample. In order to avoid substrate exhaustion and to maintain steady-state conditions, only a small portion of the substrate may be consumed in any enzymatic assay, and 20 % is commonly used.
9. Z′-factor may be calculated based on data from the control wells. Add the standard deviation of the negative control and of the positive control, multiple by 3, and divide by the difference in signal of the two control groups. Subtract this value from 1 and the result should exceed 0.5 [7, 14]. Inhibitor hit selection criteria may be based on inhibitory % response calculated from the difference in signal of the control groups, or on the standard deviation of compound wells or the negative control (for more detail *see* [9]).

Acknowledgments

This work is supported by NIH Roadmap grant # U54 HG005033 and Conrad Prebys Center for Chemical Genomics at Sanford-Burnham Medical Research Institute.

References

1. Markello TC, Pak LK, St Hilaire C et al (2011) Vascular pathology of medial arterial calcifications in NT5E deficiency: implications for the role of adenosine in pseudoxanthoma elasticum. Mol Genet Metab 103:44–50
2. Addison WN, Azari F, Sorensen ES et al (2007) Pyrophosphate inhibits mineralization of osteoblast cultures by binding to mineral, up-regulating osteopontin, and inhibiting alkaline phosphatase activity. J Biol Chem 282: 15872–15883
3. Sergienko EA, Millan JL (2010) High-throughput screening of tissue-nonspecific alkaline phosphatase for identification of effectors with diverse modes of action. Nat Protoc 5:1431–1439
4. Sergienko E, Su Y, Chan X et al (2009) Identification and characterization of novel tissue-nonspecific alkaline phosphatase inhibitors with diverse modes of action. J Biomol Screen 14:824–837
5. Chung TD, Sergienko E, Millan JL (2010) Assay format as a critical success factor for identification of novel inhibitor chemotypes of tissue-nonspecific alkaline phosphatase from high-throughput screening. Molecules 15: 3010–3037
6. Dahl R, Sergienko EA, Su Y et al (2009) Discovery and validation of a series of aryl sulfonamides as selective inhibitors of tissue-nonspecific alkaline phosphatase (TNAP). J Med Chem 52:6919–6925

7. Sidique S, Ardecky R, Su Y et al (2009) Design and synthesis of pyrazole derivatives as potent and selective inhibitors of tissue-nonspecific alkaline phosphatase (TNAP). Bioorg Med Chem Lett 19:222–225
8. Coburn SP, Mahuren JD, Jain M et al (1998) Alkaline phosphatase (EC 3.1.3.1) in serum is inhibited by physiological concentrations of inorganic phosphate. J Clin Endocrinol Metab 83:3951–3957
9. Sergienko EA (2012) Basics of HTS assay design and optimization. In: Fu H (ed) Chemical genomics. Cambridge University Press, New York, pp 159–172
10. Ramaiah SK (2007) A toxicologist guide to the diagnostic interpretation of hepatic biochemical parameters. Food Chem Toxicol 45:1551–1557
11. Babson AL, Greeley SJ, Coleman CM et al (1966) Phenolphthalein monophosphate as a substrate for serum alkaline phosphatase. Clin Chem 12:482–490
12. Morgenstern S, Kessler G, Auerbach J et al (1965) An automated p-nitrophenylphosphate serum alkaline phosphatase procedure for the AutoAnalyzer. Clin Chem 11:876–888
13. Stinson RA (1993) Kinetic parameters for the cleaved substrate, and enzyme and substrate stability, vary with the phosphoacceptor in alkaline phosphatase catalysis. Clin Chem 39:2293–2297
14. Zhang JH, Chung TD, Oldenburg KR (1999) A simple statistical parameter for use in evaluation and validation of high throughput screening assays. J Biomol Screen 4:67–73

Chapter 7

Isolation and Characteristics of Matrix Vesicles

René Buchet, Slawomir Pikula, David Magne, and Saïda Mebarek

Abstract

Mineralizing matrix vesicles (MVs) are extracellular organelles produced by chondrocytes, osteoblasts, and odontoblasts under physiological conditions and by vascular smooth muscle cells under pathological conditions. MVs are involved in the early stage of mineralization allowing calcium and phosphate to accumulate, and therefore providing an optimal environment facilitating hydroxyapatite formation. Here, we describe the isolation of MVs from osteoblasts and chondrocytes and present their main characteristics.

Key words Chondrocyte, Collagenase, Cysteine, Differential centrifugation, Osteoblast, Levamisole, Matrix vesicle, Theophylline, Trypsin

1 Introduction

Physiological mineralization is restricted to specific sites in teeth and skeletal tissues, including growth plate cartilage. Undesired or pathological mineralization can occur in several soft tissues such as articular cartilage, cardiovascular tissues, or the kidneys [1, 2]. During both physiological and pathological mineralization, extracellular matrix vesicles (MVs) seem to initiate the mineralization process [1]. 30–1,000 nm-diameter MVs are released from hypertrophic chondrocytes and osteoblasts during bone formation, from odontoblasts during teeth formation and from other cells such as smooth muscle cells [3, 4], cancer cells [1], etc., during pathological calcification. MVs are believed to initiate the mineralization by providing an optimal environment for the accumulation of calcium and phosphate inside the MV lumen facilitating the nucleation and formation of hydroxyapatite (HA) crystals. Then the preformed HA are released into the extracellular medium, where extracellular Ca^{2+} and phosphate sustain the mineralization process [5]. The role of MVs from osteoblasts and chondrocytes during mineralization is highlighted by the fact that they are enriched in enzymes controlling P_i and PP_i homeostasis, e.g., Na^+/K^+ ATPase [6, 7],

José Luis Millán (ed.), *Phosphatase Modulators*, Methods in Molecular Biology, vol. 1053,
DOI 10.1007/978-1-62703-562-0_7, © Springer Science+Business Media, LLC 2013

Ca^{2+}ATPase [7–9], inorganic pyrophosphatase 1 [8, 10], nucleoside triphosphate pyrophosphohydrolase [11], PHOSPHO1 [12, 13], and in tissue nonspecific alkaline phosphatase (TNAP) [14] and as well as by the presence of sphingomyelinase phosphodiesterase 3 [10, 15]. MVs from TNAP-deficient mice are unable to initiate mineral crystals and to self nucleate, while the failure of second-stage of mineralization may be caused by excess of an efficient inhibitor of mineralization, PP_i not sufficiently hydrolyzed due to the lack of TNAP [16]. MVs can be participants in the pathological mineralization and can be utilized for diagnosis of ectopic calcifications [1]. As reported by Wuthier, there is probably no "best" method of MV isolation and each method can serve some useful purpose [17]. There are two main methods: (1) the crude collagenase plus or minus trypsin digestion followed by differential centrifugation [8, 18] and (2) the tissue homogenization followed by subfractionation using sucrose gradient centrifugation [19, 20]. A comparative analysis of both methods has been reported [21]. The crude collagenase method using synthetic cartilage buffer [22] is the most employed method and is reported here.

2 Materials

Prepare all solutions using ultrapure water and analytical grade reagents. Store all buffers; perform ultracentrifugation and preparation of MVs at 4 °C. Treatment with collagenase is performed at 37 °C.

2.1 Materials and Equipment

1. Bradford reagent from Bio-Rad.
2. Penicillin (10,000 U/mL) and streptomycin (10 mg/mL) from Sigma.
3. McCoy's 5A (ATCC) medium from PAA.
4. Fetal bovine serum from PAA.

2.2 Assay Buffers and Solutions

Synthetic cartilage buffer. 1.42 mM $NaH_2PO_4 \cdot H_2O$, 1.83 mM $NaHCO_3$, 12.7 mM KCl, 0.57 mM $MgCl_2$, 5.55 mM D-glucose, 63.5 mM sucrose, 16.5 mM TES, 100 mM NaCl, 0.57 mM Na_2SO4. Weight 196 mg $NaH_2PO_4 \cdot H_2O$, 153.7 mg $NaHCO_3$, 947 mg KCL, 116 mg $MgCl_2$, 1 g D-glucose, 21.73 g sucrose, 3.78 g TES, 5.844 g NaCl, and 81 mg Na_2SO_4 in 1-L graduate cylinder. Add 900 mL, mix and adjust pH to 7.4. Make up to 1 L with water.

Washing buffer for growth plate and cartilage slices. Take 100 mL of synthetic cartilage buffer (see above) and dilute with 400 mL water.

Digestion buffer 1 for growth plates and cartilage slices. Add 200–500 U/g tissue of type 1 collagenase from *Clostridium histolyticum* (Sigma) and 14.7 mg $CaCl_2{\cdot}H_2O$ (1 mM) in 100 mL synthetic cartilage buffer (*see* **Notes 1** and **2**).

TNAP assay. Weight 2.19 g (10 mM) *p*-nitrophenylphosphate, 3.302 g (25 mM) glycylglycine, 1.016 g (5 mM) MgCl, 0.68 mg (5 microM) ZnCl2, and 3.976 g (25 mM) piperazine, in 0.5 L water. Adjust pH to 10.4. Make up to 1 L with water.

Mineralization assay with $Ca^{2+}$and Pi. Add 29.4 mg $CaCl_2{\cdot}2H_2O$ (2 mM) and 0 or 27.6 mg (0 or 2 mM) $NaH_2PO_4{\cdot}H_2O$ in 100 mL of Synthetic cartilage buffer. Total $CaCl_2{\cdot}2H_2O$ concentration is 2 mM and total $NaH_2PO_4{\cdot}H_2O$ concentration is 1.42 or 3.42 mM.

Mineralization assay with $Ca^{2+}$and AMP. Weight, 1.262 g AMP sodium salt monohydrate (3.42 mM), 153.7 mg $NaHCO_3$ (1.83 mM), 947 mg KCL (12.7 mM), 116 mg $MgCl_2$ (0.57 mM), 1 g D-glucose (5.55 mM), 21.73 g sucrose (63.5 mM), 3.78 g TES (16.5 mM), 5.844 g NaCl (100 mM), 81 mg Na_2SO_4 (0.57 mM) and 294 mg $CaCl_2.2H_2O$ (2 mM) in 1-L graduate cylinder. Add 900 mL, mix and adjust pH to 7.4. Make up to 1 L with water. Total $CaCl_2{\cdot}2H_2O$ concentration is 2 mM and total AMP concentration is 3.42 mM.

Stock solutions of TNAP inhibitor with $Ca^{2+}$and AMP. Prepare 10 mM inhibitor (for example: cysteine, levamisole, sinomenine, or theophylline) in 10 mL Mineralization assay with Ca^{2+} and AMP (see above).

Saos-2 cell culture growth medium. 420 mL McCoy's 5A (ATCC) supplemented with 1.5 mM glutamine, 1.1 g sodium bicarbonate (final concentration 2.2 g/L), 5 mL penicillin and streptomycin stock solution (*see* Subheading 2.1) (100 times diluted to have final 100 U/mL penicillin and 0.1 mg/mL streptomycin) and 75 mL fetal bovine serum from PAA (*see* Subheading 2.1), final concentration of bovine serum is 15 % v:v.

Mineralization medium for Saos-2 cells. Add 50 μg/mL ascorbic acid (Sigma) and 7.5 mM β-glycerophosphate (Sigma) in Saos-2 cell culture growth medium (*see* **Note 3**).

Hank's balanced salt solution (HBSS). 5.4 mM KCl, 0.3 mM $Na_2HPO_4{\cdot}2H_2O$, 0.4 mM KH_2PO_4, 0.6 mM $MgSO_4$, 137 mM NaCl, 5.6 mM D-glucose, 2.38 mM $NaHCO_3$. Weight 402.7 mg KCl, 53.4 mg $Na_2HPO_4{\cdot}2H_2O$, 54.4 mg KH_2PO_4, 122 mg $MgSO_4$, 8 g NaCl, 21.92 mg D-glucose, 199.9 mg $NaHCO_3$ in 1 L graduate cylinder. Add 900 mL, mix and adjust pH to 7.4. Make up to 1 L with water.

Digestion buffer 2 for osteoblast-like cells. Add 200 U/mL of type 1 collagenase from *Clostridium histolyticum* (Sigma) and 14.7 mg $CaCl_2{\cdot}H_2O$ (1 mM) in 100 mL HBSS buffer (*see* **Notes 1** and **2**).

3 Methods

Carry out all procedures at 4 °C except as stated.

3.1 Preparation of Matrix Vesicles from Femurs of 17-Day Chicken Embryos

1. Twenty 17-day chicken embryos are killed by decapitation.
2. Femurs were taken and put in 50 mL washing buffer for growth plate and cartilage slices (*see* Subheading 2.2.).
3. Cut 1–3 mm thick slices of growth plates and epiphyseal cartilages and wash five times with 25 mL Washing buffer for growth plate and cartilage slices.
4. Weight slices of growth plates and epiphyseal cartilages (around 3–4 g) (*see* **Note 4**).
5. Incubate in 30 mL Digestion buffer 1 for growth plates and cartilage slices (*see* Subheading 2.2) at 37 °C for 180 min by mixing continuously (*see* **Note 5**).
6. Filter through nylon filter and discard tissue debris.
7. Save filtrate around (25 mL).
8. Centrifuge filtrate at 600 × g for 15 min at 4 °C.
9. Discard debris and centrifuge supernatant at 20,000 × g for 20 min at 4 °C.
10. Discard pellet and centrifuge supernatant at 80,000 × g for 60 min at 4 °C.
11. Discard supernatant and save pellet.
12. Wash pellet with 1 mL synthetic cartilage buffer (*see* Subheading 2.2). Don't mix so that the synthetic cartilage buffer can be removed.
13. Pellet (approximately 0.1 mL) is mixed in 0.5 mL synthetic cartilage buffer. The suspension contains MVs. Mix gently and kept at 4 °C (*see* **Note 6**).

3.2 Protein Determination of Matrix Vesicles

1. Take 2 μL of MV suspension (MVs prepared as in Subheading 3.1). Add 798 μL pure water and 200 μL Bradford reagent (*see* Subheading 2.1).
2. Determine protein concentration at 595 nm. Repeat three times. Typical values of protein concentrations in MVs are 2–3 mg/mL.

3.3 TNAP Activity Determination of Matrix Vesicles

1. Take 5 μL of MV suspension (MVs prepared as in Subheading 3.1). Add 1,995 μL TNAP assay (*see* Subheading 2.2).
2. Determine activity of TNAP at 405 nm. Repeat three times. Typical values reach 15–20 μmol/min mg.

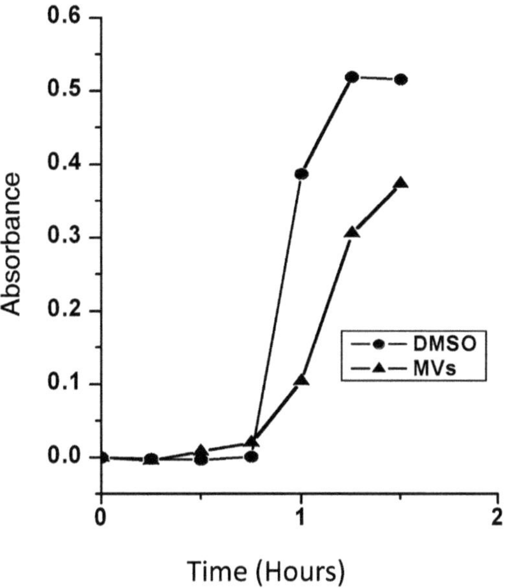

Fig. 1 Kinetics of mineral formation by MVs determined by measuring turbidity at 340 nm. MVs incubated at 37 °C in Mineralization assay with 2 mM Ca^{2+} and 3.42 mM Pi (MVs) induced mineral formation. Addition of 4 % v:v DMSO in Mineralization assay with 2 mM Ca^{2+} and 3.42 mM Pi without MVs induced mineral formation (DMSO) (positive control) while MVs in Synthetic cartilage buffer do not induce mineralization due to lack of calcium (negative control) giving a *flat line*. Taken from ref. 23

3.4 Mineralization Assay on Matrix Vesicles Supplemented with Ca^{2+} and P_i

1. Take 20 μL of MV suspension (MVs prepared as in Subheading 3.1). Add 980 μL mineralization assay (*see* Subheading 2.2). Typical MV protein concentration values are 20–30 μg/mL. Total $CaCl_2{\cdot}2H_2O$ concentration is 2 mM and total P_i concentration is 1.42–3.42 mM (*see* **Note 7**). This is the sample for mineralization assay.
2. Take 50 μL DMSO and 950 μL mineralization assay to obtain control positive (*see* **Note 8**).
3. Take 20 μL of MV suspension (MVs prepared as in Subheading 3.1). Add 980 μL Synthetic cartilage buffer (*see* Subheading 2.2) to obtain negative control. Total $CaCl_2{\cdot}2H_2O$ concentration is 0 mM and total P_i concentration is 1.42 mM. Typical MV protein concentration values are around 20–30 μg/mL.
4. Incubate sample and controls at 37 °C and determine mineralization at 340 nm at 15 min intervals. Repeat three times (Fig. 1) (*see* **Note 9**).

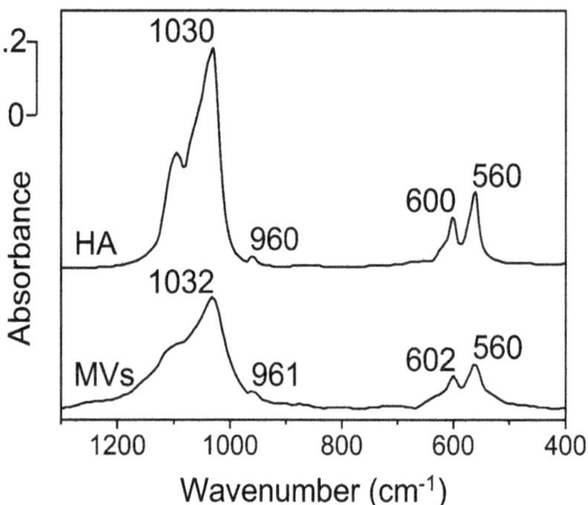

Fig. 2 Infrared spectrum of mineral obtained after 24-h incubation of MVs at 37 °C in SCl medium containing 2 mM Ca^{2+} and 3.42 mM Pi (MVs). As a control infrared spectrum of HA. Five HA characteristic bands located at 560, 600–602, 960–961, 1030–1032, and 1090–1100/cm revealed unambiguously that mineral formed by MVs is HA. Taken from ref. 24

3.5 Identification of Minerals Induced by Matrix Vesicles Supplemented with Ca^{2+} and P_i

1. Take 20 μL of MV suspension (MVs prepared as in Subheading 3.1). Add 980 μL mineralization assay (*see* Subheading 2.2). Typical MV protein concentration values are around 20–30 μg/mL. Total $CaCl_2{\cdot}2H_2O$ concentration is 2 mM and total P_i concentration is 1.42–3.42 mM (*see* **Note** 7).
2. After 24 h incubation at 37 °C, centrifuge sample containing mineral at 3,000 × *g* for 15 min.
3. Discard supernatant and wash minerals with 1 mL water.
4. Centrifuge mineral sample at 3,000 × *g* for 15 min.
5. Discard aqueous solution. Dry minerals (around 100–150 μg).
6. Mix minerals (around 100–150 μg) with 150 mg dry KBr and compress to obtain KBr pellet for IR measurements.
7. Determine IR spectra (Fig. 2).

3.6 Mineralization Assay on Matrix Vesicles Supplemented with Ca^{2+}, AMP and TNAP Inhibitor

1. Take 20 μL of MV suspension (MVs prepared as in Subheading 3.1). Add 0–400 μL of 10 mM TNAP inhibitor (e.g., cysteine, levamisole, sinomenine, or theophylline) in Mineralization assay with 2 mM Ca^{2+} and 3.42 mM AMP (*see* Subheading 2.2). Add 580 μL of Mineralization assay with Ca^{2+} and AMP (*see* Subheading 2.2). Final inhibitor concentration is from 0 to 4 mM. Typical protein concentration values are around 20–30 μg/mL (*see* **Note** 7). Total concentrations are 2 mM Ca^{2+} and 3.42 mM Pi.
2. Incubate at 37 °C each sample containing 0–4 mM of inhibitors and determine mineralization at 340 nm at 15 min intervals. Repeat three times (Fig. 3) (*see* **Notes 9** and **10**).

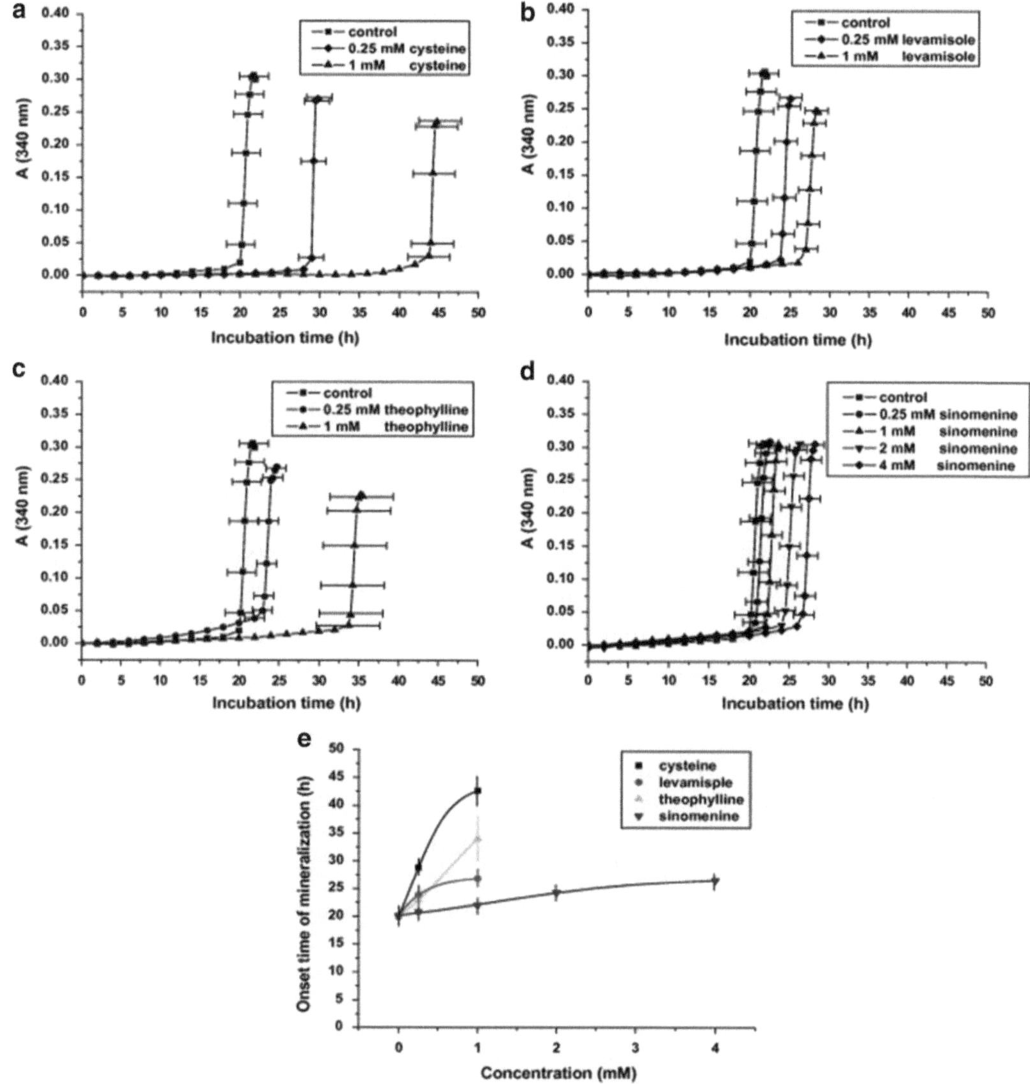

Fig. 3 Inhibition of MV-induced mineral formation by cysteine, levamisole, sinomenine, and theophylline. MVs incubated at 37 °C in Mineralization assay with 2 mM Ca^{2+} and 3.42 mM AMP, supplemented with (**a**) 0–1 mM cysteine; (**b**) 0–1 mM levamisole; (**c**) 0–1 mM theophylline; (**d**) 0–1 mM sinomenine; (**e**) onset times of mineral formation as a function of concentration of cysteine, levamisole, theophylline, or simomenine in SCL buffer containing 3.42 mM AMP and 2 mM Ca^{2+}. Increasing onset time of mineralization indicates inhibition or retardation of mineral formation. Taken from ref. 25

3.7 Preparation of Matrix Vesicles from Osteoblast-Like Saos-2 Cells

1. Culture Human osteosarcoma Saos-2 cells (ATCC HTB-85) in seven flask T75 (around 25,000 cells/cm^2) with 15 mL mineralization medium for Saos-2 cells (*see* Subheading 2.2) for each flask.
2. Incubate Saos-2 cells up to 7 days to reach confluence in a humidified incubator at 37 °C (95 % air, 5 % CO_2). Every 3 days change the mineralization medium.

3. After 7-days, change medium with 15 mL Digestion buffer 2 for osteoblast-like cells for each flask (*see* Subheading 2.2) at 37 °C for 180 min.
8. Centrifuge cells at 600 × *g* for 15 min at 4 °C
9. Discard debris and centrifuge supernatant at 20,000 × *g* for 20 min at 4 °C.
10. Discard pellet and centrifuge supernatant at 80,000 × *g* for 60 min at 4 °C.
11. Discard supernatant and save pellet.
12. Wash pellet with 1 mL synthetic cartilage buffer (*see* Subheading 2.1). Don't mix so that the synthetic cartilage buffer can be removed. Pellet (approximately 0.1 mL) is mixed in 0.5 mL synthetic cartilage buffer. Mix gently and kept at 4 °C. The suspension contains MVs (*see* **Note 6**).

3.8 Protein Determination of Matrix Vesicles

1. Take 20 μL of MV suspension. Add 780 μL pure water and 200 μL Bradford reagent (*see* Subheading 2.1).
2. Determine protein concentration. Repeat three times. Typical values around 1–2 mg/mL.

3.9 TNAP Activity Determination of Matrix Vesicles

1. Take 50 μL of MV suspension. Add 1,950 μL TNAP assay reagent (*see* Subheading 2.2).
2. Determine activity of TNAP. Repeat three times. Typical values are 15–20 μmol/min mg.

3.10 Mineralization Assay on Matrix Vesicles Supplemented with Ca^{2+} and Pi

Same procedure as described in Subheading 3.4 (*see* **Note 11**).

4 Notes

1. Collagenase quality depends from the lot and is not constant; therefore, the amount of collagenase should be adjusted.
2. Depending of the quality of collagenase, 0.1 % (w:v) trypsin can be added to facilitate the digestion.
3. The medium must be prepared freshly because ascorbic acid cannot be kept for a long time.
4. Weight the tissue to adjust the collagenase amount.
5. A rotating shaker which can kept tubes is put in oven (Thermosi).

6. MVs can be kept at 4 °C up to 5 days. Perform TNAP activity and mineralization assay to check MV integrity.
7. Perform mineralization assay supplemented with 2 mM Ca^{2+} and 0 or 2 mM $NaH_2PO_4{\cdot}H_2O$ (total concentration of $NaH_2PO_4{\cdot}H_2O$ in Synthetic cartilage buffer is 1.42 mM or 3.42 mM). Sometimes MVs do not induce nucleation and mineralization. The increase of $NaH_2PO_4{\cdot}H_2O$ concentration from 1.42 (more physiological) to 3.42 mM can boost the mineralization. Alternatively, increase MVs concentration to around 60 μg of protein/mL can induce mineralization. Mineralization assay is a more drastic test of MV integrity than TNAP activity.
8. Addition of DMSO (5 % v:v) in SCL medium supplemented with 2 mM Ca^{2+} and 0 or 2 mM $NaH_2PO_4{\cdot}H_2O$ (total concentration of $NaH_2PO_4{\cdot}H_2O$ in Synthetic cartilage buffer is 1.42 mM or 3.42 mM) induce nucleation and formation of HA without MVs. This will allow testing of the inhibition effect of candidate inhibitors directly on HA formation. For example addition of 10 μM of PP_i—a well known inhibitor of HA formation—inhibits HA formation induced by DMSO.
9. In the case where MVs do not induce mineralization, there are three critical factors that need to be checked. (1) pH of the mineralization medium should be between 7.4 and 7.8. Lower pH tends to dissolve HA. Higher pH is inhibitory. (2) Amount of collagenase should be increased and eventually 0.1 % (w:v) trypsin should be added (*see* **Notes 1** and **2**). Therefore, it is important to weight tissues to adjust collagenase amount (*see* **Note 3**). (3) Nucleational core of MVs must contain Ca^{2+}. Supplementation of 1 mM Ca^{2+} in Synthetic cartilage buffer may contribute to the preservation of the nucleational core.
10. The onset of mineral formation is delayed from 1 h (MVs in mineralization assay with 2 mM Ca^{2+} and 3.42 mM Pi) (Fig. 1) to about 20 h (MVs in mineralization assay with 2 mM Ca^{2+} and 3.42 mM AMP) (Fig. 3). This is due to the formation of Pi during AMP hydrolysis and its accumulation in the lumen of MVs.
11. Although TNAP activity is relatively high in MVs extracted from osteoblast-like cells, mineralization induced by osteoblast MVs are less effective than the mineralization induced by MVs from chondrocytes. Larger MV concentrations are needed.

Acknowledgments

This work was supported in part by CNRS grant, Polonium Grant N°27727TA and by grant N N401 140639 from the Polish Ministry of Science and Higher Education.

References

1. Anderson HC, Mulhall D, Garimella R (2010) Role of extracellular membrane vesicles in the pathogenesis of various diseases, including cancer, renal diseases, atherosclerosis, and arthritis. Lab Invest 90:1549–1557
2. Kirsch T (2006) Determinants of pathological mineralization. Curr Opin Rheumatol 18: 174–180
3. Hsu HH, Camacho NP (1999) Isolation of calcifiable vesicles from human atherosclerotic aortas. Atherosclerosis 143:353–362
4. Kapustin AN, Davies JD, Reynolds JL et al (2011) Calcium regulates key components of vascular smooth muscle cell-derived matrix vesicles to enhance mineralization. Circ Res 109:e1–e12
5. Anderson HC, Garimella R, Tague SE (2005) The role of matrix vesicles in growth plate development and biomineralization. Front Biosci 10:822–837
6. Einhorn TA, Gordon SL, Siegel SA et al (1985) Matrix vesicle enzymes in human osteoarthritis. J Orthop Res 3:160–169
7. Hsu HH, Anderson HC (1996) Evidence of the presence of a specific ATPase responsible for ATP-initiated calcification by matrix vesicles isolated from cartilage and bone. J Biol Chem 271:26383–26388
8. Ali SY, Sajdera SW, Anderson HC (1970) Isolation and characterization of calcifying matrix vesicles from epiphyseal cartilage. Proc Natl Acad Sci U S A 67:1513–1520
9. Xiao Z, Camalier CE, Nagashima K et al (2007) Analysis of the extracellular matrix vesicle proteome in mineralizing osteoblasts. J Cell Physiol 210:325–335
10. Thouverey C, Malinowska A, Balcerzak M et al (2011) Proteomic characterization of biogenesis and functions of matrix vesicles released from mineralizing human osteoblast-like cells. J Proteomics 74:1123–1134
11. Johnson K, Moffa A, Chen Y et al (1999) Matrix vesicle plasma cell membrane glycoprotein-1 regulates mineralization by murine osteoblastic MC3T3 cells. J Bone Miner Res 14:883–892
12. Roberts S, Narisawa S, Harmey D et al (2007) Functional involvement of PHOSPHO1 in matrix vesicle-mediated skeletal mineralization. J Bone Miner Res 22:617–627
13. Stewart AJ, Roberts SJ, Seawright E et al (2006) The presence of PHOSPHO1 in matrix vesicles and its developmental expression prior to skeletal mineralization. Bone 39:1000–1007
14. Register TC, McLean FM, Low MG et al (1986) Roles of alkaline phosphatase and labile internal mineral in matrix vesicle-mediated calcification. Effect of selective release of membrane-bound alkaline phosphatase and treatment with isosmotic pH 6 buffer. J Biol Chem 261:9354–9360
15. Balcerzak M, Malinowska A, Thouverey C et al (2008) Proteome analysis of matrix vesicles isolated from femurs of chicken embryo. Proteomics 8:192–205
16. Wuthier RE, Lipscomb GF (2011) Matrix vesicles: structure, composition, formation and function in calcification. Front Biosci 16: 2812–2902
17. Anderson HC, Sipe JB, Hessle L et al (2004) Impaired calcification around matrix vesicles of growth plate and bone in alkaline phosphatase-deficient mice. Am J Pathol 164:841–847
18. Majeska RJ, Wuthier RE (1975) Studies on matrix vesicles isolated from chick epiphyseal cartilage. Association of pyrophosphatase and ATPase activities with alkaline phosphatase. Biochim Biophys Acta 391:51–60
19. Watkins EL, Stillo JV, Wuthier RE (1980) Subcellular fractionation of epiphyseal cartilage: isolation of matrix vesicles and profiles of enzymes, phospholipids, calcium and phosphate. Biochim Biophys Acta 631:289–304
20. Wuthier RE, Linder RE, Warner GP et al (1978) Nonenzymatic method for isolation of matrix vesicles: characterization and initial studies on Ca and P orthophosphate metabolism. Metab Bone Dis Relat Res 1:125–136
21. Balcerzak M, Radisson J, Azzar G et al (2007) A comparative analysis of strategies for isolation of matrix vesicles. Anal Biochem 361:176–182
22. Wu LN, Yoshimori T, Genge BR et al (1993) Characterization of the nucleational core complex responsible for mineral induction by growth plate cartilage matrix vesicles. J Biol Chem 268:25084–25094
23. Li L, Buchet R, Wu Y (2008) Dimethyl sulfoxide-induced hydroxyapatite formation: a biological model of matrix vesicle nucleation to screen inhibitors of mineralization. Anal Biochem 381:123–128
24. Thouverey C, Bechkoff G, Pikula S et al (2009) Inorganic pyrophosphate as a regulator of hydroxyapatite or calcium pyrophosphate dihydrate mineral deposition by matrix vesicles. Osteoarthritis Cartilage 17:64–72
25. Li L, Buchet R, Wu Y (2010) Sinomenine, theophylline, cysteine, and levamisole: comparisons of their kinetic effects on mineral formation induced by matrix vesicles. J Inorg Biochem 104:446–454

Chapter 8

The Use of Tissue-Nonspecific Alkaline Phosphatase (TNAP) and PHOSPHO1 Inhibitors to Affect Mineralization by Cultured Cells

Tina Kiffer-Moreira and Sonoko Narisawa

Abstract

Here, we describe methods to evaluate the ability of small molecules inhibitors of TNAP and PHOSPHO1 in preventing mineralization of primary cultures of murine vascular smooth muscle cells. The procedures are also applicable to primary cultures of calvarial osteoblasts. These cell-based assays are used to complement kinetic testing during structure–activity relationship studies aimed at improving scaffolds in the generation of pharmaceuticals for the treatment for medial vascular calcification.

Key words TNAP, PHOSPHO1, Vascular smooth muscle cells, High-throughput screening, Small-molecules, Pharmacological inhibitors, Kinetic studies, Cell culture

1 Introduction

The calcification of the medial elastic layer of the arteries occurs in several pathologies and it is referred to as medial vascular calcification (MVC). MVC is common in patients with chronic kidney disease, obesity, aging and in rare pediatric conditions that herald generalized arterial calcification of infancy (GACI). MVC develops through an actively regulated process that resembles skeletal mineralization, resulting from chondro-osteogenic transformation of vascular smooth muscle cells (VSMCs) [1].

It has been hypothesized that phosphate accelerates this chondro-osteogenic conversion by inducing expression of runt-related transcription factor 2 (RUNX2), osteocalcin and tissue-nonspecific alkaline phosphatase (TNAP) [2]. Elevated TNAP further favors vascular calcification by hydrolyzing the calcification inhibitor inorganic pyrophosphate (PP_i) [3–6]. In the *Enpp1*$^{-/-}$ mouse model of GACI, inhibition of TNAP can restore PP_i to sufficient levels to maintain normal mineralization [3].

José Luis Millán (ed.), *Phosphatase Modulators*, Methods in Molecular Biology, vol. 1053,
DOI 10.1007/978-1-62703-562-0_8, © Springer Science+Business Media, LLC 2013

The phosphatase PHOSPHO1, first identified in the chicken as a member of the haloacid dehalogenase superfamily of Mg^{2+}-dependent hydrolases, [7–10] is expressed in mineralizing cartilage at levels 120-fold higher than in non-mineralizing tissues [11]. PHOSPHO1 shows high phosphohydrolase activity towards phosphoethanolamine (P-Etn) and phosphocholine (P-Cho), both of which are key phospholipid components of the matrix vesicle (MV) [12]. Ultrastructural studies have identified hydroxyapatite-containing MVs in human aorta, which indicates that these structures may provide the nidus for vascular calcification [13, 14].

Using natural PHOSPHO1 substrates, potent and specific inhibitors of PHOSPHO1 were identified via high-throughput screening and mechanistic analysis, and two were selected for further analysis, MLS-0390838 and MLS-0263839 [15]. Their therapeutic potential for preventing VSMC calcification by targeting PHOSPHO1 was assessed, both alone and in combination with the potent TNAP inhibitor MLS-0038949 [3, 15]. That study revealed that TNAP and PHOSPHO1 play a critical role in VSMC mineralization and that "phosphatase inhibition" may be a useful therapeutic strategy to reduce MVC.

2 Materials

Prepare all solutions using Milli-Q water and analytical grade reagents. Prepare and store all reagents at room temperature (unless indicated otherwise).

2.1 Materials and Equipment

In case material/equipment not described in this section, the specifications can be found in the text body.

1. Biomol Green reagent (Biomol International, Plymouth Meeting, PA, USA).
2. Origin Scientific plotting software (Northampton, MA, USA).
3. 1,2-dioleoyl-*sn*-glycero-3-phosphoethanolamine (Sigma-Adrich, P-Eth); catalog number P0503.
4. Phosphocholine chloride calcium salt tetrahydrate (Sigma-Aldrich, P-Chol); catalog number P0378.

2.2 Preparation of Recombinant PHOSPHO1

Preparation of recombinant PHOSPHO1.

2.2.1 Characteristics

pBAD-Phosopho1.

Molecular Weight about 36.8 KD with C-Term His Tag.
Theoretical pI = 6.73.
Extinction Coefficient = 0.78.

2.2.2 Expression

1. Grow 500 ml overnight culture in Top10 cells with Ampicillin.
2. Inoculate TB/amp with 100 ml overnight culture.
3. Grow until OD_{600} = 0.6–1.0. Induce with 0.2 % arabinose. Transfer to lower temperature.
4. Grow overnight at 28 °C.
5. Harvest. Resuspend pellets in 10 ml 1× His binding buffer. freeze in liquid nitrogen (*see* **Note 1**).

2.2.3 Purification

1. Quick thaw cell pellets. Add proteinase inhibitor cocktail. Lyse cells with homogenizer.
2. Spin down at 16 K, 30 min.
3. Prepare Ni-NTA beads with binding buffer for a 10 ml column.
4. Load supernatant on column. Save flow through for analysis.
5. Wash 100 ml binding buffer.
6. Wash 50 ml wash buffer.
7. Elute with 300 mM Imidazole buffer.
8. Immediately dialyze into Dialysis Buffer 1 (*see* **Note 2**).

2.3 Assay Buffers and Solutions

1× Binding Buffer (High Salt). 5 mM Imidazole, 500 mM NaCl, 20 mM Tris–HCl, pH 7.9.

Elution Buffer (Med Salt). 300 mM Imidazole, 250 mM Nacl, 20 mM Tris–HCl, pH 7.9.

Wash Buffer. 30 mM Imidazole, 250 mM NaCl, 20 mM Tris–HCl pH 7.9.

Dialysis Buffer 1. 20 mM Tris–HCl, pH 7.9, 250 mM NaCl.

Reaction Mix. 20 mM MES-NaOH, pH 6.7, 0.01 % (w/v) BSA, 0.0125 % (v/v) Tween-20, 2 mmol/L $MgCl_2$, 62.5 μmol/L P-Etn or P-Cho (Sigma-Aldrich, St. Louis, MO, USA).

Alizarin Red solution. Sigma A5533, 40 mM, Dissolve 2.738 g in ~150 ml distilled water (ddw), adjust pH 5.0 with drops of diluted ×1/10 NH_4OH (*see* **Note 3**). Final 200 ml with ddw.

Cetylpyridinium chloride. Sigma C 5460, 10 % (w/v) in 10 mM phosphate buffer (pH 7), 10 g powder + 1 ml 1 M PB (pH 7.0–7.4)/bring up to final 100 ml with ddw.

DNA lyses solution. 10 mg/ml Proteinase K added into 50 mM Tris–HCl (pH 8.0), 100 mM EDTA (pH 8.0), 1 % SDS-100 mM NaCl (1:20 vol).

1 M PB (phosphate buffer). pH ~7–7.4, $Na_2HPO_4 \cdot 12H_2O$, 28.7 g (0.08 mol), $Na_2HPO_4 \cdot 2H_2O$, 3.3 g (0.021 mol), H_2O final 100 ml.

3 Methods

HTS of 55,000 compounds from the MLSMR compound collection was conducted using a colorimetric assay based on the ability of PHOSPHO1 to liberate phosphate from P-Etn and its reaction with the Biomol Green reagent (Biomol International, Plymouth Meeting, PA, USA). HTS provided approximately 5,000 compounds that showed greater than 50 % activity in the single point assay, a hit rate of 3 %. Subsequent hit follow up and validation in dose response identified sub-micromolar inhibitors of PHOSPHO1 (*see* PubChem BioAssay AID 1666 for details). Initial HTS was performed in duplicate at a concentration of 20 μmol/L with dose–response assays using a 10-point twofold serial dilution of the hit compounds in DMSO. Hit confirmation was performed using the Biomol Green colorimetric assay (described below) to verify inhibitory activity against PHOSPHO1 in dose–response mode (*see* Fig. 1 for the dose–response curves for the inhibition of PHOSPHO1 by the nine best inhibitors using PHOSPHO1 natural substrates P-Etn and P-Cho).

VSMC isolated from mice were used for in vitro calcification studies in the presence of PHOSPHO1 inhibitors as well as the potent TNAP inhibitor MLS-0038949 [16]. The PHOSPHO1 and TNAP inhibitors alone or in combination were able to significant reduce calcification in VSMC culture and alizarin red staining was used for the quantification of mineralization reduction (Fig. 2).

3.1 Biomol Green Colorimetric Assay

Reactions were measured in triplicate in 96-well plates (Evergreen scientific) with a final volume of 50 μL. Blanks containing DMSO only were discounted (*see* **Note 4**).

1. Dissolve test compound in DMSO (Sigma Aldrich catalog number D8418) ranging from 0.0143 nmol/L to 300 μmol/L (*see* **Note 5**).
2. Mix with purified recombinant PHOSPHO1 (0.3 nmol/L) in the reaction mix for a 50 μL final volume.
3. Allow reaction to proceed for 60 min at room temperature.
4. Stop reaction with 100 μL of Biomol Green reagent and measure absorbance at 620 nm.
5. Calculate the inhibitory effect as a percent of control.
6. The IC_{50} was calculated from plots of residual enzyme activity against inhibitor concentration, using Origin Scientific plotting software (Northampton, MA, USA).

3.2 Isolation of Smooth Muscle Cells from Aorta

(*see* **Note 6**)

VSMCs were isolated from excised aortas using a collagenase digestion method and the smooth muscle phenotype was confirmed by RT-PCR analysis for smooth muscle α-actin as before [9]. One mouse aorta provided an average of 5×10^5 VSMCs (*see* **Note 7**).

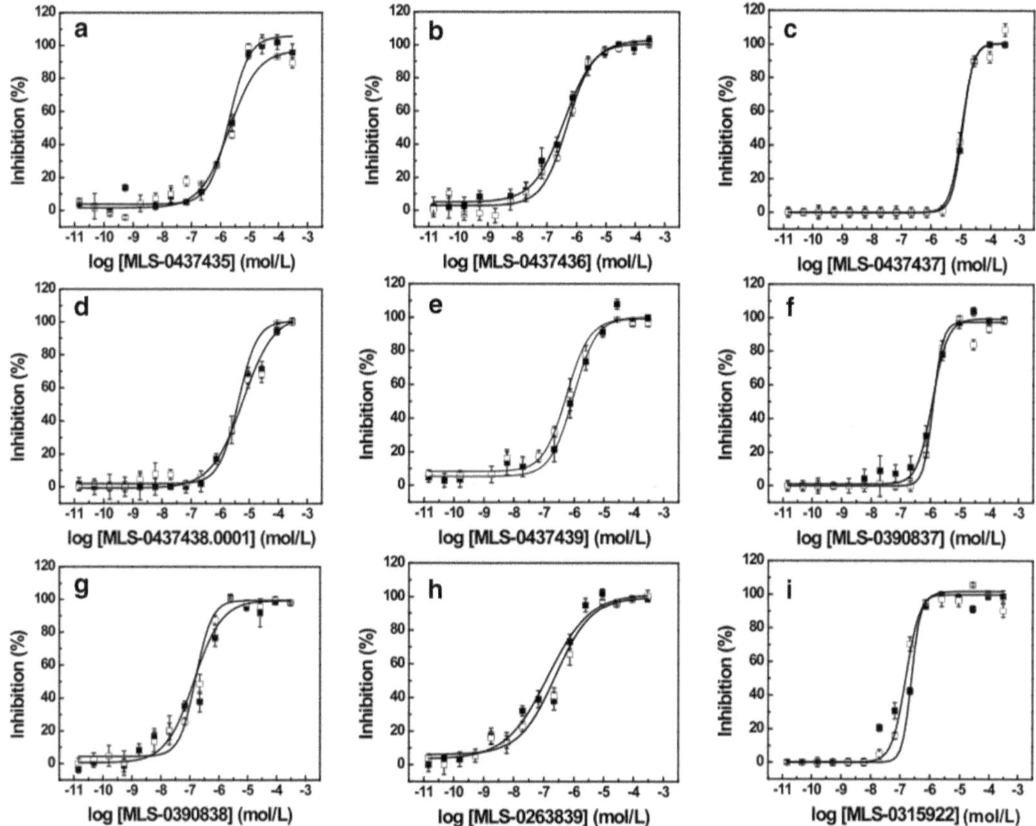

Fig. 1 Characterization of PHOSPHO1 inhibitors. Degree of PHOSPHO1 inhibition measured with the substrates P-Etn (*filled square*) or P-Cho (*open square*) as a function of PHOSPHO1 inhibitor concentration: (**a**), MLS-0437435.001; (**b**) MLS-0437436.001 (**c**), MLS-0437437.001; (**d**), MLS-0437438.001; (**e**), MLS-0437439.001; (**f**), MLS-0390837; (**g**), MLS-0390838; (**h**), MLS-0263839; (**i**), MLS-0315922. Origin Scientific plotting software (Northampton, MA, USA). Results are presented as mean ± SEM

1. Anesthetize a mouse with intraperitoneal injection with Avertin (*see* **Note 8**).
2. Open up abdominal cavity and open diaphragm.
3. Pull internal organs (intestines, liver, spleen etc.) to the side.
4. Open diaphragm to see the heart.
5. Make a cut on the left vena cava by a needle or scissors. Blood will come out immediately.
6. Perfuse from the left ventricle with 10 ml PBS at room temperature.
7. Remove ribcage to expose thoracic cavity.
8. Cut esophagus at upper position. Lower intestines should not be damaged to avoid bacteria contamination.

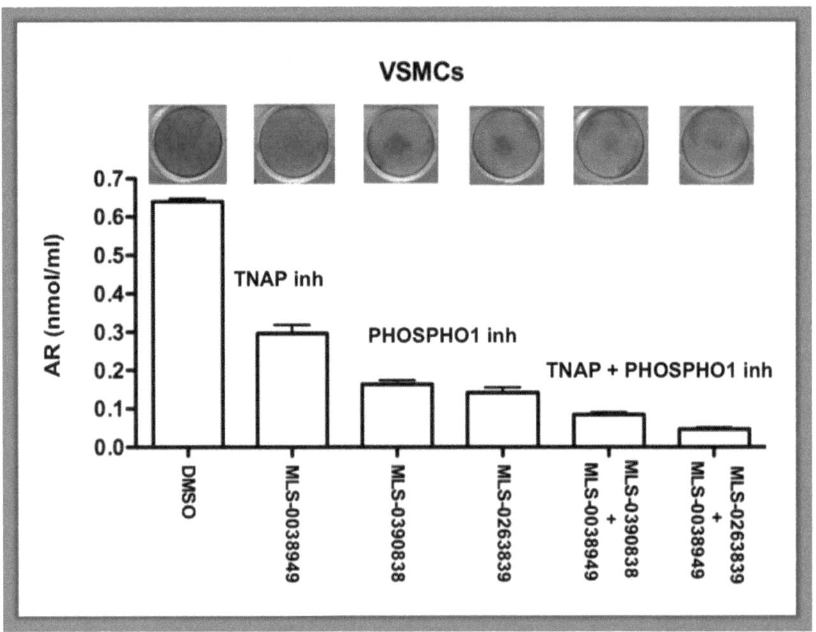

Fig. 2 Inhibition of matrix mineralization. Alizarin Red-staining ($A_{570\ nm}$) of 21 day cultures of WT VSMC in the presence of β-glycerophosphate. The cells were cultured with 30 μmol/L of the TNAP inhibitor MLS-0038949 and the PHOSPHO1 inhibitors MLS-0263839 and MLS-0390838, either alone or combined, as indicated. Calcium deposition in 21-day VSMC cultures was evaluated by staining cell layers with Alizarin Red

9. Find and hold common iliac artery as recognized as downwards Y shape, down the kidney and trim adhered tissue around the aorta and go up by cutting off artery branches.
10. Remove diaphragm and continue to go up till the aortic arc separating from yellowish adipose tissue.
11. Remove esophagus and lungs.
12. Pull the heart gently and trim out fat and connective tissue suspending the heart (*see* **Note 9**).
13. Pull the heart and trim out fat and connective tissue around the aorta.
14. Trim out fat around the aorta as much as possible.
15. Dissect and hold the end near the bottom end (just above the Y shape area).
16. Open up the aorta longitudinal direction all the way up the heart.
17. Cut to disconnect from the left ventricle.
18. Transfer to a dish containing serum free media. (Trim out fat as much as possible if necessary). Keep at 4 °C to finish all mice. (Under tissue culture hood).

19. Tools, tubes, 3.5 cm dish, PBS, Trypsin, collagenase, 37 °C water bath should be ready. 1 mg/ml trypsin-PBS solution [Sigma T-47999] (5 ml per a mouse × 3 runs). 2 mg/ml collagenase/PBS solution [Worthington type2, 4176] (1 ml per a mouse × 2 runs). (both prepared and sterilized by 0.2 μM filter on the day).
20. Wash aorta in ~2 ml PBS in 3.5 dish.
21. Transfer to round bottom tube containing 5 ml trypsin solution.
22. Incubate 10 min in 37 °C water bath under mild shaking.
23. Remove spent trypsin solution.
24. Repeat (total three changes of trypsin solution).
25. Remove the third trypsin solution.
26. Add 6–7 ml PBS and remove the PBS (this is to reduce carry over of trypsin).
27. Add 1 ml collagenase solution.
28. Incubate for 20 min in 37 °C water bath under rotation.
29. Transfer the spent collagenase solution into a new conical tube.
30. Repeat (total two times collagenase).
31. Combine the two changes of collagenase solution containing digested cells.
32. Centrifuge (500 *g* × min) and remove supernatant.
33. Resuspend cells in 2.5 ml 10 % FCS-αMEM (with nucleosides, gln, pyruvate, Invitrogen 12571063).
34. Filtrate with 40 μm strainer.
35. Plate onto one well of 24-well plate/aorta from single mouse and incubate 37 °C 5–7 % CO_2.
36. Next day, partial change of the media. (Remove 2 ml, add fresh 2 ml).
37. Once confluent (3–4 days), trypsinize and expand into 1 of 35 mm dish.
38. Once confluent trypsinize and expand into 1 of 25 cm^2 flask/head.
39. Once confluent, seed at 2×10^4/well/24-well plate for differentiation assay.
40. Next day, add mineralizing media, 10 % FCS αMEM containing 10 mM β-glycerophosphate-100 μg/ml ascorbic acid and 30 μM of freshly dissolved PHOSPHO1 inhibitors in DMSO (βGP, AsA Stocks stored in −20 °C) and renew the mineralizing media every second day till desired point (Day 15–21). Controls contain DMSO only.

3.3 Calcium Measurement with Alizarin Red Staining

Culture (Be careful to keep the cells adhered to the dish).

1. Rinse with PBS briefly, decant. ~1 ml PBS/well.
2. Add a few drop 50 %/PBS slowly. Add more 50 %/PBS.
3. Suction, leave only ~1 ml/well. Add more 50 %/PBS. Suction/decant.
4. Add ~1.5 ml/well 50 % EtOH/PBS, wait a few minutes. Suction/decant.
5. Add ~1.5 ml/well. 70 % EtOH/water.
   Suction/decant.
6. Add ~1.8 ml/well 100 % EtOH (can be stored in −20°).
7. Discard the EtOH. Wash with ddw or dry completely.
8. Add 300 μl/well Alizarin Red solution. Incubate 30 min at room temp under mild rotation. Wash with ddw several times.
9. Wash with PBS until the color of PBS gets very pale (even among all the wells). Scan bottom side if necessary.
10. Suction PBS completely.
11. Add exactly 500 μl/well cetylpyridinium chloride solution.
12. Incubate 30–60 min under rotation at room temp.
13. Collect the dye extracts.
14. Wash wells with ddw once.
15. Add DNA lyses solution.
16. Extract genomic DNA with the phenol chloroform method (SST tube), High salt method or other methods.
17. For measurement, mix 10 μl samples (standards) + 90 μl 10 mM phosphate buffer. (PB can be 40 μl if samples have low OD or 190 μl if samples have high OD).
18. Measure OD562. Triplicate is desirable. Use serially diluted AR solutions as standards.
19. Obtain values of AR/μg DNA of samples.

4 Notes

1. Run SDS PAGE to check expression.
2. Run SDS PAGE to check expression.
3. Be very cautious since slow change occurs.
4. Each reaction contained a final concentration of 1 % (v/v) DMSO, which was determined to be insufficient to interfere with PHOSPHO1 activity.
5. The compounds can be diluted in DMSO 100 % and storage as frozen aliquots at −20°. This procedure will guarantee reproducibility of kinetics measurements and culture effects.

6. All aseptic procedure. Sterilize all tools with 70 % EtOH.
7. The in vitro experiments that utilize cultures of vascular smooth muscle cells (VSMC) require initial isolation of primary aortic VSMC from adult mice (5–8 weeks old). Because adult vascular cell cultures undergo significant phenotypic modulation with prolonged culture it is necessary to continuously generate and maintain low-passage cultures (≤6).
8. All mice need to be housed in a temperature and climate controlled facility. The food and bedding need to be autoclaved to maintain the cleanest possible setting. The animals need to be observed daily for any IACUC endpoint suggesting discomfort, distress, pain, and injury. Urine volume, and food and water intake also need to be observed. In all cases, discomfort, distress, pain, and injury will be limited to that which is unavoidable in the conduct of scientifically sound research. Any animals showing signs of distress need to receive either buprenorphine or Meloxicam, or if necessary, euthanized.
9. Because the experiment is performed with VSMC primary culture the cleaning of aorta trimming out fat as much as possible is extremely necessary to avoid contamination of another type of cells in the culture.

Acknowledgments

This work was supported in part by grant R01AR053102 and R01AR047908 from the National Institute of Arthritis and Musculoskeletal Diseases (NIAMS), and American Recovery and Reinvestment Act (ARRA) Challenge grant RC1HL10899 from the National Heart, Lung, and Blood Institute (NHLB), National Institutes of Health (NIH), USA, and Institute Strategic Program Grant Funding from the Biotechnology and Biological Sciences Research Council.

References

1. Towler DA (2008) Vascular calcification: a perspective on an imminent disease epidemic. IBMS Bonekey 5:41–58
2. O'Neill WC, Sigrist MK, McIntyre CW (2010) Plasma pyrophosphate and vascular calcification in chronic kidney disease. Nephrol Dial Transplant 25:187–191
3. Narisawa S, Harmey D, Yadav MC et al (2007) Novel inhibitors of alkaline phosphatase suppress vascular smooth muscle cell calcification. J Bone Miner Res 22:1700–1710
4. Lomashvili KA, Garg P, Narisawa S et al (2008) Upregulation of alkaline phosphatase and pyrophosphate hydrolysis: potential mechanism for uremic vascular calcification. Kidney Int 73:1024–1030
5. Villa-Bellosta R, Wang X, Millán JL et al (2011) Extracellular pyrophosphate metabolism and calcification in vascular smooth muscle. Am J Physiol Heart Circ Physiol 301:H61–H68
6. Hessle L, Johnson KA, Anderson HC et al (2002) Tissue-nonspecific alkaline phosphatase and plasma cell membrane glycoprotein-1 are central antagonistic regulators of bone mineralization. Proc Natl Acad Sci U S A 99: 9445–9449

7. Houston B, Seawright E, Jefferies D et al (1999) Identification and cloning of a novel phosphatase expressed at high levels in differentiating growth plate chondrocytes. Biochim Biophys Acta 1448:500–506
8. Stewart AJ, Schmid R, Blindauer CA et al (2003) Comparative modelling of human PHOSPHO1 reveals a new group of phosphatases within the haloacid dehalogenase superfamily. Protein Eng 16:889–895
9. Houston B, Stewart AJ, Farquharson C (2004) PHOSPHO1-A novel phosphatase specifically expressed at sites of mineralisation in bone and cartilage. Bone 34:629–637
10. Stewart AJ, Roberts SJ, Seawright E et al (2006) The presence of PHOSPHO1 in matrix vesicles and its developmental expression prior to skeletal mineralization. Bone 39:1000–1007
11. Roberts S, Narisawa S, Harmey D et al (2007) Functional involvement of PHOSPHO1 in matrix vesicle-mediated skeletal mineralization. J Bone Miner Res 22:617–627
12. Roberts SJ, Stewart AJ, Sadler PJ et al (2004) Human PHOSPHO1 exhibits high specific phosphoethanolamine and phosphocholine phosphatase activities. Biochem J 382:59–65
13. Hsu HH, Camacho NP (1999) Isolation of calcifiable vesicles from human atherosclerotic aortas. Atherosclerosis 143:353–362
14. Kapustin AN, Davies JD, Reynolds JL et al (2011) Calcium regulates key components of vascular smooth muscle cell-derived matrix vesicles to enhance mineralization. Circ Res 109:e1–e12
15. Kiffer-Moreira T, Yadav M, Zhu D et al (2013) Pharmacological inhibition of PHOSPHO1 suppresses vascular smooth muscle cell calcification. J Bone Miner Res 28(1):81–91
16. Dahl R, Sergienko EA, Su Y et al (2009) Discovery and validation of a series of aryl sulfonamides as selective inhibitors of tissue-nonspecific alkaline phosphatase (TNAP). J Med Chem 52:6919–6925

Chapter 9

Modulators of Intestinal Alkaline Phosphatase

Ekaterina V. Bobkova, Tina Kiffer-Moreira, and Eduard A. Sergienko

Abstract

Small molecule modulators of phosphatases can lead to clinically useful drugs and serve as invaluable tools to study functional roles of various phosphatases in vivo. Here, we describe lead discovery strategies for identification of inhibitors and activators of intestinal alkaline phosphatases. To identify isozyme-selective inhibitors and activators of the human and mouse intestinal alkaline phosphatases, ultrahigh throughput chemiluminescent assays, utilizing CDP-Star as a substrate, were developed for murine intestinal alkaline phosphatase (mIAP), human intestinal alkaline phosphatase (hIAP), human placental alkaline phosphatase (PLAP), and human tissue-nonspecific alkaline phosphatase (TNAP) isozymes. Using these 1,536-well assays, concurrent HTS screens of the MLSMR library of 323,000 compounds were conducted for human and mouse IAP isozymes monitoring both inhibition and activation. This parallel screening approach led to identification of a novel inhibitory scaffold selective for murine intestinal alkaline phosphatase. SAR efforts based on parallel testing of analogs against different AP isozymes generated a potent inhibitor of the murine IAP with IC50 of 540 nM, at least 65-fold selectivity against human TNAP, and >185 selectivity against human PLAP.

Key words Alkaline phosphatase, Intestinal alkaline phosphatase, Tissue-nonspecific alkaline phosphatase, Placental alkaline phosphatase, High throughput assay, Chemiluminescence, CDP-star, Inhibition, Activation, Structure–activity relationship

1 Introduction

Alkaline phosphatase (AP; orthophosphoric monoester phosphohydrolase, EC 3.1.3.1) isozymes are present in a wide range of species from bacteria to man and are capable of dephosphorylation and transphosphorylation of a broad spectrum of substrates in vitro [1]. Their broad substrate specificity and localization on the outside leaf of the cytoplasmic membrane suggests potential involvement in numerous extracellular processes. In humans, four isozymes of APs have been identified. One of them, tissue-nonspecific alkaline phosphatase (TNAP), is ubiquitously expressed, demonstrating especially high level of expression in bone, liver and kidney tissues. Three other isozymes demonstrate tissue-specific

José Luis Millán (ed.), *Phosphatase Modulators*, Methods in Molecular Biology, vol. 1053,
DOI 10.1007/978-1-62703-562-0_9, © Springer Science+Business Media, LLC 2013

character of expression; they are named according to the tissue of their predominant expression: intestinal (IAP), placental (PLAP), and germ cell (GCAP) alkaline phosphatases.

IAP isozyme is expressed in the brush border of the intestinal epithelium. Although the IAP isozyme was discovered more than half a century ago, its physiologic function is still not fully understood. Most of our understanding of IAP potential functions comes from mouse IAP (*Akp3*) knockout studies, which suggest that IAP is involved in fat absorption [2], in the maintenance of healthy gut barrier function [3–5] and in the establishment of the gut microbiota [6]. It is also believed that IAP promotes an adequate detoxification of bacterial products such lipopolysaccharides (LPS), reducing excessive intestinal inflammation [7]. In a proof of concept study, duodenal delivery of calf IAP to ulcerative colitis patients resulted in improvement in clinical and serological measures [8].

Mouse IAP (*Akp3*) KO models were extremely informative and helpful for outlining the physiological processes involving IAP (see Chapter 3); on the other hand, full interpretation of these data and translation of this knowledge onto humans is not straightforward, since multiple AP isozymes are present in mouse gut, while only one is expressed in human intestines [9]. Availability of small-molecule modulators of human and mouse IAP with exquisite selectivity against other isozymes would be extremely helpful in further exploration of physiologic and pathophysiologic functions of IAP, similarly to other isozymes of AP and other phosphatases [10–12]. Previously, we identified highly selective small-molecule inhibitors and activators of human TNAP and PLAP [13–16]. Pertinent to the current study, we also discovered inhibitors of human IAP with moderate selectivity over human TNAP and PLAP isozymes [17, 18].

Interestingly, all these compounds are inactive against mouse IAPs, therefore, unsuitable for studies of IAP function in mouse models. The objectives of the current studies were the identification of small-molecule modulators of mouse IAP and, potentially, identification of novel selective scaffolds of human IAP. To this end, we applied and optimized the chemiluminescent assays that were successfully utilized for screening human IAP, TNAP, and PLAP and developed a novel assay for mouse IAP, analogous to the assays of other isozymes. These assays utilize CDP-star, a substrate of alkaline phosphatases specifically invented for and commonly utilized in blotting techniques [19, 20]. Development and utilization of the prototype plate-reader enzymatic assay for TNAP isozyme with CDP-star substrate was previously described in detail elsewhere [21].

2 Materials

Prepare all solutions using ultrapure water and analytical grade reagents. Prepare and store all reagents at room temperature (unless indicated otherwise).

2.1 Materials and Equipment

1. CDP-Star reagent (New England Biolabs # N7001S).
2. 1,536-well white untreated plate (Corning, #3725).
3. Echo550 (Labcyte).
4. MultiDrop Combi dispenser (Thermo Scientific).
5. Centrifuge 5810 (Eppendorf).
6. EnVision (Perkin Elmer).

2.2 Preparation of Phosphatases

FLAG-Tagged isozymes of murine IAP, human IAP, human TNAP, and human PLAP were used in the ultrahigh throughput chemiluminescent assays. The enzymes in which the FLAG sequence is substituted for the GPI anchor were transiently transfected into COS-1 cells. The transfected cells cultured in OPTI-MEM serum-free medium, and conditioned media (containing secreted enzymes) were collected 60 h after transfection and submitted to dialysis in 1× Tris buffered saline containing 1 mM $MgCl_2$ and 20 μM $ZnAc_2$ for 48 h prior to use. The transient expression using COS-1 cells was performed as followed:

1. Expansion of COS-1 cells: grow stock cells in 75 cm^2 flask using 10 % FCS growth media* until reaching confluence.
2. Trypsinization: remove spent medium (20 ml) and wash plate with ~20 ml PBS followed by addition of 2 ml Trypsin–EDTA ×1 solution. Incubate for 5–15 min pipetting to disperse cells. Add 8 ml growth media to stop trypsinization and transfer cell suspension to 15 ml tube. Centrifuge (1,000 × 5 min) and resuspend cells in serum free media** for cell counting.
3. Transfection (described for one kind of plasmid/ one electroporation): for cell count combine 0.2 ml cell suspension and 0.6 ml Trypan blue (×1/4 dilution) and mix. Count numbers (use 1–2 counting chambers ($=10^{-3}$ ml)). If the number is 200 (from all ten squares), the cell concentration is $200 \times 10^3 \times 4$ dilution = 8.00×10^5 cells/ml. Then, you have 8×10^6 cells (8.00×10^5 cells/ml × 10 ml). Wash cells with serum free medium one more time and spin down cell. To cells ($0.7–1 \times 10^7$ cells) add 0.8 ml HBS. First, place DNA solution (exactly 10 mg) into the electroporation cuvette (0.4 cm distance). Second, suspend cells containing HBS with 1 ml pipet and add to the cuvette (cells and DNA will be mixed). Electroporate at 220 V 960 mF and immediately place cuvette on ice, let stand 20 min. Release into fresh 30 ml growth media in 144 cm^2

dish. Incubate at 37 °C 5 % CO_2 for 24 h and change media to Opti-MEM. After 60 h further incubation collect media and filter using filter device (0.2 μm).

**10 % FCS growth media*

DMEM high glucose (w/o L-glutamine, w/o Na-pyruvate), 500 ml/bottle.

Add.

10 ml 200 mM L-glutamine (defrost, keep 4 °C).

5 ml 100 mM Na pyruvate (4 °C).

5 ml 100 mM MEM nonessential amino acids (4 °C).

53 ml FCS (56 °C 30 min treated, stored in 4 °C).

***Serum-free media*

DMEM high glucose (w/o L-glutamine, w/o Na-pyruvate), 500 ml/bottle.

Add.

10 ml 200 mM L-glutamine (defrost, keep 4 °C).

5 ml 100 mM Na pyruvate (4 °C).

5 ml 100 mM MEM nonessential amino acids (4 °C).

2.3 Assay Buffers and Solutions

2× Assay Buffer used for all isozymes: 200 mM DEA, pH 9.8, 0.04 mM $ZnCl_2$, 2 mM $MgCl_2$.

3 Methods

CDP-Star, phenylphosphate substituted-1,2-dioxetane, is a chemiluminescent substrate for alkaline phosphatases routinely utilized in immunoblotting approaches with AP as a reporter [19, 20]. Previously, we successfully developed a luminescent assay for TNAP using this substrate [21], later expanding this approach to IAP and PLAP isozymes [14, 17]. This substrate is readily dephosphorylated by APs, leading to formation of high-energy unstable dephosphorylated form that decomposes with production of light. CDP-Star provides high sensitivity and speed of detection which makes it compatible with a wide range of microplate readers (*see* Fig. 1) [21]. All procedures are carried out at room temperature unless otherwise specified. The assays were developed and optimized according to the principles described elsewhere [22].

3.1 mIAP 1,536-Well High-Throughput Assay

1. Using Echo550, transfer 40 nL DMSO to columns 1–4 (positive and negative control wells), and test compounds to columns 4–48 of a 1,536 well assay plate to achieve the desired

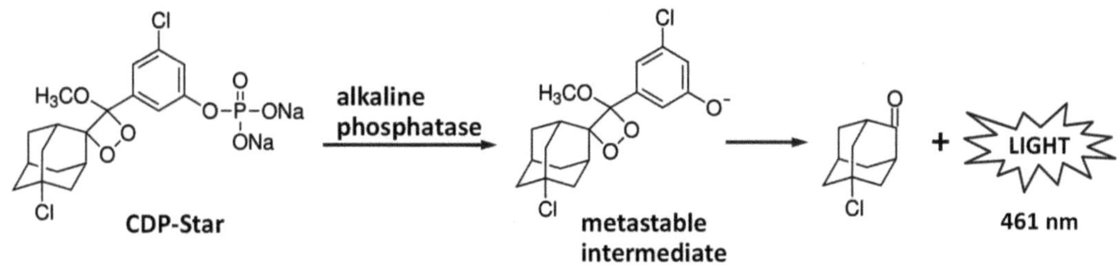

Fig. 1 Detection of alkaline phosphatase activity using CDP-Star substrate

test concentrations. Compounds are transferred from a 10 mM stock to give the final concentration. Test compound wells in the assay plate are back-filled with DMSO to equalize final volume to 40 nL.

2. Prepare 400 μM CDP-star solution in MQ water.
3. Prepare 2× mIAP Solution: Dilute the enzyme 125-fold by aliquoting 53.6 μL of the mIAP stock solution into 6.7 mL of 2× Assay Buffer. Mix slowly by inverting the container. Store on ice until used.
4. Using MultiDrop Combi dispenser add 2 μL/well of 2× Assay Buffer to columns 1 and 2 (positive control wells).
5. Using MultiDrop Combi Dispenser add 2 μL/well of 2× mIAP Solution to columns 3 through 48.
6. Pre-incubate compounds with enzyme for 30 min.
7. Start reactions by adding 2 μL/well of CDP-Star (400 μM in MQ water) to all wells.
8. Using Eppendorf 5810 centrifuge, spin the plates down at 182 RCF for 5 min. to eliminate air bubbles and to level the liquid surface.
9. Cover the plate and incubate at room T for 30 min.
10. Read the plate on Perkin Elmer EnVision 0.2 s/well at 0.5 mm above well top, using US-Luminescence mode.

3.2 hIAP 1,536-Well High Throughput Assay

1. Using Echo550, transfer 40 nL DMSO to columns 1–4 (positive and negative control wells), and test compounds to columns 4–48 of a 1,536 well assay plate to achieve the desired test concentrations. Compounds are transferred from a 10 mM stock to give the final concentration. Test compound wells in the assay plate are back-filled with DMSO to equalize final volume to 40 nL.
2. Prepare 400 μM CDP-star solution in MQ water.
3. Prepare 2× hIAP Solution: Dilute the enzyme 200-fold by aliquoting 33.5 μL of the stock solution of hIAP into 6.7 mL of 2× Assay Buffer. Mix slowly by inverting the container. Store on ice until used.

4. Using MultiDrop Combi dispenser add 2 μL/well of 2× Assay Buffer to columns 1 and 2 (positive control wells).
5. Using MultiDrop Combi Dispenser add 2 μL/well of 2× mIAP Solution to columns 3 through 48.
6. Pre-incubate compounds with enzyme for 30 min.
7. Start reactions by adding 2 μL/well of CDP-Star (354 μM in MQ water) to all wells.
8. Using Eppendorf 5810 centrifuge, spin the plates down at 182 RCF for 5 min. to eliminate air bubbles and to level the liquid surface.
9. Cover the plate and incubate at room T for 30 min.
10. Read the plate on Perkin Elmer EnVision 0.2 s/well at 0.5 mm above well top, using US-Luminescence mode.

3.3 PLAP 1,536-Well High Throughput Assay

1. Using Echo550, transfer 40 nL DMSO to columns 1–4 (positive and negative control wells), and test compounds to columns 4–48 of a 1,536-well assay plate to achieve the desired test concentrations. Compounds are transferred from a 10 mM stock to give the final concentration. Test compound wells in the assay plate are back-filled with DMSO to equalize final volume to 40 nL.
2. Prepare 500 μM CDP-star solution in MQ water.
3. Prepare 2× hIAP Solution: Dilute the enzyme 500-fold by aliquoting 13.4 μL of the stock solution of PLAP into 6.7 mL of 2× Assay Buffer. Mix slowly by inverting the container. Store on ice until used.
4. Using MultiDrop Combi dispenser add 2 μL/well of 2× Assay Buffer to columns 1 and 2 (positive control wells).
5. Using MultiDrop Combi Dispenser add 2 μL/well of 2× mIAP Solution to columns 3 through 48.
6. Pre-incubate compounds with enzyme for 30 min.
7. Start reactions by adding 2 μL/well of CDP-Star (500 μM in MQ water) to all wells.
8. Using Eppendorf 5810 centrifuge, spin the plates down at 182 RCF for 5 min. to eliminate air bubbles and to level the liquid surface.
9. Cover the plate and incubate at room T for 30 min.
10. Read the plate on Perkin Elmer EnVision 0.2 s/well at 0.5 mm above well top, using US-Luminescence mode.

3.4 TNAP 1,536-Well High Throughput Assay

1. Using Echo550, transfer 40 nL DMSO to columns 1–4 (positive and negative control wells), and test compounds to columns 4–48 of a 1,536-well assay plate to achieve the desired

test concentrations. Compounds are transferred from a 10 mM stock to give the final concentration. Test compound wells in the assay plate are back-filled with DMSO to equalize final volume to 40 nL.

2. Prepare 500 μM CDP-star solution in MQ water.
3. Prepare 2× hIAP Solution: Dilute the enzyme 1,000-fold by aliquoting aliquoting 6.7 μL of the stock solution of TNAP into 6.7 mL of 2× Assay Buffer. Mix slowly by inverting the container. Store on ice until used.
4. Using MultiDrop Combi dispenser add 2 μL/well of 2× Assay Buffer to columns 1 and 2 (positive control wells).
5. Using MultiDrop Combi Dispenser add 2 μL/well of 2× mIAP Solution to columns 3 through 48.
6. Pre-incubate compounds with enzyme for 30 min.
7. Start reactions by adding 2 μL/well of CDP-Star (500 μM in MQ water) to all wells.
8. Using Eppendorf 5810 centrifuge, spin the plates down at 182 RCF for 5 min. to eliminate air bubbles and to level the liquid surface.
9. Cover the plate and incubate at room *T* for 30 min.
10. Read the plate on Perkin Elmer EnVision 0.2 s/well at 0.5 mm above well top, using US-Luminescence mode.

3.5 High Throughput Screens, Hit Profiling, and Lead Optimization

The HTS screening results, hit confirmation, and potency profiling against the primary target are summarized in Fig. 2. About 330,000 compounds comprising the MLSMR small molecule library were screened at 10 μM against the mIAP and hIAP assays, yielding 781 and 706 initial hits, respectively, with some of these overlapping. Resulting 1,377 unique hits were cherry-picked from the MLSMR library and retested at a single concentration of 10 μM in duplicate against both IAP enzymes. 227 modulators of mIAP were confirmed, with 55 activators and 172 inhibitors, and 256 modulators were confirmed for hIAP, comprising 86 activators and 170 inhibitors. 586 compounds inhibited both IAP isozymes. To establish the ranking by potency, all active compounds were evaluated in 10 pt dose response studies in duplicate. In parallel, all active hits were also profiled in PLAP and TNAP selectivity assays. This strategy led to identification of multiple inhibitory and activating scaffolds for both IAP isozymes. Best hits and their analogs were reordered from dry powders, with 525 compounds advanced to the full profiling in 16 pt dose response screens in mIAP and hIAP assays, as well as in TNAP and PLAP screens to evaluate selectivity. Most of the confirmed DR actives were eliminated due to poor or non-selectivity against both PLAP and TNAP. After an inspection of the actual scaffolds and their level of selectivity, two mIAP inhibitory scaffolds

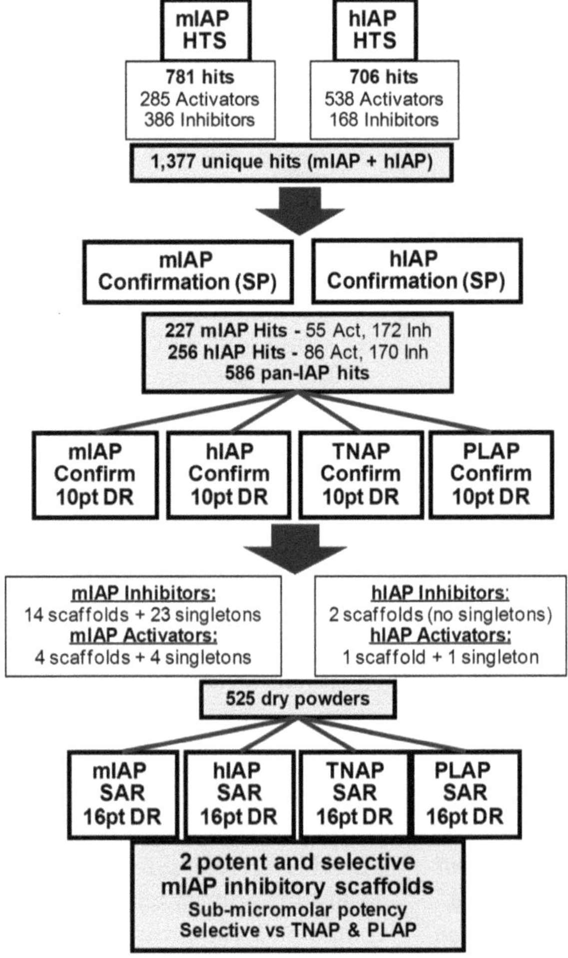

Fig. 2 Lead discovery and optimization for modulators of intestinal alkaline phosphatases

emerged as being of sufficient merit based on the initial hit validation criteria. Structure–Activity Relationship studies generated a potent and selective inhibitor of mIAP, molecular probe ML260, with potency against mIAP of 540 nM and 65-fold selectivity against TNAP and the more than 185-fold selectivity against PLAP and hIAP (PubChem AID 2818). Replicate dose response titrations of the molecular probe compound ML260 against mIAP, TNAP and PLAP are shown in Fig. 3.

4 Notes

1. Alkaline phosphatase stock solutions are usually stable at 4 °C for a few weeks. For long-term storage it is recommended to flash freeze small aliquots of enzymes in liquid nitrogen and store at −80 °C.

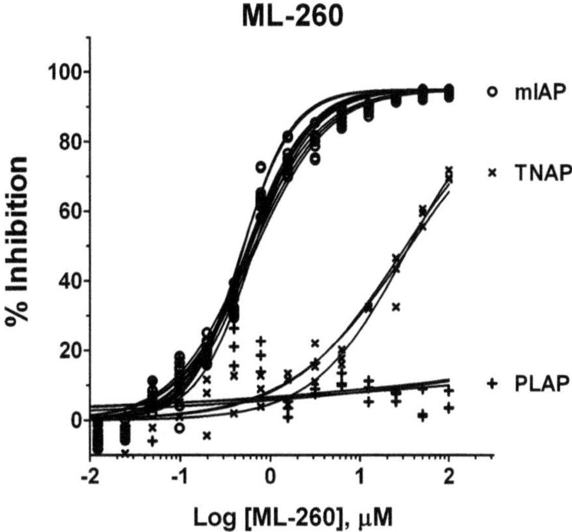

Fig. 3 Potency and selectivity of ML-260 mIAP inhibitor. The potency of the inhibitor against mIAP (o), PLAP (+) and TNAP (x) was assessed by dose–response testing in 1,536-well format in duplicate in 1× Assay Buffer, containing 100 mM DEA, pH 9.8, 0.02 mM $ZnCl_2$, 1 mM $MgCl_2$, and 1:250, 1:1,000, and 1:2,000 diluted enzymes, respectively, in a total volume of 4 μL/well. The reactions were started by addition of the CDP-star substrate to the final concentrations of 200 μM for the mIAP reaction, and 250 μM for the PLAP and TNAP reactions, with the luminescence intensity measured after 30 min incubation at room *T*

2. Due to highly miniaturized microliter reaction volumes, care should be taken to seal the plates during the prolonged incubations—we used gasket-sealed Kalypsys lids, but usual inexpensive sealing films for microtiter plates should work fine as well.
3. To ensure sensitivity of the assay towards modulators, the level of the enzyme in the reaction mix should be optimized for every new batch of protein by performing the titration curve under various enzyme dilutions – the enzyme concentration from the linear range of the curve should be selected as described elsewhere [22].
4. For selectivity studies, to directly compare potencies of modulators using IC_{50} values against different enzymes, all assays were performed at the substrate concentration of $1 \times K_m$ for the corresponding isozyme.

Acknowledgments

This work was performed at Sanford-Burnham Medical Research Institute and was supported by the NIH Roadmap grant U54 HG005033 and grant # X01-MH077602-01.

References

1. Millán JL (2006) Mammalian alkaline phosphatases: from biology to applications in medicine and biotechnology. Wiley-VCH Verlag GmbH, Weinheim, Germany
2. Narisawa S, Huang L, Iwasaki A et al (2003) Accelerated fat absorption in intestinal alkaline phosphatase knockout mice. Mol Cell Biol 23:7525–7530
3. Chen KT, Malo MS, Beasley-Topliffe LK et al (2011) A role for intestinal alkaline phosphatase in the maintenance of local gut immunity. Dig Dis Sci 56:1020–1027
4. Goldberg RF, Austen WG Jr, Zhang X et al (2008) Intestinal alkaline phosphatase is a gut mucosal defense factor maintained by enteral nutrition. Proc Natl Acad Sci U S A 105:3551–3556
5. Ramasamy S, Nguyen DD, Eston MA et al (2011) Intestinal alkaline phosphatase has beneficial effects in mouse models of chronic colitis. Inflamm Bowel Dis 17:532–542
6. Malo MS, Alam SN, Mostafa G et al (2010) Intestinal alkaline phosphatase preserves the normal homeostasis of gut microbiota. Gut 59:1476–1484
7. Poelstra K, Bakker WW, Klok PA et al (1997) A physiologic function for alkaline phosphatase: endotoxin detoxification. Lab Invest 76:319–327
8. Lukas M, Drastich P, Konecny M et al (2010) Exogenous alkaline phosphatase for the treatment of patients with moderate to severe ulcerative colitis. Inflamm Bowel Dis 16:1180–1186
9. Narisawa S, Hoylaerts MF, Doctor KS et al (2007) A novel phosphatase upregulated in *Akp3* knockout mice. Am J Physiol Gastrointest Liver Physiol 293:G1068–1077
10. Kiffer-Moreira T, Yadav M, Zhu D et al (2012) Pharmacological inhibition of PHOSPHO1 suppresses vascular smooth muscle cell calcification. J Bone Miner Res 28(1):81–91
11. Narisawa S, Harmey D, Yadav MC et al (2007) Novel inhibitors of alkaline phosphatase suppress vascular smooth muscle cell calcification. J Bone Miner Res 22:1700–1710
12. Villa-Bellosta R, Wang X, Millán JL et al (2011) Extracellular pyrophosphate metabolism and calcification in vascular smooth muscle. Am J Physiol Heart Circ Physiol 301:H61–68
13. Dahl R, Sergienko EA, Su Y et al (2009) Discovery and validation of a series of aryl sulfonamides as selective inhibitors of tissue-nonspecific alkaline phosphatase (TNAP). J Med Chem 52:6919–6925
14. Lanier M, Sergienko E, Simão AM et al (2010) Design and synthesis of selective inhibitors of placental alkaline phosphatase. Bioorg Med Chem 18:573–579
15. Sergienko E, Su Y, Chan X et al (2009) Identification and characterization of novel tissue-nonspecific alkaline phosphatase inhibitors with diverse modes of action. J Biomol Screen 14:824–837
16. Sidique S, Ardecky R, Su Y et al (2009) Design and synthesis of pyrazole derivatives as potent and selective inhibitors of tissue-nonspecific alkaline phosphatase (TNAP). Bioorg Med Chem Lett 19:222–225
17. Chung TD, Sergienko E, Millán JL (2010) Assay format as a critical success factor for identification of novel inhibitor chemotypes of tissue-nonspecific alkaline phosphatase from high-throughput screening. Molecules 15:3010–3037
18. PubChem AID 2574 (2011) Summary assay for identification of inhibitors of human intestinal alkaline phosphatase. http://pubchem.ncbi.nlm.nih.gov/assay/assay.cgi?aid=2574&loc=ea_ras
19. Bronstein I, Brooks E, Rouh-Rong J, inventors (1994) Chemiluminescent 3-(substituted adamant-2′-ylidene) 1,2-dioxetanes. United States patent 5,326,882
20. Brooks E, Voyta JC, inventors (1990) Purification of stable water-soluble dioxetanes. United States patent 4,931,569
21. Sergienko EA, Millán JL (2010) High-throughput screening of tissue-nonspecific alkaline phosphatase for identification of effectors with diverse modes of action. Nat Protoc 5:1431–1439
22. Sergienko E (2012) Basics of HTS assay design and optimization. Chemical genomics. Cambridge University Press, New York, pp 159–172

Chapter 10

New Activity Assays for ENPP1 with Physiological Substrates ATP and ADP

Chen-Ting Ma and Eduard A. Sergienko

Abstract

Existing assays monitoring ENPP1 activity are either not physiologically relevant or not suitable for high-throughput screening (HTS). Here, we describe the development and implementation of two new ENPP1 activity assays that address these drawbacks. These assays employ physiological substrates of ENPP1, ATP and ADP. They rely on detection of inorganic phosphate using a special modification of the malachite green-molybdate colorimetric procedure that ensures stability of acid-labile compounds, such as the ones containing phosphodiester bonds. The pyrophosphate generated in ENPP1 reaction is converted to inorganic phosphate in the presence of inorganic phosphatase; whereas, omission of this coupling enzyme enables detection of the inorganic phosphate generated by ENPP1. These new ENPP1 assays were miniaturized into high-density microplate formats. With minimal requirement for ENPP1 enzyme, low micromolar phosphate detection sensitivity, and simple protocol involving three to four simple liquid handling steps, these robust assays are suitable for HTS.

Key words ENPP1, ATP, ADP, Pyrophosphatase, Phosphodiesterase, Phosphate detection, PiColorLock

1 Introduction

Ectonucleotide pyrophosphatase/phosphodiesterase 1 (ENPP1), also known as plasma cell membrane glycoprotein 1 (PC-1), is a member of the NPP family of transmembrane enzymes that catalyzes the hydrolysis of extracellular nucleotides. Different family members demonstrate preference for distinct substrates, but share ability to hydrolyze phosphodiester and pyrophosphatase bonds [1]. ENPP1 appears to have preference for ATP, but also demonstrates pyrophosphatase and phosphodiesterase activities toward pyrophosphate and ADP [1, 2].

ENPP1 is expressed on the extracellular membrane of osteoblasts and chondrocytes and on the surface of their matrix vesicles. Its localization and substrate specificity seem to implicate the enzyme is playing an important role in regulation of mineratization [1, 2].

José Luis Millán (ed.), *Phosphatase Modulators*, Methods in Molecular Biology, vol. 1053,
DOI 10.1007/978-1-62703-562-0_10, © Springer Science+Business Media, LLC 2013

Pyrophosphohydrolase activity of ENPP1 converts ATP into AMP and pyrophosphate, supplying the reservoir of extracellular pyrophosphate, a known inhibitor of hydroxyapatite formation. Another ectoenzyme expressed by osteoblasts and chondrocytes, tissue-nonspecific alkaline phosphatase (TNAP), co-localizes with ENPP1 and is known to hydrolyze pyrophosphate, providing the second arm of mineralization regulation. Supporting the major role of ENPP1 in mineralization, ENPP1 knockout cells demonstrate excessive calcification that could be corrected by transfecting in ENPP1, but not another NPP isozyme [3].

Human single nucleotide polymorphism (SNP) and genome-wide association studies (GWAS) indicate that ENPP1 is also involved in insulin signaling and is a predisposition factor for type 2 diabetes, obesity and diabetic neuropathy [4, 5]. Intriguingly, the insulin signaling effect of ENPP1 appears to be mediated through a direct interaction between ENPP1 and insulin receptor (IR). SNP responsible for K121Q mutation of human ENPP1 protein linked to insulin resistance and susceptibility to obesity results in ENPP1 protein variant that has 2- to 3-fold higher affinity to insulin receptor and is more potent in inhibiting IR [6]. Thus, identification of small molecule modulators of ENPP1 that could specifically affect either ENPP1 catalytic activity or IR binding interface is of great benefit for further delineation of ENPP1 physiological functions.

Currently available activity assays measuring ENPP1 activity require a substantial amount of enzyme [7], special instrumentations [2, 7], or radioactive materials [8]. Some of them utilize nitrophenyl-based substrates [8, 9], which are useful to study the enzyme reaction, but do not provide sufficient assay sensitivity or relevance to physiology. Here, we describe two new ENPP1 activity assays that use two biologically relevant substrates, ATP and ADP; easy stepwise procedures lead through the development and optimization of these assays. The latter is an important step for adopting the assay for every new batch of the ENPP1 enzyme produced in-house or purchased commercially to ensure assays robustness and sensitivity.

2 Materials

Prepare all solutions using ultrapure water (prepared by purifying deionized water to attain resistivity greater than 18.0 MΩ—cm at 25 °C) and analytical grade reagents. Prepare and store all reagents at room temperature (unless indicated otherwise). Diligently follow all waste disposal regulations when disposing waste materials.

1. 1 M Tris pH 7.5. Add 150 ml water to glass beaker. Weigh 24.2 g Tris and add to beaker. Dissolve Tris powder and pH to

7.5 with concentrated HCl. Add water to bring volume up to 200 ml. Pass through 0.22 μm filtration unit. Store at 4 °C.

2. 1 M $MgCl_2$. Add 75 ml water to glass beaker. Weigh 20.3 g $MgCl_2$ hexahydrate and add to beaker. Dissolve $MgCl_2$ powder and add water to bring volume up to 100 ml. Pass through 0.22 μm filtration unit. Store at 4 °C.
3. 10 % v/v Tween 20 detergent. Add 9 ml water to 15 ml disposable conical tube. Take 1 ml pipette tip and remove the tip with scissors. Pipette 1 ml Tween 20 to conical tube and mix thoroughly by vortexing. Pass the solution through a 0.22 μm syringe filter into a dark 15 ml conical tube. Store at 4 °C for no more than 2 weeks.
4. 0.6 mg/ml or 6 μm enzyme stock solution, amino acid 85–905 of mouse origin, expressed in 293 cells. Store aliquots at −80 °C. Store thawed aliquot on ice and discard at the end of assay day.
5. 40 mM ATP. Weigh 22 mg ATP powder (*see* **Note 1**) and add to 1.6 ml Eppendorf tube. Add 1 ml of water and mix by vortexing. Do not pH ATP solution. Store at −20 °C in smaller aliquots. Store thawed aliquot on ice and discard at the end of assay day.
6. 2,000 U/ml inorganic pyrophosphatase from Baker's yeast (Sigma-Aldrich, St. Louis, MO, USA). Resuspend 500 U lyophilized enzyme and buffer salts in 250 μl water. Store at −80 °C in 5 μl aliquots. Store thawed aliquot on ice and discard at the end of assay day.
7. 40 mM ADP. Weigh 17 mg ADP powder (Sigma-Aldrich) and add to 1.6 ml Eppendorf tube. Add 1 ml of water and mix by vortexing. Do not pH ADP solution. Store at −20 °C in smaller aliquots. Store thawed aliquot on ice and discard at the end of assay day.
8. PiColorLock Phosphate Detection System (Innova Biosciences, Cambridge, UK): 100 μM phosphate standard, Gold reagent solution, accelerant solution, and stabilizer solution. Store at 4 °C or on ice during assay.
9. Agilent Bravo Automated Liquid Handling Platform (Agilent Technologies, Santa Clara, CA, USA).
10. P30 pipette tips in 384-well racks (Velocity11, Santa Clara, CA, USA).
11. Multidrop Combi Reagent Dispenser (Thermo Scientific, Waltham, MA, USA).
12. Small tube metal/plastic tip dispensing cassettes for Combi (Thermo Scientific).
13. Viaflow 12.5 μl electronic multichannel pipette (Integra Biosciences, Hudson, NH, USA).

14. PHERAstar plate reader (BMG Labtech, Cary, NC, USA).
15. Echo 555 liquid handler (Labcyte, Sunnyvale, CA, USA).
16. 384-Well polypropylene small volume intermediate plate (catalog# 784201) (Greiner Bio-One, Monroe, NC, USA).
17. 1,536-Well clear polystyrene, HiBase, flat-bottom, square well assay plate (Catalog# 782101) (Greiner Bio-One).

3 Methods

Carry out all procedures at room temperature unless otherwise specified. Assay Buffer refers to buffer mixture described in Subheading 3.1, **step 1**. Prime all Combi tubing with the appropriate assay reagent, empty the tubing back into the conical tube holding the reagent, and prime again to ensure adequate mixing.

3.1 Optimization of Enzyme and Substrate Concentrations

1. Mix 1,000 μl of 1 M Tris, pH 7.5, 40 μl of 1 M $MgCl_2$ solution, 10 μl of 10 % Tween 20 solution, and 18.95 ml water in a 50 ml disposable conical tube, followed by thorough mixing by vortexer. Prepare this Assay Buffer fresh for every assay day.
2. Prepare 0.8 U/ml diluted pyrophosphatase solution by diluting 4 μl of 2,000 U/ml pyrophosphatase solution in 10 ml Assay Buffer in 15 ml conical tube. Mix by pipetting.
3. Prepare 200 μM diluted ATP solution by diluting 4 μl of 40 mM ATP solution in 800 μl 0.8 U/ml diluted pyrophosphatase solution in a 1.6 ml Eppendorf tube. Mix by pipetting and then transfer 400 μl of the mixture to a fresh tube with 400 μl diluted pyrophosphatase solution for twofold serial dilutions to reach the following concentrations: 200, 100, 50, 25, 12.5, 6.25, 3.13, 0 μM. Evenly distribute the content of each vial into 12 wells in a new 384-well intermediate plate, each ATP concentration occupying a column of the plate at the upper left corner. Repeat this step for ADP.
4. Prepare 12.5 nM diluted ENPP1 by diluting 3.55 μl of 6 μM ENPP1 enzyme stock in 1,697 μl Assay Buffer in a 2 ml Eppendorf tube. Perform twofold serial dilutions to reach the following concentrations: 12.5, 6.25, 3.13, 1.56, 0.78, 0.39, 0.20, 0 nM. Evenly distribute the content of each vial into 16 wells in a new 384-well intermediate plate, each enzyme concentration occupying a row of the plate at the upper left corner. 0 nM enzyme (assay buffer only, unless otherwise specified) should occupy at least five rows to ensure sufficient sample size for subsequent data analysis.
5. Spin down both the enzyme and the substrate intermediate plates at 182 × *g* for 1 min. Use Bravo liquid handler and two unused boxes of P30 tips to aspirate 13 μl from the enzyme

and substrate assay plates and dispense 3 μl into quadruplicate wells in the 1,536-well assay plate (*see* **Note 2**). Cross-titration of ENPP1 enzyme and ATP substrate has been achieved. Spin down the assay plate and cover with adhesive aluminum seal. Incubate for 110 min.

6. During the incubation, prepare ColorLock Gold working solution by mixing 30 μl of the accelerant solution and 3 ml Gold reagent in a 15 ml round-bottom Falcon tube. Mix by vortexing. Prepare Stabilizer working solution by mixing 1.5 ml stabilizer solution and 1 ml water in a 15 ml round-bottom Falcon tube. Mix by vortexing. Increase volume proportionally if screening a large number of compounds.
7. Near the end of incubation, prepare serial dilutions of phosphate standard from the ColorLock reagent kit, in water. Dispense 6 μl of each concentration (100, 50, 25, 12.5, 6.25, 3.13, 1.56, 0 μM) in quadruplicate into the 1,536-well assay plate holding the assay mixture under incubation. For 0 μM prepare 16–32 replicate wells for sufficient sample size in subsequent data analysis.
8. At the end of incubation, set up the Combi dispenser with plastic tip cassette, and dispense 1.5 μl ColorLock Gold working solution to all occupied wells in the 1,536-well assay plate (*see* **Note 3**). Spin down the plate and wait 5 min before dispensing 1 μl stabilizer working solution with a different plastic tip cassette to all occupied wells. Spin down again and read on the PHERAstar plate reader under OD_{635}. Seal the plate to prevent evaporation if the reader is currently unavailable.
9. Start the data analysis by plotting OD of phosphate standard against phosphate concentration and fit the linear region to a straight line (Fig. 1). The limit of detection (*LOD*) of the ColorLock kit can be determined by taking the standard deviation of the blank/background control (0 μM phosphate, only water) and multiply by 3 and divide by the slope of the linear fit [10]. This is a measurement of the minimal signal detectable above background noise by this assay. If the *LOD* is above 2 μM, the slope of the fit is below 0.012, or the background signal is above 0.2 OD, then the ColorLock reagent may be degraded or one of the assay reagents may be contaminated with phosphate (*see* **Note 4**).
10. Continue the data analysis by plotting the change in OD against ATP concentrations, the data separated into groups by the amount of enzyme present (Fig. 2). Fit the data to the Michaelis–Menten equation for K_m value [10]. Repeat for ADP. If the ATP K_m value is below 15 μM, use 15 μM in the final screening assay and proceed to the next step. If the K_m value is above 15 μM, use K_m value for the final screening assay (*see* **Note 5**).

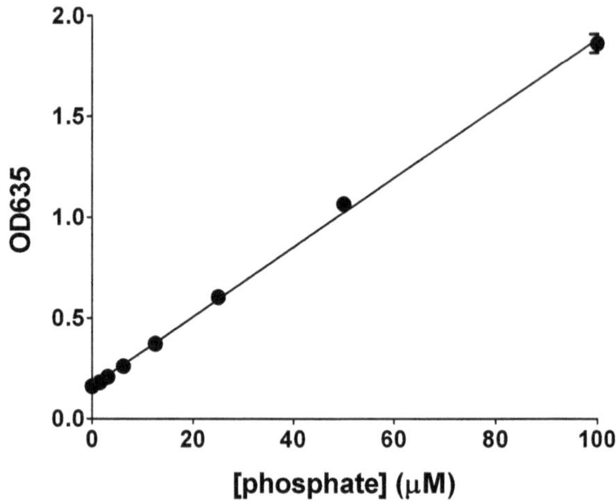

Fig. 1 Sample Phosphate standard calibration curve. OD_{635} values are plotted against phosphate concentration in 6 μl assay volume, and the data fitted linearly. Slope is 0.017 ± 0.0002, standard deviation of the background control is 0.0077, and the limit of detection (LOD) is 1.4 μM

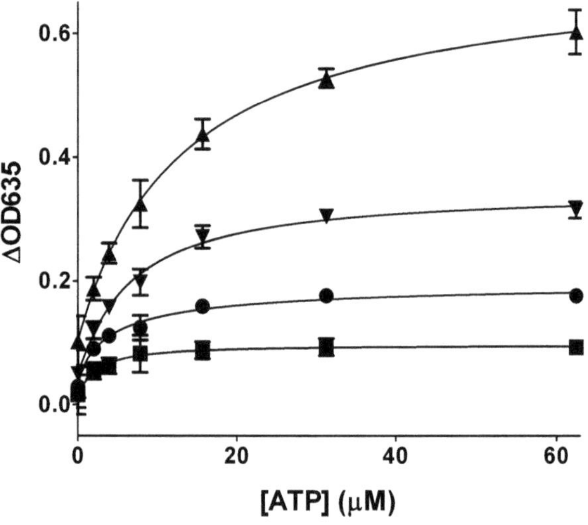

Fig. 2 Sample ATP titration curve. Change in OD_{635} values are plotted against initial ATP concentration in 6 μl assay volume, and the data fitted to Michaelis–Menten equation. K_m values are 13.3 ± 1.1, 6.7 ± 1.4, 4.4 ± 1.8, and 2.4 ± 0.4 μM for 0.78 (*filled upright triangle*), 0.39 (*filled inverted triangle*), 0.195 (*filled circle*), and 0.098 nM (*filled square*) ENPP1, in the presence of 0.4 U/ml pyrophosphatase

11. Finish the data analysis by plotting the change in OD against ENPP1 concentrations (Fig. 3), at a fixed ATP or ADP concentration selected in **step 10**. Fit the linear region to a straight line and calculate *LOD*. If 10× *LOD* in enzyme concentration is far beyond the linear region, select the next higher substrate value and check again (*see* **Note 6**).

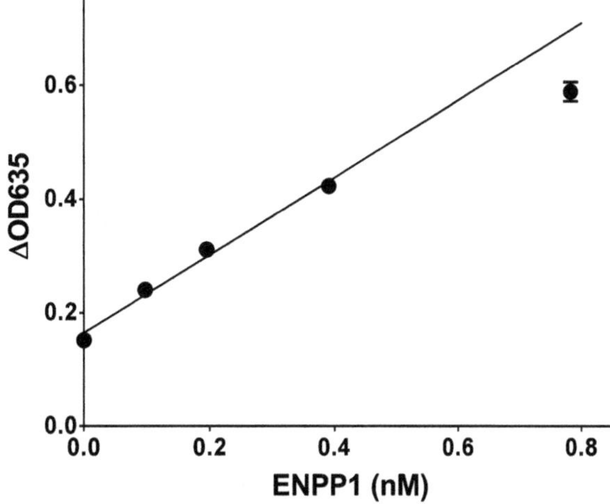

Fig. 3 Sample ENPP1 enzyme dependence. Change in OD635 values are plotted against ENPP1 concentration in 6 μl assay volume, and the data fitted linearly. Slope is 0.68 ± 0.05, standard deviation of the background control is 0.0071, and the limit of detection (LOD) is 0.031 nM

3.2 Screening Assay for ENPP1 Using ATP as Substrate

1. (Optional) Perform phosphate standard calibration by repeating **steps 6–9**, Subheading 3.1 to ensure stability of the ColorLock reagents.
2. Prepare Assay buffer and diluted pyrophosphatase solution similar to Subheading 3.1, **steps 1** and **2**.
3. Prepare 30 μM diluted ATP solution by diluting 4.95 μl of 40 mM ATP solution in 6,595 μl 0.8 U/ml diluted pyrophosphatase solution in a 15 ml round-bottom Falcon tube (*see* **Note 2**).
4. Prepare 0.8 nM diluted ENPP1 by diluting 2.6 μl of 6 μM ENPP1 enzyme stock in 19.8 ml Assay Buffer in a 50 ml conical tube.
5. With the Echo liquid dispenser, dispense nanoliter amounts of compounds resuspended in DMSO into empty 1,536-well assay plate, followed by an equal amount of DMSO into columns reserved for no-enzyme positive control wells and vehicle negative control wells.
6. With the Multidrop Combi dispenser and small metal tip cassette, dispense 3 μl Assay Buffer in columns 1–2 for Positive Control, and 3 μl diluted ENPP1 in columns 3–48. With a different cassette, dispense 3 μl diluted ATP solution to all assay wells. Spin down the plate and seal with aluminum adhesive seal. Incubate for 110 min.
7. Repeat **steps 6–8**, Subheading 3.1 to dispense ColorLock reagents and read the plate for signal (*see* **Note 7**).

3.3 Screening Assay for ENPP1 Using ADP as Substrate

1. (Optional) Perform phosphate standard calibration by repeating **steps 6–9,** Subheading 3.1 to ensure stability of the ColorLock reagents.
2. Prepare Assay buffer and diluted pyrophosphatase solution similar to Subheading 3.1, **steps 1** and **2**.
3. Prepare 30 μM diluted ADP solution by diluting 4.95 μl of 40 mM ADP solution in 6,595 μl 0.8 U/ml diluted pyrophosphatase solution in a 15 ml round-bottom Falcon tube (*see* **Note 2**).
4. Prepare 11.2 nM diluted ENPP1 by diluting 12.3 μl of 6 μM ENPP1 enzyme stock in 6.6 ml Assay Buffer in a 50 ml conical tube.
5. With the Echo liquid dispenser, dispense nanoliter amounts of compounds resuspended in DMSO into empty 1,536-well assay plate, followed by an equal amount of DMSO into columns reserved for no-enzyme positive control and vehicle negative controls.
6. With the Multidrop Combi dispenser and small metal tip cassette, dispense 3 μl Assay Buffer in columns 1–2 for Positive Control, and 3 μl diluted ENPP1 in columns 3–48. With a different cassette, dispense 3 μl diluted ADP solution to all assay wells. Spin down the plate and seal with aluminum adhesive seal. Incubate for 110 min.
7. Repeat **steps 6–8**, Subheading 3.1 to dispense ColorLock reagents and read the plate for signal (*see* **Note** 7).

4 Notes

1. The phosphate detection assay is very sensitive to phosphate contamination in any reagent, for example, the ATP solution, and the best way to lower background signal is to purchase ATP powder from a supplier that provides an analysis report of the level of phosphate impurities. Sigma-Aldrich (catalog #: A7699) supplies a BioXtra grade ATP suitable for this purpose. *See* **Note 4** for additional details.
2. The final concentration in the assay plate, after mixing by spinning, is always half the concentration in the intermediate plates. Thus, enzyme titration starts at 6.25 nM, and substrate titration starts at 100 μM.
3. ColorLock Gold reagent contains strong acid and, therefore, is slightly corrosive and should be handled with caution. Due to its tendency to stick to Combi tubings, pre-dispense into 4–24 columns to ensure accurate dispensing. Do not use metal tip

Combi cassette. The ColorLock Gold assay kit is based on malachite green-molybdate colorimetric assay, and the yellow reagent turns green in the presence of phosphate. The stabilizer stabilizes acid-labile substrates and the molybdate complex to maintain signal stability.

4. The most likely source of phosphate contamination would be ATP, ADP, or the enzyme. ATP and ADP may be unstable and undergo hydrolysis in aqueous solution, over time. The enzyme stock buffer must not contain phosphate, excluding common buffer selections such as PBS. If necessary, test individual component of the assay at appropriately diluted assay-level concentrations in ColorLock reagents for phosphate contamination. *See* **Note 1** for additional details.

5. 15 μM pyrophosphate generated by consumption of 15 μM ATP or ADP would yield 30 μM phosphate in the presence of saturating pyrophosphatase. Generally, for a steady-state kinetic assay, one should aim for 10–30 % consumption of the substrate to avoid substrate supply exhaustion and gradually declining enzymatic rate. Generation of 3–9 μM phosphate should yield a signal that is up to 10-fold higher than *LOD*, sufficient for inhibitor-oriented screening and kinetic characterization of the ENPP1 enzyme. Under steady-state kinetic conditions the enzymatic rate does not change, and changes in OD635 are directly proportional to the enzymatic rate and can be used directly in kinetic calculations and enzyme characterization.

6. Steady-state kinetics presumes catalytic amount of enzyme and a large supply of substrate, such that increase in enzyme concentration would be followed by linear increase in enzymatic rate, without any concern of substrate supply exhaustion. If enzyme increase is not followed by the proportional increase in assay signal, then steady-state assumption is no longer true and the assay is no longer optimal. *LOD* is calculated from no-enzyme background controls in this case.

7. Z′ factor may be calculated based on data from the control wells. Add the standard deviation of the negative control and of the positive control, multiple by 3, and divide by the difference in signal of the two control groups. Subtract this value from 1 and the result should exceed 0.5 [11]. Inhibitor hit selection criteria may be based on inhibitory % response calculated from the difference in signal of the control groups (commonly > 40–50 %), or on the standard deviation of the negative control (commonly > 3–7 standard deviations).

Acknowledgments

We are grateful to Dr. José Luis Millán and his lab for providing ENPP1 enzyme for these studies. This work is supported by NIH Roadmap grant # U54 HG005033 and Conrad Prebys Center for Chemical Genomics at Sanford-Burnham Medical Research Institute.

References

1. Simao AM, Yadav MC, Narisawa S et al (2010) Proteoliposomes harboring alkaline phosphatase and nucleotide pyrophosphatase as matrix vesicle biomimetics. J Biol Chem 285: 7598–7609
2. Ciancaglini P, Yadav MC, Simao AM et al (2010) Kinetic analysis of substrate utilization by native and TNAP-, NPP1-, or PHOSPHO1-deficient matrix vesicles. J Bone Miner Res 25: 716–723
3. Johnson K, Goding J, Van Etten D et al (2003) Linked deficiencies in extracellular PP(i) and osteopontin mediate pathologic calcification associated with defective PC-1 and ANK expression. J Bone Miner Res 18:994–1004
4. Lin CC, Wu CT, Wu LS (2011) Ectonucleotide pyrophosphatase/phosphodiesterase 1 K173Q polymorphism is associated with diabetic nephropathy in the Taiwanese population. Genet Test Mol Biomarkers 15:239–242
5. Wang R, Zhou D, Xi B et al (2011) ENPP1/PC-1 gene K121Q polymorphism is associated with obesity in European adult populations: evidence from a meta-analysis involving 24,324 subjects. Biomed Environ Sci 24:200–206
6. Goldfine ID, Maddux BA, Youngren JF et al (2008) The role of membrane glycoprotein plasma cell antigen 1/ectonucleotide pyrophosphatase phosphodiesterase 1 in the pathogenesis of insulin resistance and related abnormalities. Endocr Rev 29:62–75
7. Iqbal J, Levesque SA, Sevigny J et al (2008) A highly sensitive CE-UV method with dynamic coating of silica-fused capillaries for monitoring of nucleotide pyrophosphatase/phosphodiesterase reactions. Electrophoresis 29: 3685–3693
8. Cimpean A, Stefan C, Gijsbers R et al (2004) Substrate-specifying determinants of the nucleotide pyrophosphatases/phosphodiesterases NPP1 and NPP2. Biochem J 381:71–77
9. Zalatan JG, Fenn TD, Brunger AT et al (2006) Structural and functional comparisons of nucleotide pyrophosphatase/phosphodiesterase and alkaline phosphatase: implications for mechanism and evolution. Biochemistry 45: 9788–9803
10. Sergienko EA, Millan JL (2010) High-throughput screening of tissue-nonspecific alkaline phosphatase for identification of effectors with diverse modes of action. Nat Protoc 5:1431–1439
11. Zhang JH, Chung TD, Oldenburg KR (1999) A Simple Statistical Parameter for Use in Evaluation and Validation of High Throughput Screening Assays. J Biomol Screen 4:67–73

Chapter 11

Structure of Acid Phosphatases

César L. Araujo and Pirkko T. Vihko

Abstract

Acid phosphatases are enzymes that have been studied extensively due to the fact that their dysregulation is associated with pathophysiological conditions. This characteristic has been exploited for the development of diagnostic and therapeutic methods. As an example, prostatic acid phosphatase was the first marker for metastatic prostate cancer diagnosis and the dysregulation of tartrate resistant acid phosphatase is associated with abnormal bone resorption linked to osteoporosis.

The pioneering crystallization studies on prostatic acid phosphatase and mammalian tartrate-resistant acid phosphatase conformed significant milestones towards the elucidation of the mechanisms followed by these enzymes (Schneider et al., EMBO J 12:2609–2615, 1993). Acid phosphatases are also found in nonmammalian species such as bacteria, fungi, parasites, and plants, and most of them share structural similarities with mammalian acid phosphatase enzymes.

Acid phosphatase (EC 3.1.3.2) enzymes catalyze the hydrolysis of phosphate monoesters following the general equation (1).

$$\text{Phosphate monoester} + H_2O \Leftrightarrow \text{alcohol} + \text{phosphate} \quad (1)$$

The general classification "acid phosphatase" relies only on the optimum acidic pH for the enzymatic activity in assay conditions using non-physiological substrates. These enzymes accept a wide range of substrates in vitro, ranging from small organic molecules to phosphoproteins, constituting a heterogeneous group of enzymes from the structural point of view. These structural differences account for the divergence in cofactor dependences and behavior against substrates, inhibitors, and activators. In this group only the tartrate-resistant acid phosphatase is a metallo-enzyme whereas the other members do not require metal-ion binding for their catalytic activity. In addition, tartrate-resistant acid phosphatase and erythrocytic acid phosphatase are not inhibited by L-(+)-tartrate ion while the prostatic acid phosphatase is tartrate-sensitive. This is an important difference that can be exploited in in vitro assays to differentiate between different kinds of phosphatase activity. The search for more sensitive and specific methods of detection in clinical laboratory applications led to the development of radioimmunoassays (RIA) for determination of prostatic acid phosphatase in serum. These methods permit the direct quantification of the enzyme regardless of its activity status. Therefore, an independent structural classification exists that helps to group these enzymes according to their structural features and mechanisms. Based on this we can distinguish the histidine acid phosphatases (Van Etten, Ann N Y Acad Sci 390:27–51, 1982), the low molecular weight protein tyrosine acid phosphatases and the metal-ion dependent phosphatases.

A note of caution is worthwhile mentioning here. The nomenclature of acid phosphatases has not been particularly easy for those new to the subject. Unfortunately, the acronym PAP is very common in

José Luis Millán (ed.), *Phosphatase Modulators*, Methods in Molecular Biology, vol. 1053,
DOI 10.1007/978-1-62703-562-0_11, © Springer Science+Business Media, LLC 2013

the literature about purple acid phosphatases and prostatic acid phosphatase. In addition, LPAP is the acronym chosen to refer to the lysophosphatidic acid phosphatase which is a different enzyme. It is important to bear in mind this distinction while reviewing the literature to avoid confusion.

Key words Prostate cancer, Bone resorption, Osteoporosis, Metallo-enzymes, Tartrate-resistant

1 Histidine Acid Phosphatases

The histidine acid phosphatase family comprises those acid phosphatases with α/β domains (parallel β sheets in α/β/α units) which use a histidine residue as a nucleophile during the mechanism for the removal of the phosphate group from the substrate. According to the folding of these proteins, they belong to the phosphoglycerate mutase-like superfamily (Fig. 1). At the sequence level, the members of the histidine acid phosphatase family are characterized by the general signature motif RHGXRXP. This motif can be found in prostatic acid phosphatase, lysosomal acid phosphatase, lysophosphatidic acid phosphatase and the testicular acid phosphatase. The original crystallization studies on rat prostatic acid phosphatase carried out by Vihko et al. provided fundamental information to confirm the enzymatic mechanism of a histidine acid phosphatase [4–6]. The active enzyme is a dimer presenting positive cooperativity on substrate binding [7]. Our crystallization studies revealed that each subunit is composed of two domains, an α/β domain and an α-helical domain. The active site lies in an open cleft between these two structural domains with the active site in each subunit of the dimeric form far apart from

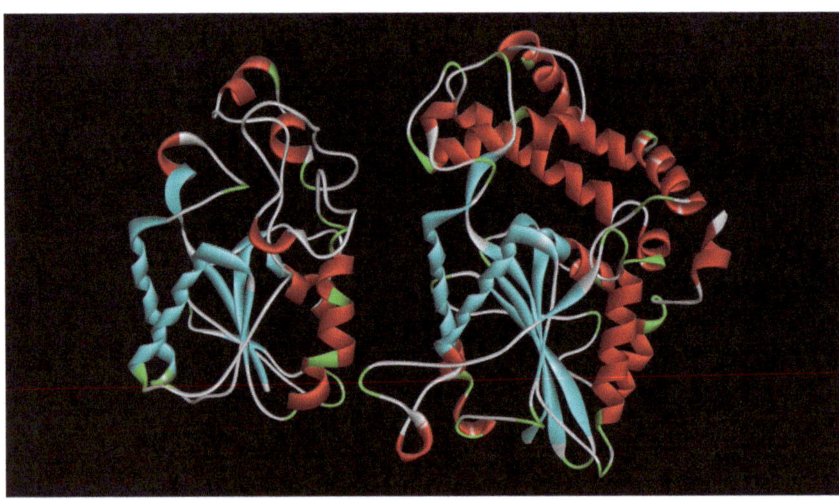

Fig. 1 Three-dimensional representation of phosphoglycerate mutase (*left*) [47] and rat prostatic acid phosphatase (*right*). Based on Schneider et al. [1]

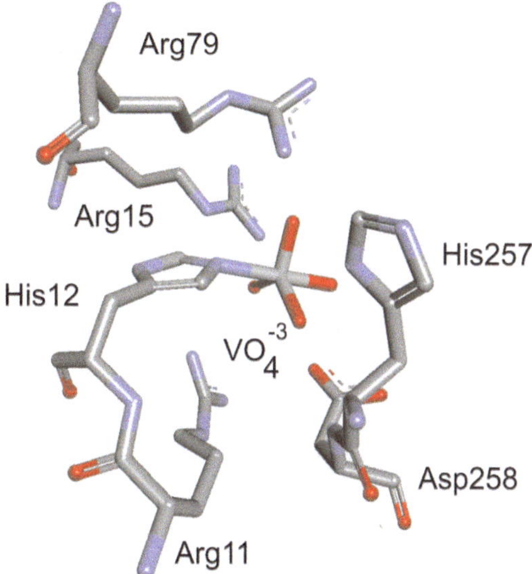

Fig. 2 Three-dimensional representation of the active site of rat prostatic acid phosphatase complexed with transition-state analogue vanadate ion (1rpt). Based on Lindqvist et al. [4]

each other (ca. 34 Å) [1]. The active site is composed of arginine residues (Arg11 and Arg15 from the histidine acid phosphatase signature motif and Arg79), which provides the positive environment contributing to the stabilization and positioning of the negative phosphate group; the nucleophile histidine residue (His12), the general acid–base catalyst aspartic acid residue (Asp258) and a histidine residue (His257), which provides extra stabilization by hydrogen bonding to the substrate [1]. Figure 2 depicts the active site residues from the rat prostatic acid phosphatase complexed with transition state analogue vanadate (VO_4^{-3}) [4].

In addition, our studies showed that the activity of histidine acid phosphatases relies in the general acid–base catalysis provided by an acidic residue in the active site. This is a two-step mechanism in which the first step consists of the formation of a phospho-histidine intermediate by the nucleophilic attack of an active site histidine to the phosphate group of the substrate by following a concerted SN_2 mechanism with release of an alcohol molecule. In the second step (considered to be the rate-limiting step) the phospho-histidine intermediate is hydrolyzed by an active site water molecule (or another phosphate acceptor molecule) assisted by the general acid–base catalyst residue. In the human prostatic acid phosphatase, the aspartic acid present in the binding pocket (Asp258) plays the role of general acid–base catalyst [6]. Figure 3

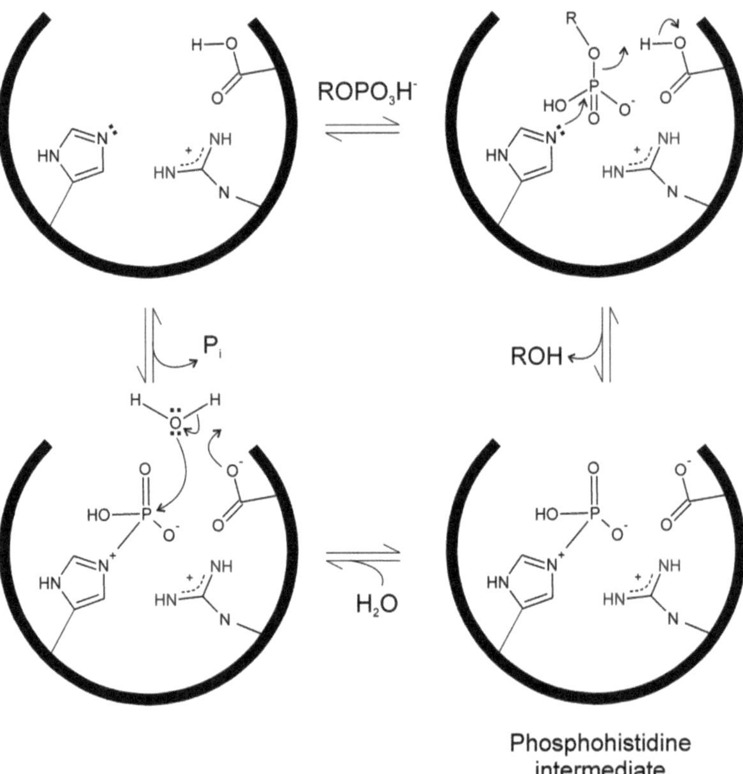

Fig. 3 Schematic representation of a typical mechanism for a histidine acid phosphatase enzyme. Based on Lindqvist et al. [4]

depicts the proposed mechanism for a histidine acid phosphatase based on these findings.

In addition, the crystal structure of the histidine acid phosphatase from the pathogen *Francisella tularensis* (FtHAP) has been solved showing that the three-dimensional structure is similar to that of prostatic acid phosphatase [8]. Moreover, a mutation of the Asp261 (the equivalent to Asp258 in prostatic acid phosphatase) traps the substrate 3′-AMP (3′-adenosine monophosphate) which provides supporting evidence to the importance of this residue in the catalytic mechanism [8].

The lysosomal acid phosphatase enzyme played a key role in the discovery of lysosomes by de Duve in 1963 and is widely used as a lysosomal marker. This enzyme shows a high degree of sequence similarity (ca. 49 % identity) with prostatic acid phosphatase [9] and both are inhibited by L-(+)-tartrate ion [10].

In 2007 we characterized a novel transmembrane type-I isoform of the prostatic acid phosphatase enzyme (TMPAP) as the product of a splice variant of the same gene encoding the secreted form (sPAP). This transmembrane type-I isoform contains a tyrosine-based lysosomal targeting (YxxΦ) motif at the

C-terminus and is widely expressed across different tissues [11]. This tyrosine-based targeting motif is also found in the lysosomal acid phosphatase isozyme which suffers a sequential proteolytic cleavage by a cytoplasmic thiol proteinase and a lysosomal aspartyl proteinase releasing the enzyme into the lumen of the lysosomal organelle [12]. Differences in the amino acid composition required for the recognition by these proteases lead the transmembrane prostatic acid phosphatase enzyme to the plasma membrane-endosomal-lysosomal recycling pathway. Topologically, the N-terminal catalytic domain of transmembrane prostatic acid phosphatase is extracellular while it is present in plasma membrane and intraluminal when is bound to vesicular membranes. Intracellularly, the transmembrane prostatic acid phosphatase enzyme traffics from the endoplasmic reticulum where it is synthesized to the plasma membrane via the trans-Golgi network. The enzyme present at the plasma membrane may be endocytosed in a clathrin-dependent manner through the endosomal-lysosomal pathway co-localizing with multivesicular endosomes and lysosomal markers [11].

We showed that prostatic acid phosphatase also exerts in vivo catalytic activity on 5′-AMP in mouse dorsal root ganglia resulting in pain relief effect up to eight-times greater than morphine [13]. In addition, the thiamine monophosphatase activity, which is a classic histochemical marker of dorsal root ganglia (DRG) neurons, is in effect due to the prostatic acid phosphatase activity [13]. The low molecular weight protein tyrosine acid phosphatase from erythrocyte does not present activity against the thiamine monophosphate substrate [13]. Recent works showed that the antinociceptive effect of benfotiamine, which is a thiamine derivative used in the treatment of thiamine deficiency, relies entirely on prostatic acid phosphatase activity [14]. In vitro, prostatic acid phosphatase can dephosphorylate phosphotyrosine based peptides [15] which led to the postulation of the hypothesis that the enzyme could dephosphorylate epidermal growth factor receptor in vivo via a cytosolic cellular form as has been suggested in the literature [16, 17]. However, such an isoform has been never cloned and no direct mechanism can be devised considering the topological arrangement of these proteins.

All these significant physiological findings must be taken into account during the evaluation of inhibitors and activators of prostatic acid phosphatase enzyme for therapeutic use.

2 Metal-Ion Dependent Acid Phosphatases

Metallo-dependent acid phosphatases (a.k.a. purple acid phosphatases) contain a binuclear ion center [18, 19] in the active site which confers a characteristic purple color in solution. The color is produced by a charge transfer mechanism between a conserved tyrosine residue and a Fe^{+3} ion in the active site [20]. These enzymes contain α+β domains (parallel β sheets in segregated α and β regions). One important feature of these enzymes is their resistance to tartrate inhibition.

Still there is no official nomenclature for this class of enzymes but there is a consensus in the literature by which the term *purple acid phosphatase* refers to those metallo-dependent acid phosphatases found in nonmammalian species while the term *tartrate-resistant acid phosphatase* is kept for those enzymes found in mammals.

The nature of the metal-ions in the active site also varies between species. Whereas the purple acid phosphatase isolated from red kidney beans (rkbPAP) contains Fe^{+3} and Zn^{+2}, the tartrate-resistant acid phosphatase isolated from rat osteoclasts (TRAcP) contains two iron atoms in different oxidation states, an stabilized Fe^{+3} ion and a redox-active Fe^{+2} ion. In this way, the ability of the ferrous ion to act as an electron donor confers to the enzyme an alternative function as generator of reactive oxygen species (ROS) [20, 21]. The enzyme may appear in an inactive purple form when the redox-active iron is oxidized to the ferric state, or it can be in an active pink form where the redox-active iron is reduced to the ferrous state [22]. In particular, the tartrate-resistant acid phosphatase isolated from osteoclasts is synthetized as a precursor which is activated by cysteine proteinases resulting in an active two subunit enzyme [23].

Lam et al., showed the existence of two distinct isoforms in human serum [24]. They were named according to their electrophoretic mobility as TRAcP 5a and TRAcP 5b isoforms and the difference in size was attributed to the additional content of sialic acid in the carbohydrate chain of the TRAcP 5a isoform. The osteoclasts secrete only the TRAcP 5b isoform whereas activated macrophages express both isoforms but only secrete the TRAcP 5a isoform into the circulation [25].

In vitro studies have shown that TRAcP 5b isoform remains intracellularly in macrophages and dendritic cells which only secrete the TRAcP 5a isoform [25]. Therefore, for clinical determinations it is assumed that the tartrate-resistant acid phosphatase activity associated with bone resorbing osteoclasts derives only from the TRAcP 5b isoform.

The tartrate-resistant acid phosphatases showed preferential activity towards esters of aromatic compounds and less towards phosphoanhydrides such as nucleotide triphosphate and diphosphate and phosphotyrosine but not against phosphoserine, phosphothreonine [26], or aliphatic alcohols [27].

The activity of the tartrate-resistant acid phosphatase enzyme is affected by the strength of reducing agents. Whereas mild-reducing agents, such as ascorbate or β-mercaptoethanol, lead to the activation of the enzyme, strong reducing agents, such as dithionite anion, elicit the inhibition of enzyme activity by removal of the iron ions from the catalytic site [28]. Also other species affect the enzymatic activity, such as mercurial ions that produce a reversible inhibition while manganese and magnesium ions were found to activate the enzyme to a small extent [29].

In 1995, Sträter and his coworkers determined the three-dimensional structure of kidney bean purple acid phosphatase at 2.9 Å resolution providing insights on the groups involved in the catalytic mechanism. However, due to the low degree of sequence homology between mammalian tartrate-resistant acid phosphatases and plant purple acid phosphatases, we decided to determine the three-dimensional structure of tartrate-resistant acid phosphatase from rat bone [30]. For this purpose we first set up a high-scale production of recombinant rat bone TRAcP using a baculovirus expression vector system. This provided us with suitable quantities of protein for crystallization studies. In addition, we generated specific monoclonal antibodies against rat bone TRAcP resulting in a valuable research tool to study drug effects on an experimental animal model of osteoporosis [31]. Based on our results from the analysis of the three-dimensional structure, we proposed a mechanism for the mammalian enzyme [30]. In this mechanism, the substrate binds to the enzyme by coordination with the ferrous ion. In this way, the hydroxyl group, which forms the μ-hydroxo bridge between the two iron centers, locates in a suitable position to attack the phosphorus atom from the opposite site to the leaving group producing an inversion of the configuration on the phosphorus atom and hence providing evidence against the formation of a phospho-enzyme intermediate [31].

Further mechanistic studies during the last years focused on purple acid phosphatases and proposed a more detailed mechanism with up to eight steps [32].

3 Protein Tyrosine Acid Phosphatases

Protein tyrosine phosphatases (PTPs) of high and low molecular weight contain the highly conserved motif CX_5R and employ a common catalytic mechanism [33]. The low molecular weight protein tyrosine phosphatase found in erythrocytes belongs to this group of enzymes. These enzymes show a high degree of sequence homology and are extensively found among prokaryotes and eukaryotes. The mechanism involves a conserved active site cysteine residue acting as a nucleophile on the substrate to form a cysteinyl-phosphate intermediate. Figure 4 show a schematic representation of the mechanism for this type of enzymes.

In humans, acid phosphatase from erythrocytes is polymorphic with several alleles expressing the enzyme. This particular

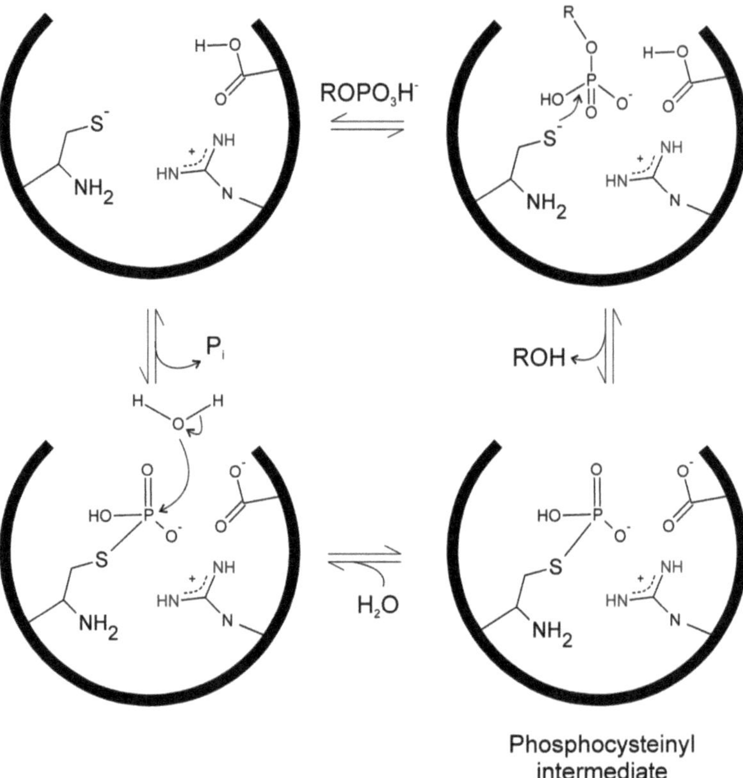

Fig. 4 Schematic representation of the mechanism for a low molecular weight phosphotyrosine acid phosphatase. Adapted from Lindqvist et al. [4]

characteristic has been exploited in forensic medicine for paternity tests along with other red cell systems [34]. This enzyme is identified as two electrophoretic bands representing the *fast* and the *slow* forms. Although first identified in red cells, this cytosolic acid phosphatase is found in all human tissues [35]. These isoforms vary in an alternative spliced sequence known as the variable loop which confers a differential substrate specificity and inhibitor/activator responses [36].

The low molecular weight acid phosphatase has been implicated in the insulin resistance pathway producing effects that are independent from the high molecular weight protein tyrosine phosphatase (PTP1B) [37, 38]. This renders the low molecular weight protein tyrosine acid phosphatase an interesting target for the treatment of insulin resistance and type 2 diabetes.

However the development of inhibitors for PTPs with suitable drug-like features of selectivity, absorption, distribution, metabolism, excretion and toxicity (ADMET) has proven to be challenging in part due to the highly conserved and charged active site [39]. A potent inhibitor of human low molecular weight protein tyrosine phosphatase is a 3-phenoxyphenyl derivative of an 5-arylidene-2,4-thiazolidinedione [40].

4 Inhibitors and Activators of Acid Phosphatases

Since disturbed acid phosphatase activity has been associated with pathological conditions, the research has focused on the development of diagnostic methods for detection of activity as a marker for the onset of the disease, and in some extent to the development of inhibitors rather than activators to treat those conditions in which the increase in enzyme activity has a direct effect on the evolution of the disease. In particular, only the development of bisphosphonate derivatives as inhibitors for tartrate-resistant acid phosphatase found their way to the market for treatment of osteoporosis [41]. Typical inhibition of phosphatase activity includes anionic species such as L-(+)-tartrate, phosphate, vanadate, molybdate, and fluoride and neutral molecules such as formaldehyde. Vanadate ion, VO_4^{-3}, is a competitive unspecific inhibitor for acid phosphatases by forming transition state analogs. Other oxoanions such as molybdate and tungstate also show inhibitory effects on

Table 1
List of human acid phosphatases and their corresponding official and alternative names and gene symbols

Official name (as approved by HGNC[a])	Official gene symbol	Gene ID	Other names	Other symbols
Acid phosphatase 1, soluble	*ACP1*	52	Acid phosphatase of erythrocyte; Low molecular weight phosphotyrosine protein phosphatase	HAAP
Acid phosphatase 2, lysosomal	*ACP2*	53		LAP
Acid phosphatase, prostate	*ACPP*	55	Prostatic acid phosphatase	PAP, sPAP (secreted isoform), TMPAP (transmembrane isoform), ACP3
Acid phosphatase 5, tartrate resistant	*ACP5*	54	Tartrate-resistant acid phosphatase (in mammalian species), Purple acid phosphatase (in nonmammalian species)	TRAP, TRAcP, PAP, SPENCDI
Acid phosphatase 6, lysophosphatidic	*ACP6*	51205	Lysphosphatidic acid phosphatase	LPAP, ACPL1, PACPL1
Acid phosphatase, testicular	*ACPT*	93650	Testicular acid phosphatase	

[a]HUGO Gene Nomenclature Committee

Table 2
Summary of inhibitors and activators of acid phosphatases

Enzyme	Inhibitors	Activators
Low molecular weight acid phosphatase	Formaldehyde, derivatives of 5-arylidene-2, 4-thiazolidinedione, 1,4-butanediol	Aliphatic alcohols
Prostatic acid phosphatase	Phosphate (PO_4^{-3}), vanadate (VO_4^{-3}), arsenate (AsO_4^{-3}), molybdate (MoO_4^{-2}), L-(+)-tartrate, benzylaminophosphonic acid and benzylaminobenzylphosphonic acid, *N*-propyltartramate	Aliphatic alcohols
Lysosomal acid phosphatase	L-(+)-tartrate	
Lysophosphatidic acid phosphatase	N/A	N/A
Tartrate-resistant acid phosphatase	Fluoride (F^-), strong reducing agents such as dithionite anion ($S_2O_4^{-2}$), mercurial ions (Hg^{+2}), bisphosphonates (such as zoledronate and alendronate), oxidizing agents such as hydrogen peroxide	Mild-reducing agents: β-mercaptoethanol, ascorbate, dithiothreitol.
Testicular acid phosphatase	N/A	N/A

phosphatase activity. Formaldehyde inhibits the erythrocyte acid phosphatase but not the prostatic acid phosphatase enzyme [42]. The fluoride ion inhibits the purple acid phosphatase and the tartrate-resistant acid phosphatase activity in osteoclasts. Some studies on specific inhibitors for prostatic acid phosphatase focused on benzylaminophosphonic [43] and benzylaminobenzylphosphonic acids [44]. However, since the physiological mechanisms in which prostatic acid phosphatase is involved in are still poorly understood, their therapeutic use is limited. One particular important aspect of inhibition relies on the differential enzymatic activity between prostatic acid phosphatase and tartrate-resistant acid phosphatase in the presence of L-(+)-tartrate ion. This property is exploited in research and clinical assays of phosphatase activity to differentiate between these two kinds of enzymes. It also constitutes the basis of a purification method of prostatic acid phosphatase [8]. The tartrate-resistant acid phosphatase activity in serum not only includes the acid phosphatase activity from osteoclasts (TRAcP) but also the acid phosphatase activity derived from erythrocytes (ACP1) [45]. Therefore, the tartrate-sensitive acid phosphatase activity is thought to derive exclusively from the prostatic acid phosphatase enzyme and this is the basis for the colorimetric determination in clinical assays.

In general, the addition of aliphatic alcohols to the reaction media may enhance the activity of acid phosphatases by acting as phosphate acceptors in a transphosphorylation reaction [46].

A summary of the nomenclature for acid phosphatases is described in Table 1, and their inhibitors and activators that can be found in the literature are listed in Table 2.

References

1. Schneider G, Lindqvist Y, Vihko P (1993) Three-dimensional structure of rat acid phosphatase. EMBO J 12:2609–2615
2. Vihko P (1978) Human prostatic acid phosphatase and its radioimmunoassay. Acta Universitatis Ouluensis, Series D Medica 33, Clinica Chemica 1:1–78
3. Van Etten RL (1982) Human prostatic acid phosphatase: a histidine phosphatase. Ann N Y Acad Sci 390:27–51
4. Lindqvist Y, Schneider G, Vihko P (1994) Crystal structures of rat acid phosphatase complexed with the transition-state analogs vanadate and molybdate. Implications for the reaction mechanism. Eur J Biochem 221:139–142
5. Lindqvist Y, Schneider G, Vihko P (1993) Three-dimensional structure of rat acid phosphatase in complex with L(+)-tartrate. J Biol Chem 268:20744–20746
6. Porvari KS, Herrala AM, Kurkela RM et al (1994) Site-directed mutagenesis of prostatic acid phosphatase. Catalytically important aspartic acid 258, substrate specificity, and oligomerization. J Biol Chem 269:22642–22646
7. Luchter-Wasylewska E (2001) Cooperative kinetics of human prostatic acid phosphatase. Biochim Biophys Acta 1548:257–264
8. Singh H, Felts RL, Schuermann JP et al (2009) Crystal Structures of the histidine acid phosphatase from Francisella tularensis provide insight into substrate recognition. J Mol Biol 394:893–904
9. Peters C, Geier C, Pohlmann R et al (1989) High degree of homology between primary structure of human lysosomal acid phosphatase and human prostatic acid phosphatase. Biol Chem Hoppe Seyler 370:177–181
10. Pohlmann R, Krentler C, Schmidt B et al (1988) Human lysosomal acid phosphatase: cloning, expression and chromosomal assignment. EMBO J 7:2343–2350
11. Quintero IB, Araujo CL, Pulkka AE et al (2007) Prostatic acid phosphatase is not a prostate specific target. Cancer Res 67:6549–6554
12. Gottschalk S, Waheed A, Schmidt B et al (1989) Sequential processing of lysosomal acid phosphatase by a cytoplasmic thiol proteinase and a lysosomal aspartyl proteinase. EMBO J 8:3215–3219
13. Zylka MJ, Sowa NA, Taylor-Blake B et al (2008) Prostatic acid phosphatase is an ecto-nucleotidase and suppresses pain by generating adenosine. Neuron 60:111–122
14. Hurt JK, Coleman JL, Fitzpatrick BJ et al (2012) Prostatic acid phosphatase is required for the antinociceptive effects of thiamine and benfotiamine. PLoS One 7:e48562
15. Li HC, Chernoff J, Chen LB et al (1984) A phosphotyrosyl-protein phosphatase activity associated with acid phosphatase from human prostate gland. Eur J Biochem 138:45–51
16. Lin MF, DaVolio J, Garcia-Arenas R (1992) Expression of human prostatic acid phosphatase activity and the growth of prostate carcinoma cells. Cancer Res 52:4600–4607
17. Veeramani S, Chou YW, Lin FC et al (2012) Reactive oxygen species induced by p66Shc longevity protein mediate nongenomic androgen action via tyrosine phosphorylation signaling to enhance tumorigenicity of prostate cancer cells. Free Radic Biol Med 53:95–108
18. Campbell HD, Dionysius DA, Keough DT et al (1978) Iron-containing acid phosphatases: comparison of the enzymes from beef spleen and pig allantoic fluid. Biochem Biophys Res Commun 82:615–620
19. Davis JC, Averill BA (1982) Evidence for a spin-coupled binuclear iron unit at the active site of the purple acid phosphatase from beef spleen. Proc Natl Acad Sci USA 79:4623–4627
20. Oddie GW, Schenk G, Angel NZ et al (2000) Structure, function, and regulation of tartrate-resistant acid phosphatase. Bone 27:575–584
21. Halleen JM, Raisanen S, Salo JJ et al (1999) Intracellular fragmentation of bone resorption products by reactive oxygen species generated by osteoclastic tartrate-resistant acid phosphatase. J Biol Chem 274:22907–22910
22. Antanaitis BC, Aisen P, Lilienthal HR (1983) Physical characterization of two-iron uteroferrin. Evidence for a spin-coupled binuclear iron cluster. J Biol Chem 258:3166–3172

23. Ljusberg J, Ek-Rylander B, Andersson G (1999) Tartrate-resistant purple acid phosphatase is synthesized as a latent proenzyme and activated by cysteine proteinases. Biochem J 343(Pt 1):63–69
24. Lam WK, Eastlund DT, Li CY et al (1978) Biochemical properties of tartrate-resistant acid phosphatase in serum of adults and children. Clin Chem 24:1105–1108
25. Janckila AJ, Nakasato YR, Neustadt DH et al (2003) Disease-specific expression of tartrate-resistant acid phosphatase isoforms. J Bone Miner Res 18:1916–1919
26. Andersson G, Lindunger A, Ek-Rylander B (1989) Isolation and characterization of skeletal acid ATPase–a new osteoclast marker? Connect Tissue Res 20:151–158
27. Lam KW, Yam LT (1977) Biochemical characterization of the tartrate-resistant acid phosphatase of human spleen with leukemic reticuloendotheliosis as a pyrophosphatase. Clin Chem 23:89–94
28. Schlosnagle DC, Bazer FW, Tsibris JC et al (1974) An iron-containing phosphatase induced by progesterone in the uterine fluids of pigs. J Biol Chem 249:7574–7579
29. Hayman AR, Warburton MJ, Pringle JA et al (1989) Purification and characterization of a tartrate-resistant acid phosphatase from human osteoclastomas. Biochem J 261:601–609
30. Lindqvist Y, Johansson E, Kaija H et al (1999) Three-dimensional structure of a mammalian purple acid phosphatase at 2.2 A resolution with a mu-(hydr)oxo bridged di-iron center. J Mol Biol 291:135–147
31. Kaija H (2002) Tartrate-resistant acid phosphatase: three-dimensional structure and structure-based functional studies. Oulu University Press, Oulu
32. Schenk G, Elliott TW, Leung E et al (2008) Crystal structures of a purple acid phosphatase, representing different steps of this enzyme's catalytic cycle. BMC Struct Biol 8:6
33. Barford D, Das AK, Egloff MP (1998) The structure and mechanism of protein phosphatases: insights into catalysis and regulation. Annu Rev Biophys Biomol Struct 27:133–164
34. Bull H, Murray PG, Thomas D et al (2002) Acid phosphatases. Mol Pathol 55:65–72
35. Wo YY, McCormack AL, Shabanowitz J et al (1992) Sequencing, cloning, and expression of human red cell-type acid phosphatase, a cytoplasmic phosphotyrosyl protein phosphatase. J Biol Chem 267:10856–10865
36. Zhang M, Stauffacher CV, Lin D et al (1998) Crystal structure of a human low molecular weight phosphotyrosyl phosphatase. Implications for substrate specificity. J Biol Chem 273:21714–21720
37. Pandey SK, Yu XX, Watts LM et al (2007) Reduction of low molecular weight protein-tyrosine phosphatase expression improves hyperglycemia and insulin sensitivity in obese mice. J Biol Chem 282:14291–14299
38. Chiarugi P, Cirri P, Marra F et al (1997) LMW-PTP is a negative regulator of insulin-mediated mitotic and metabolic signalling. Biochem Biophys Res Commun 238:676–682
39. Maccari R, Ottana R, Ciurleo R et al (2009) Structure-based optimization of benzoic acids as inhibitors of protein tyrosine phosphatase 1B and low molecular weight protein tyrosine phosphatase. ChemMedChem 4:957–962
40. Maccari R, Paoli P, Ottana R et al (2007) 5-Arylidene-2,4-thiazolidinediones as inhibitors of protein tyrosine phosphatases. Bioorg Med Chem 15:5137–5149
41. Watts NB (2003) Bisphosphonate treatment of osteoporosis. Clin Geriatr Med 19:395–414
42. Abul-Fadl MA, King EJ (1948) The inhibition of acid phosphatases by formaldehyde and its clinical application for the determination of serum acid phosphatases. J Clin Pathol 1:80–90
43. Beers SA, Schwender CF, Loughney DA et al (1996) Phosphatase inhibitors–III. Benzylaminophosphonic acids as potent inhibitors of human prostatic acid phosphatase. Bioorg Med Chem 4:1693–1701
44. Ortlund E, LaCount MW, Lebioda L (2003) Crystal structures of human prostatic acid phosphatase in complex with a phosphate ion and alpha-benzylaminobenzylphosphonic acid update the mechanistic picture and offer new insights into inhibitor design. Biochemistry 42:383–389
45. Abul-Fadl MA, King EJ (1949) Properties of the acid phosphatases of erythrocytes and of the human prostate gland. Biochem J 45:51–60
46. Valcour AA, Bowers GN Jr, McComb RB (1989) Evaluation of a kinetic method for prostatic acid phosphatase with use of self-indicating substrate, 2,6-dichloro-4-nitrophenyl phosphate. Clin Chem 35:939–945
47. Winn SI, Watson HC, Harkins RN et al (1981) Structure and activity of phosphoglycerate mutase. Philos Trans R Soc Lond B Biol Sci 293:121–130

Chapter 12

Purification of Prostatic Acid Phosphatase (PAP) for Structural and Functional Studies

Annakaisa M. Herrala, Ileana B. Quintero, and Pirkko T. Vihko

Abstract

High-scale purification methods are required for several protein studies such as crystallography, mass spectrometry, circular dichroism, and function. Here we describe a purification method for PAP based on anion exchange, L-(+)-tartrate affinity, and gel filtration chromatographies. Acid phosphatase activity and protein concentration were measured for each purification step, and to collect the fractions with the highest acid phosphatase activity the *p*-nitrophenyl phosphate method was used. The purified protein obtained by the procedure described here was used for the determination of the first reported three-dimensional structure of prostatic acid phosphatase.

Key words Prostatic acid phosphatase, Recombinant protein, High-scale purification, Affinity chromatography, L-(+)-tartrate, Acid phosphatase activity, *p*-Nitrophenyl phosphate

1 Introduction

Prostatic acid phosphatase (EC 3.1.3.2) can be purified from seminal fluid, prostate tissue, or as a recombinant protein. High-scale purification methods are important in order to obtain mass amounts of homogeneous, purified protein required by structural and functional studies such as inhibitor and activator analyses. Acid phosphatases can be divided into two groups according to their sensitivity or resistance to L-(+)-tartrate inhibition. The activity of these enzymes can be analyzed by the *p*-nitrophenyl phosphate method in presence [1] or absence of L-(+)-tartrate [2]. PAP belongs to the group of L-(+)-tartrate sensitive acid phosphatases, and this feature is used as the basis of the purification method described in this chapter.

Structural and functional studies require high amounts of homogeneous protein, requirements difficult to meet by the purification of protein from prostate tissue or seminal fluid. The production of PAP as a recombinant protein enables obtaining mass amounts of homogeneous, glycosylated protein. In our case, we

José Luis Millán (ed.), *Phosphatase Modulators*, Methods in Molecular Biology, vol. 1053,
DOI 10.1007/978-1-62703-562-0_12,

have produced recombinant PAP [3] in mass-scale using a baculovirus expression system in *Spodoptera frugiperda* 9 (Sf9) insect cells. Briefly, Sf9 insect cell cultures were co-transfected with the transfer vector rPAP-pVL1392 and *Autographa californica* nuclear polyhedrosis virus (AcNPV) [4] to generate the rPAP-AcNPV. Sf9 insect cell cultures were scaled from spinner flasks via 2 L bioreactor to 30 L bioreactor and when the cell concentration reached 2×10^6 cells/mL, the cultures were infected with rPAP-AcNPV, and cells were harvested 5 days post-infection. Cells were removed by centrifugation and recombinant PAP was purified from the culture medium.

We adapted this purification method (Fig. 1) from Vihko et al. [5] where human PAP was purified from seminal fluid or prostate tissue. In addition, we describe the changes required for small-scale purification method.

2 Materials

Prepare all solutions using ultrapure water (deionized water to attain a resistivity of 18 MΩ at room temperature (RT = 22 °C)) and analytical grade reagents. Prepare reagents at RT, and store in indicated temperatures. Sodium azide has to be added to dialysis buffers to avoid bacterial growth. Always check the safety sheets of all the reagents before starting and follow the good laboratory practice. Follow the waste disposal regulations of your laboratory.

1. Dialysis membrane, flat width 40 mm, MWCO 6–8 kDa.
2. Dialysis membrane, flat width 20 mm, MWCO 6–8 kDa.
3. Ultrafiltration device, Pellicon, MWCO 10 kDa.
4. Ultrafiltration cell device, Amicon, MWCO 10 kDa.
5. Ultra concentrators, Amicon, MWCO 10 kDa.
6. Q-Sepharose HP anion-exchange chromatography column, 2.6 mL × 10 cm, GE Healthcare Lifesciences.
7. Automated fast flow liquid chromatography system, Äkta™ purification system, GE Healthcare Lifesciences.
8. Aminohexyl agarose, AH-Sepharose 4B, GE Healthcare Lifesciences.
9. Standard Econo column, Bio-Rad Laboratories.
10. Sephacryl S-200 column, 2.6 × 80 cm, GE Healthcare Lifesciences.
11. EDC, 1-*ethyl*-(3-dimethylaminopropyl)-carboniimide hydrochloride, MW: 155.24 g/mol.

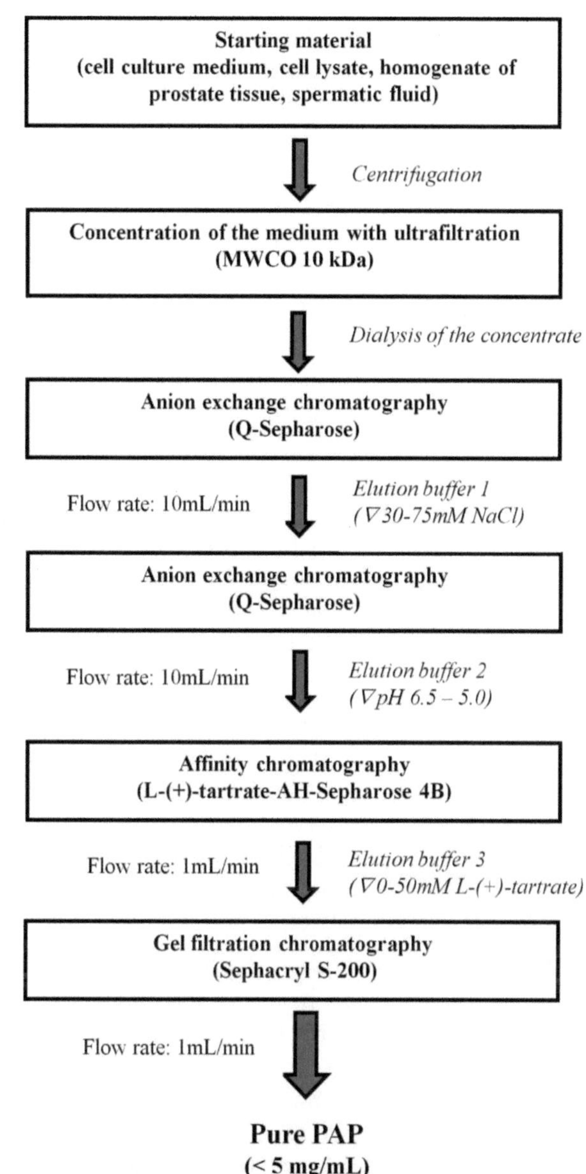

Fig. 1 Flow-chart of PAP purification (Linear gradient = *inverted open triangle*) [3]

12. Other reagents: *p*-nitrophenyl phosphate, *p*-nitrophenol, NaCl, NaOH, $NaHCO_3$, Na-acetate, Bis-Tris, HCl, L-(+)-tartrate, citric acid, Na-citrate.
13. pH-meter, spectrophotometer, magnetic stirrer, rotating mixer, centrifuge, water bath.

2.1 Preparation of L-(+)-Tartrate Affinity Column

1. *0.5 M NaCl*, dissolve 58.44 g NaCl in 1,800 mL water in a graduated cylinder and adjust volume up to 2 L.
2. *Water, pH 4.5*, Adjust the pH of 2 L water with weak HCl.

3. *Wash buffer 1*, 0.1 M $NaHCO_3$, pH 9.0. Add 200 mL water to a 2 L graduated cylinder. Weigh 16.80 g $NaHCO_3$ (MW 84 g/mol) and transfer to the cylinder. Mix on a magnetic stirrer and adjust pH with 1 M NaOH. Adjust volume up to 2 L with water. Filtrate and store at RT.
4. *Wash buffer 2*, 0.1 M Na-acetate, pH 4.0. Weigh 16.40 g Na-acetate and make the buffer as described above and adjust pH 4.0 with glacial acetic acid. Adjust volume up to 2 L with water. Filtrate and store at RT.
5. *Wash buffer 3*, 0.1 M Na-acetate, pH 4.0, 1 M NaCl. Make as wash buffer 2, but weigh also 116.88 g NaCl and add to the solution before adjusting the pH. Adjust volume up to 2 L with water. Filtrate and store at RT.

2.2 Purification of Prostatic Acid Phosphatase

1. *Dialysis buffer 1*, 25 mM Bis-Tris/HCl, pH 6.5, 0.05 % NaN_3. Add 1 L water to a 5-L plastic beaker. Weigh 26.16 g Bis-Tris (MW 209.24 g/mol) and transfer to the beaker. Mix on magnetic stirrer, add water to a volume of 4.5 L and adjust the pH with HCl. Adjust volume up to 5 L with water add 0.25 g NaN_3. Store at +4 °C.
2. *Loading buffer for anion exchange chromatography (I, II)*, 25 mM Bis-Tris/HCl, pH 6.5. Add 200 mL water to a 2 L glass beaker. Weigh 10.46 g Bis-Tris and transfer to the glass beaker. Mix on magnetic stirrer, add water to a volume 1.9 L and adjust the pH with HCl. Adjust volume up to 2 L with water. Filter through 0.22 μm filter membrane and store at +4 °C.
3. *Elution buffer 1* for anion exchange chromatography (I), 25 mM Bis-Tris/HCl, pH 6.5, 0.5 M NaCl. Add 200 mL water to a 2 L graduated cylinder or beaker. Weigh 10.46 g Bis-Tris and 58.44 g NaCl and transfer both to the beaker. Mix on magnetic stirrer, add water to a volume 1.9 L and adjust the pH with HCl. Adjust volume up to 2 L with water. Filter through 0.22 μm filter membrane and store at +4 °C.
4. *Elution buffer 2* for anion exchange chromatography (II), 25 mM Bis-Tris/HCl, pH 5.0. Add 200 mL water to a 2 L graduated cylinder or beaker. Weigh 10.46 g Bis-Tris and transfer to the beaker. Mix on magnetic stirrer, add water to a volume 1.9 L and adjust the pH with HCl. Adjust volume up to 2 L with water. Filter through 0.22 μm filter membrane and store at +4 °C.
5. *Dialysis buffer 2*, 50 mM Na-acetate, pH 6.0, 0.05 % NaN_3. Add 1 L water to a 5-L plastic beaker. Weigh 20.51 g Na-acetate (mw 82.03 g/mol) and transfer to the beaker. Mix on magnetic stirrer, add water to a volume 4.5 L and adjust the pH with glacial acetic acid. Adjust volume up to 5 L with water and add 0.25 g NaN_3. Store at +4 °C.

6. *Loading buffer for* L*-(+)-tartrate affinity chromatography*, 50 mM Na-acetate, pH 6.0. Add 200 mL water to 2 L graduated cylinder. Weigh 8.20 g Na-acetate and transfer to the cylinder. Mix on a magnetic stirrer, add water to a volume of 1.8 L and adjust the pH with glacial acetic acid. Adjust volume up to 2 L with water. Filter through 0.22 μm filter membrane and store at +4 °C.
7. *Elution buffer 3*, 50 mM Na-acetate, pH 6.0, 50 mM L-(+)-tartrate. Add 100 mL water to 1 L graduated cylinder. Weigh 4.51 g Na-acetate and 9.50 g L-(+)-tartrate and transfer them to the cylinder. Mix on a magnetic stirrer, add water to a volume 900 mL and adjust the pH with glacial acetic acid. Adjust volume up to 1 L with water. Filter through 0.22 μm filter membrane and store at +4 °C.
8. *Buffer for gel filtration chromatography*, 50 mM Na-acetate, pH 5.0, 150 mM NaCl. Add 200 mL water to 2 L graduated cylinder. Weigh 8.20 g Na-acetate and 17.53 g NaCl and transfer both to the cylinder. Mix on a magnetic stirrer, add water to a volume 1,900 mL and adjust the pH with glacial acetic acid. Adjust volume up to 2 L with water. Filter through 0.22 μm filter membrane and store at +4 °C.
9. *Dialysis buffer for storing pure PAP*, 50 mM Na-acetate, pH 5.5. Add 200 mL water to 2-L graduated cylinder. Weigh 8.20 g Na-acetate and transfer to the cylinder. Mix on a magnetic stirrer, add water to a volume 1.8 L and adjust the pH with glacial acetic acid. Make up to 2 L with water store at +4 °C. No NaN_3 is added.
10. *Regeneration buffer for anion exchange column*, 0.25 M Bis-Tris, 1 M NaCl, pH 6.5. Add 200 mL water to a 2 L graduated cylinder or beaker. Weigh 10.46 g Bis-Tris and 116.88 g NaCl and transfer both to the beaker. Mix on magnetic stirrer, add water to a volume 1.9 L and adjust the pH with HCl. Adjust volume up to 2 L with water. Filter through 0.22 μm filter membrane and store at +4 °C.

2.3 Acid Phosphatase Activity Measurement

1. *Buffer 1*, 50 mM Sodium citrate buffer, pH 4.8. Prepare the following solutions. Solution A: dissolve 7.35 g sodium citrate dihydrate ($C_6H_5Na_3O_7{\cdot}2H_2O$, FW 294.1 g/mol) in 500 mL water and Solution B: dissolve 5.35 g citric acid monohydrate ($C_6H_8O_7{\cdot}H_2O$, MW 210.14 g/mol) in 500 mL water. Adjust the pH 4.8 by mixing solution A and solution B (*see* **Note 1**).
2. *Buffer 2*, 50 mM Sodium Tartrate in 50 mM Sodium citrate buffer, pH 4.8. Prepare the following solutions. Solution C: Dissolve 7.35 g Sodium citrate dihydrate ($C_6H_5Na_3O_7{\cdot}2H_2O$, FW 294.1 g/mol) and 7.05 g potassium sodium tartrate tetrahydrate ($C_4H_4KNaO_6{\cdot}4H_2O$, FW: 282.23 g/mol) in

500 mL water and Solution D: Dissolve 5.35 g Citric acid monohydrate ($C_6H_8O_7{\cdot}H_2O$, MW 210.14 g/mol) and 7.05 g potassium sodium tartrate tetrahydrate ($C_4H_4KNaO_6{\cdot}4H_2O$, FW: 282.23 g/mol) in 500 mL water. Adjust the pH mixing solution C and solution D (*see* **Note 1**).

3. *Substrate solution 1*, 5.5 mM *p*-nitrophenyl phosphate. Weigh 14.5 mg *p*-nitrophenyl phosphate disodium salt and dissolve it in 10 mL *Buffer 1* (*see* **Note 2**).
4. *Substrate solution 2*, 5.5 mM *p*-nitrophenyl phosphate. Weigh 14.5 mg of *p*-nitrophenyl phosphate disodium salt and dissolve it in 10 mL *Buffer 2* (*see* **Note 2**).

3 Methods

3.1 Preparation of L-(+)-Tartrate Affinity Column

Couple L-(+)-tartrate to aminohexyl agarose (AH-Sepharose 4B), according to manufacturer's instructions and Vihko et al. [5] as follows:

1. Prepare 50 g wet gel, 1 g of aminohexyl agarose will be 4 mL swollen gel, by dissolving 50 g aminohexyl agarose to 0.5 M NaCl solution (200 mL/g freeze-dried powder) and wash on a sintered glass filter for 15 min in order to remove any additives.
2. To remove NaCl wash the gel with distilled water, pH 4.5 (*see* **Note 3**).
3. Combine two parts of gel and one part of water pH 4.5 to make a slurry for stirring, $V_{final} = 300$ mL.
4. Add 0.5 g L-(+)-tartrate to 50 g of wet gel. Stir the mixture gently at RT. Do not use a magnetic stirrer (*see* **Note 4**).
5. Add EDC to the mixture to final concentration of 0.1 M (4.66 g in 300 mL) to form amine bonds (*see* **Note 5**).
6. Continue stirring and maintain the pH at 4.5–6.0 for 1 h by addition of diluted NaOH. After that the changes of pH are small.
7. Continue stirring for 24 h at RT
8. Wash the gel thoroughly with *Wash buffer 1* using a sintered glass filter. Continue washing with *Wash buffer 2* and *Wash buffer 3* alternately the buffer four times. Finally wash with distilled water (*see* **Note 6**).
9. Pack the gel into standard Econo column.

3.2 Purification of Prostatic Acid Phosphatase

Carry out all the procedures at RT unless otherwise specified (*see* **Note** 7 for small-scale purification).

1. Harvest the recombinant PAP production media from the insect cell culture and remove cells by centrifugation 2,000 × *g* for 10 min.

2. Take samples from each step for acid phosphatase activity and protein concentration measurements (*see* **Note 8**).
3. Concentrate (*see* **Note 9**) the centrifuged media with a MWCO 10 kDa membrane (e.g. Pellicon cassette system).
4. Dialyze the concentrate in *Dialysis buffer 1* overnight. Volume of the concentrate 50–100 mL, volume of the dialysis buffer 5 L (*see* **Note 10**).
5. Filtrate the dialyzed concentrate through 0.22 μm membrane. Take a sample of the concentrate for enzyme activity measurement.
6. Balance the anion-exchange Q-Sepharose HP column connect to Äkta™ system with *Loading buffer for anion exchange chromatography* with 3–4 times column volume, 10 mL/min.
7. Load the filtrated concentrate to the anion-exchange column, 10 mL/min. Collect the flow through and wash unbound material until the UV baseline is zero. Take a sample of the flow through and the unbound material for enzyme activity measurement to confirm that PAP is bound to the column.
8. Anion exchange chromatography (I): elute the column with linear salt gradient from 30 to 75 mM NaCl mixing *Loading buffer* and *Elution buffer 1*. Collect 10 mL fractions and measure acid phosphatase activity from the fractions. Pool the fractions with highest activity, concentrate with concentrator (MWCO 10 kDa) to ca. 40 mL and dialyze in *Dialysis buffer 2*. Measure acid phosphatase and protein concentration to calculate specific activity. Regenerate the column (*see* **Note 11**).
9. Balance the anion-exchange Q-Sepharose HP column with *Loading buffer* and load the dialyzed pool from the previous step to the column.
10. Anion exchange chromatography (II): elute the column with linear pH gradient from 6.5 to 5.0 mixing *Loading buffer* and *Elution buffer 2*. Collect 10 mL fractions (*see* **Note 11**).
11. Measure acid phosphatase activity from all collected fractions. Pool fractions with highest activity and concentrate them with an ultrafiltration cell device (MWCO 10 kDa, *see* **Note 12**) to ca. 10 mL.
12. Dialyze the pooled fractions in *Dialysis buffer 2* and take a sample for acid phosphatase activity and protein concentration measurements to calculate specific activity.
13. Balance the L-(+)-tartrate affinity column (1.0 × 12 cm) connect to Äkta™ system with 3–4 times column volume 1.0 mL/min) with *Loading buffer for L-(+)-tartrate affinity chromatography*.
14. Load the dialyzed fraction pool to the L-(+)-tartrate affinity column, 1 mL/min. Collect the flow through and wash the

unbound material with *Loading buffer for L-(+)-tartrate affinity chromatography*. Take a sample for acid phosphatase activity measurement to confirm that PAP is bound to the column.

15. Elute the PAP with a linear L-(+)-tartrate gradient mixing *Loading buffer for L-(+)-tartrate affinity chromatography and Elution buffer 3*, 1 mL/min. Collect 1 mL fractions and measure acid phosphatase activity from each fraction. Pool the fractions with the highest activity (*see* **Note 13**).
16. Concentrate the pool with ultrafiltration cell as described previously to 2 mL for gel filtration chromatography. Take a sample for acid phosphatase activity and protein concentration measurements to calculate specific activity.
17. Balance the Sephacryl S-200 gel filtration column connected to Äkta™ system with *Buffer for gel filtration chromatography*, 1 mL/min.
18. Load the pool from L-(+)-tartrate affinity column to Sephacryl S-200 column and elute the pure PAP with *Buffer for gel filtration chromatography*, 1 mL/min. Collect 1 mL fractions. Measure acid phosphatase activity from fractions and pool the fractions with the highest activity.
19. Dialyze the pooled fractions *in Dialysis buffer for storing pure PAP* as previously, 2 L dialysis buffer is enough.
20. Concentrate the pool with ultracentrifugal concentrator at +4 °C (*see* **Note 14**).
21. Measure acid phosphatase activity and protein concentration to calculate specific activity.
22. Check the purity of the purified protein with native and SDS/PAGE gel electrophoresis with silver staining detection (Fig. 2a), followed by Western-blot and immunodetection of the product (Fig. 2b).

3.3 Acid Phosphatase Activity Measurement

1. Pipette into three labeled tubes the following solutions. Tube 1: 200 μl *substrate solution 1*, Tube 2: 200 μl *substrate solution 1*, Tube 3: 200 μl *substrate solution 2*.
2. Incubate the tubes in 37 °C water bath for 5 min.
3. Add 20 μl sample to tubes 2 and 3 (*see* **Note 15**).
4. Incubate all the tubes in a 37 °C water bath for exactly 30 min.
5. Stop the reaction by adding 2 mL 0.02 N NaOH (*see* **Note 16**).
6. Add 20 μl sample to tube 1 (*see* **Note 17**).
7. Measure the absorbance of all samples at λ 400–420 nm using water as reference.
8. Calculations: The difference in absorbance between tube 2 and tube 1 is the total acid phosphatase activity, and the difference between tube 2 and tube 3 is the prostatic acid phosphatase activity.

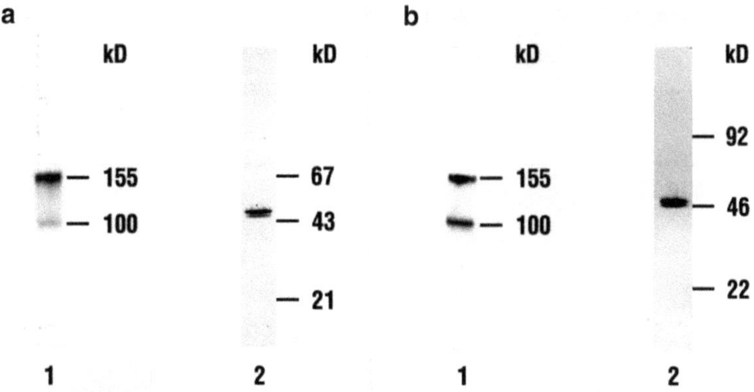

Fig. 2 (**a**) Silver stain of gel electrophoresis of pure recombinant rat PAP (size in kDa). *Lane 1*: native PAGE. *Lane 2*: SDS-PAGE. (**b**) Immunodetection (size in kDa) of purified recombinant rat PAP by Western-blotting. *Lane 1*: native PAGE. *Lane 2*: SDS-PAGE. Immunodetection with rabbit polyclonal antibody against an N-terminal peptide of rat PAP (in house antibody) [3]

9. Prepare the calibration curve by diluting 0.5 mL *p-nitrophenol* standard solution (10 μmol/mL) to 100 mL with 0.02 N NaOH (*see* **Note 18**). Pipette in every numbered tube the following volumes:

Tube No.	Diluted *p*-nitrophenol solution (mL)	0.02 N NaOH (mL)	Amount of *p*-nitrophenol (μmol)
1	1.0	10.0	0.05
2	2.0	9.0	0.10
3	4.0	7.0	0.20
4	6.0	5.0	0.30
5	8.0	3.0	0.40
6	10.0	1.0	0.50

10. Read absorbance at λ 400–400 nm. Use 0.02 N NaOH solution as reference.
11. Plot the absorbance versus the amount of *p*-nitrophenol (μmol). Interpolate the absorbance value from the sample to determine the amount of *p*-nitrophenol produced by the reaction in 30 min (*see* **Note 19**). Calculate the enzyme activity of the sample.
12. Measure the protein concentration of the sample and calculate the specific activity of the sample.

    Specific activity (μmol/min mg) = Enzyme activity/protein concentration.

13. The purification factor is calculated as follows:
    Purification factor for step X = Specific activity in X/Specific activity in starting material.

4 Notes

1. To adjust the pH of *Buffer 1* place a magnetic stir bar at the bottom of a glass beaker and add the solution A. Introduce the pH-meter electrode and allow to stabilize the pH measurement (remember to calibrate the pH-meter before starting). Add small amounts of solution B to the glass beaker allowing the pH to stabilize. Continue adding Solution B until reach the required pH. Follow the same procedure for *Buffer 2.* Citrate buffer at pH 4.8 is optimal for the acid phosphatase activity test of PAP using *p*-nitrophenyl phosphate as substrate.
2. *p*-Nitrophenyl phosphate reagent is stored at −20 °C, allow the substrate to warm to room temperature before opening the container to avoid absorption of moisture. Prepare fresh *p*-nitrophenyl phosphate solutions (with and without tartrate) in the moment of use.
3. To check that all the NaCl has been removed from the gel matrix the conductivity of the washing solution can be tested.
4. Place the gel mixture into a plastic bottle and attach it to a rotating mixer. Do not use a magnetic stirrer, since a magnetic bar will break the gel matrix.
5. Make sure that there are no lumps in the EDC before adding it to the mixture.
6. If the gel is not used immediately store it in buffer with a neutral or slightly acidic pH and a high concentration of salt (1 M NaCl) at 4–8 °C in the presence of a suitable bacteriostatic agent, e.g. 0.05 % NaN_3.
7. Small-scale protein purification protocol is suitable when seminal fluid or prostate tissue is used as starting material. Perform all the steps for small-scale protein purification in +4 °C (or cold room). Same buffers than in mass-scale purification are used. Prepare your own columns using strong anion exchanger matrix such as QAE Sephadex A-25, and for gel filtration matrix use e.g. Sephacryl S-200 HR. L-(+)-tartrate affinity column is prepared in the same way than for mass-scale purification method. Use peristaltic pumps, gradient mixer and fraction collector in +4 °C.
8. Take samples for acid phosphatase activity measurement from each step. Save all flow throughs and washing pools until acid

phosphatase activity measurements have been performed to confirm that PAP is not lost during the purification steps. Measure protein concentration with a suitable method such as Lowry [6]. The relation between acid phosphatase activity and protein concentration is defined as the specific activity, which is an indicator of the purification level of the enzyme.

9. Perform the concentration of cell culture media in a cold room. If not possible, keep the media and concentrate on ice in order to avoid the heating of the material and precipitation of proteins.
10. Perform all dialyses at +4 °C on a magnetic stirrer.
11. Regenerate the anion exchange column with running 0.25 M Bis-Tris, 1 M NaCl, pH 6.5 100 mL through the column to remove all bound material before the next run.
12. Perform the concentration at +4 °C to avoid the protein denaturation.
13. L-(+)-tartrate is a competitive inhibitor of PAP.
14. Final protein concentration should be <5 mg/mL to avoid precipitation.
15. Samples for this assay may be cell supernatant from cell culture experiments, cell lysates, tissue lysates, or biological fluids such as serum or plasma. High degree of hemolysis in serum or blood in tissue samples interferes with the measurement due to the presence of erythrocyte acid phosphatase.
16. Alkaline medium stops the reaction and induces the development of color. *p*-Nitrophenol is the result compound of the dephosphorylation of *p*-nitrophenyl phosphate by the phosphatase and it turns a yellow color in alkaline medium.
17. The addition of the sample after stopping the reaction serves two purposes. First, it makes all volumes equal for the measurement, and secondly it acts as the blank for the reaction.
18. A new calibration curve has to be done for every assay to avoid variations due to the equipment. For the calibration curve, make a fresh solution every time.
19. The enzyme activity is defined as the moles of substrate or product converted in a reaction per time unit. The enzyme activity depends on the reaction conditions, which should be specified. In the International system (SI) the unit is the katal (*1 katal = 1 mol/s*), but a more practical and commonly used value is *1 enzyme unit (U) = 1 μmol/min*. One enzyme unit corresponds to 16.67 nanokatals [7].

References

1. Jacobsson K (1960) The determination of tartrate-inhibited phosphatase in serum. Scand J Clin Lab Invest 12:367–380
2. Andersch MA, Szczypinski AJ (1947) Use of p-nitrophenylphosphate as the substrate in determination of serum acid phosphatase. Am J Clin Pathol 17:571–574
3. Vihko P, Kurkela R, Porvari K et al (1993) Rat acid phosphatase: overexpression of active, secreted enzyme by recombinant baculovirus-infected insect cells, molecular properties, and crystallization. Proc Natl Acad Sci USA 90:799–803
4. Summers M, Smith G (1987) Manual of methods for baculovirus vector and insect cell culture procedures (bul.no. 1555). Texas Agric. Exp. Stn. and Texas A&M University, College Station, TX
5. Vihko P, Kontturi M, Korhonen LK (1978) Purification of human prostatic acid phosphatase by affinity chromatography and isoelectric focusing. Part I. Clin Chem 24:466–470
6. Lowry OH, Rosebrough NJ, Farr AL et al (1951) Protein measurement with the Folin phenol reagent. J Biol Chem 193:265–275
7. Nomenclature Committee of the International Union of Biochemistry (NC-IUB) (1979) Units of enzyme activity. recommendations 1979. Eur J Biochem 97:319–320

Chapter 13

Protein Tyrosine Phosphatases: Structure, Function, and Implication in Human Disease

Lutz Tautz, David A. Critton, and Stefan Grotegut

Abstract

Protein tyrosine phosphorylation is a key regulatory mechanism in eukaryotic cell physiology. Aberrant expression or function of protein tyrosine kinases and protein tyrosine phosphatases can lead to serious human diseases, including cancer, diabetes, as well as cardiovascular, infectious, autoimmune, and neuropsychiatric disorders. Here, we give an overview of the protein tyrosine phosphatase superfamily with its over 100 members in humans. We review their structure, function, and implications in human diseases, and discuss their potential as novel drug targets, as well as current challenges and possible solutions to developing therapeutics based on these enzymes.

Key words Tyrosine phosphorylation, Phosphatases, PTP, Structure, Function, Substrate recognition, Specificity, Human disease, PTP1B, SHP2, Inhibitors

1 Introduction

Protein phosphorylation is a fundamental regulatory mechanism for numerous important aspects of cell physiology throughout all kingdoms of life [1–6]. Protein phosphorylation is a rapidly reversible posttranslational modification, controlled by the opposing activities of protein kinases and protein phosphatases. A change in phosphorylation state can be the result of a change in the activity or access of either enzyme. In eukaryotes, about one third of the entire proteome is phosphorylated, predominantly on serine, threonine, and tyrosine [7]. While phosphorylation on serine/threonine is most abundant (~98 %), tyrosine phosphorylation [8], which accounts for less than 2 % of the total phosphoproteome, is a key regulatory mechanism in numerous important aspects of eukaryotic cell physiology [9–17]. The importance of tyrosine phosphorylation in normal cell physiology is highlighted by the fact that many human diseases are the result of aberrant protein tyrosine kinase (PTK) [18] or

José Luis Millán (ed.), *Phosphatase Modulators*, Methods in Molecular Biology, vol. 1053,
DOI 10.1007/978-1-62703-562-0_13, © Springer Science+Business Media, LLC 2013

protein tyrosine phosphatase (PTP) function [9, 13, 19]. Historically, research had long been focused on the role of PTKs in signaling, as it was generally believed that PTPs functioned merely as indiscriminate "housekeeping" enzymes with broad specificities. Now, we know that PTPs are highly specific and tightly regulated, both in space and time [9, 13]. Here, we review the PTP family, describe their classification into subfamilies, discuss their common structural features and shared catalytic mechanism, analyze their substrate recognition and specificity, and finally review their implication in human disease and highlight their potential as novel drug targets and challenges that need to be overcome in order to develop therapeutics based on these pivotal enzymes.

2 The Human PTP Superfamily

PTPs constitute the largest family of phosphatase genes and are defined by their active-site signature motif $\mathbf{C}(X)_5\mathbf{R}$. The catalytic mechanism of PTPs (as discussed in more detail below) is based on a nucleophilic cysteine residue that is part of this motif. The complement of human PTP genes can be divided into three distinct families based on the sequence of their PTP domains: class I PTPs (100 genes), class II PTPs (1 gene), and class III PTPs (3 genes) (Table 1) [9]. The class I PTPs constitute the vast majority of human PTP genes and can be divided into the classical, phosphotyrosine (pTyr)-specific PTP, and the VH1-like or dual specificity protein phosphatase (DSP, DUSP) subfamilies. The pTyr-specific PTPs (37 genes) are further divided into the transmembrane receptor-like PTPs (RPTPs) and the intracellular nonreceptor-like PTPs (NRPTPs) [9, 13, 20]. The DUSP subfamily (63 genes) is the most diverse group in terms of substrate specificity. This group includes the phosphothreonine (pThr)/pTyr-specific mitogen-activated protein (MAP) kinase phosphatases (MKPs), the pThr/pTyr- or mRNA-specific atypical DSPs, the phosphoserine (pSer)-specific slingshots, the pTyr-specific phosphatases of regenerating liver (PRLs), the pSer/pThr-specific CDC14s, the PTENs, which are specific for the 3'-phosphate of the inositol ring in phosphatidylinositol-(3,4,5)-triphosphate, the myotubularins, which dephosphorylate phosphatidylinositol-3-phosphate and phosphatidylinositol-(3,5)-bisphosphate, as well as the inositol 4-phosphatases, which catalyze the hydrolysis of the 4'-phosphate of phosphatidylinositol-(3,4)-bisphosphate, inositol-(1,3,4)-trisphosphate, and inositol(3,4)-bisphosphate [9, 21, 22]. All class I PTPs appear to have evolved from a common ancestor, based on their similar structural folds [20]. The class II PTP family is represented by only a single gene, which encodes a pTyr-specific low molecular weight PTP (LMPTP) [23] that is

Table 1
The human PTPome

	Gene	Protein, synonymes	UniProt	1	2	3	4	5	6	7	8	9	10	11	12	13	14	15	16	17	18	19	20	21	22	23	24	25	26	27	28	29	30	31	32	33	34	35
Transmembrane receptor-like PTPs (RPTPs)	*PTPRA*	RPTPα	P18433	X																							X											
	PTPRB	RPTPβ	B2RU80	X		X																																
	PTPRC	CD45, LCA	P08575	X	X	X																																
	PTPRD	RPTPδ	P23468	X	X	X				X																												
	PTPRE	RPTPε	P23469	X																							X											
	PTPRF	LAR	P10586	X	X	X				X																												
	PTPRG	RPTPγ	P23470	X	X	X																	X															
	PTPRH	SAP1	Q9HD43	X		X																																
	PTPRJ	DEP1, CD148, RPTPη	Q12913	X		X																																
	PTPRK	RPTPκ	Q15262	X	X	X				X					X	X																						
	PTPRM	RPTPμ	P28827	X	X	X				X					X	X																						
	PTPRN	IA-2, Islet cell antigen 512	Q16849		X																									X								
	PTPRN2	PTPRP, RPTPπ, IA-2β, phogrin	Q92932		X																									X								
	PTPRO	GLEPP1, PTP-U2, PTPROτ isoforms A/B/C	Q16827	X		X																																
	PTPRQ	PTPS31	Q9UMZ3	X		X																																
	PTPRR	PTP-SL, PCPTP, PTPBR7, PC12-PTP1	Q15256	X																X																		
	PTPRS	RPTPσ	Q13332	X	X	X				X																												
	PTPRT	RPTPρ	Q99M80	X	X	X				X					X	X																						
	PTPRU	PTPJ/PTP-U1/ PTPRomicron isoforms 1/2/3	Q92729	X	X	X				X					X	X																						
	PTPRZ	RPTPζ	P23471	X	X	X																	X															

(continued)

Table 1
(continued)

	Gene	Protein, synonymes	UniProt	1	2	3	4	5	6	7	8	9	10	11	12	13	14	15	16	17	18	19	20	21	22	23	24	25	26	27	28	29	30	31	32	33	34	35
Intracellular nonreceptor-like PTPs (NRPTPs)	*PTPN1*	PTP1B	P18031	X																																		
	PTPN2	TCPTP, MPTP, PTP-S	P17706	X																																		
	PTPN3	PTPH1	P26045	X								X									X																	
	PTPN4	PTP-MEG1, TEP	P29074	X								X									X																	
	PTPN5[a]	STEP	P54829	X																X																		
	PTPN6	SHP1, PTP1C, SH-PTP1, HCP	P29350	X										X																								
	PTPN7	HePTP, LCPTP	P35236	X																X																		
	PTPN9	PTP-MEG2	P43378	X																																		X
	PTPN11	SHP2, SH-PTP2, Syp, PTP1D, PTP2C, SH-PTP3	Q06124	X										X																								
	PTPN12	PTP-PEST, PTP-P19, PTPG1	Q05209	X							X																											
	PTPN13	PTP-BAS, FAP-1, PTP1E, RIP, PTPL1, PTP-BL	Q12923	X								X									X																	
	PTPN14	PTP36, PEZ, PTPD2	Q15678	X								X																										
	PTPN18	PTP-HSCF, PTP20, BDP	Q99952	X							X																											
	PTPN20	TypPTP	Q4JDL3	X							X																											
	PTPN21	PTPD1, PTP2E, PTP-RL10	Q16825	X								X																										
	PTPN22	LYP	Q9Y2R2	X							X																											
	PTPN23	HD-PTP, HDPTP, PTP-TD14, KIAA1471	Q9H3S7	X							X																				X					X		

Map kinase phosphatases (MKPs)	*DUSP1*	MKP-1, 3CH134, PTPN10, erp, CL100/ HVH1	P28562	X		X			
	DUSP2	PAC-1	Q05923	X		X			
	DUSP4	MKP-2, hVH2/TYP1	Q13115	X		X			
	DUSP5	hVH3/B23	Q16690	X		X			
	DUSP6	MKP-3, PYST1	Q16828	X		X			
	DUSP7	MKP-X, PYST2, B59	Q16829	X		X			
	DUSP8	hVH5, M3/6, HB5	Q13202	X		X	X		
	DUSP9	MKP-4, Pyst3	Q99956	X		X			
	DUSP10	MKP-5	Q9Y6W6	X		X			
	DUSP16	MKP-7, MKP-M	Q9BY84	X		X	X		
	DUSP24	MK-STYX, MKSTYX, DUSP24, STYXL1	Q9Y6J8		X	X			
Atypical dual-specificity phosphatases	*DUSP3*	VHR, T-DSP11	P51452	X					
	DUSP11	PIR1	O75319	X					
	DUSP12	HYVH1, GKAP, LMW-DSP4	Q9UNI6	X					X
	DUSP13A[b]	BEDP	Q6B8I1	X					
	DUSP13B[b]	TMDP, TS-DSP6	Q9UII6	X					
	DUSP14	MKP6, MKP-L	O95147	X					
	DUSP15	VHY, Q9H1R2	Q9H1R2	X				X	
	DUSP18	DUSP20, LMW-DSP20	Q8NEJ0	X					
	DUSP19	DUSP17, SKRP1, LDP-2, TS-DSP1	Q8WTR2	X					
	DUSP21	LMW-DSP21, BJ-HCC-26 tumor antigen	Q9H596	X					
	DUSP22	VHX, MKPX, JSP1, LMW-DSP2, TS-DSP2, JKAP	Q9NRW4	X				X	
	DUSP23	VHZ, DUSP25, FLJ20442, LMW-DSP3	Q9BVJ7	X					

(continued)

Table 1
(continued)

	Gene	Protein, synonymes	UniProt	1	2	3	4	5	6	7	8	9	10	11	12	13	14	15	16	17	18	19	20	21	22	23	24	25	26	27	28	29	30	31	32	33	34	35
	DUSP26	MKP-8, DUSP24, LDP4, NATA1, SKRP3	Q9BV47	X																																		
	PTPMT1	MOSP, PLIP	Q8WUK0	X																																		
	DUSP27[c]	FMDSP, DUPD1, 'similar to cyclophilin',	Q5VZP5	X																																		
	DUSP28	VHP, DUSP26	Q4G0W2	X																																		
	EPM2A	Laforin	O95278	X																													X					
	RNGTT	mRNA capping enzyme	O60942	X																															X			
	STYX	STYX	Q8WUJ0		X																																	
Slingshots	*SSH1*	SSH1, slingshot 1	Q8WYL5	X																		X																
	SSH2	SSH2, slingshot 2	Q76I76	X																		X																
	SSH3	SSH3, slingshot 3	Q8TE77	X																		X																
PRLs	*PTP4A1*	PRL-1	Q93096	X															X																			
	PTP4A2	PRL-2, OV-1	Q12974	X															X																			
	PTP4A3	PRL-3	O75365	X															X																			
CDC14s	*CDC14A*	CDC14A	Q9UNH5	X																					X													
	CDC14B	CDC14B	O60729	X																					X													
	CDKN3	KAP	Q16667	X																																		
	PTPDC	PTPDC, PTP9Q22	A2A3K4	X																																		

PTENs	*PTEN*	PTEN, MMAC1, TEP1	P60484	X						X	X					
	TPIP	TPIPα, TPTE and PTEN homologous	Q6XPS3	X							X			X		
	TPTE	PTEN-like, PTEN2	P56180	X							X			X		
	TNS1	Tensin-1, TNS	Q9HBL0		X			X	X							
	TENC1	C1-TEN, TENC1, KIAA1075, TNS2	Q63HR2		X			X	X						X	
	TNS3	Tensin 3	Q68CZ2		X			X	X							
Myotubularins	*MTM1*	Myotubularin	Q13496	X		X	X			X						
	MTMR1	MTMR1	Q13613	X		X	X			X						
	MTMR2	MTMR2	Q13614	X		X	X			X						
	MTMR3	MTMR3, FYVE-DSP1	Q13615	X		X	X						X			
	MTMR4	MTMR4, FYVE-DSP2	Q9NYA4	X		X	X						X			
	MTMR5	MTMR5, SBF1	O95248		X	X	X					X				X
	MTMR6	MTMR6	Q9Y217	X		X	X									
	MTMR7	MTMR7	Q9Y216	X		X	X									
	MTMR8	MTMR8	Q96EF0	X		X	X									
	MTMR9	MTMR9, LIP-STYX	Q96QG7		X	X	X									
	MTMR10	MTMR10	Q9NXD2		X		X									
	MTMR11	MTMP11, CRA α/β	A4FU01		X	X	X									
	MTMR12	MTMR12, 3-PAP	Q9C0I1		X	X		X								
	MTMR13	MTMR13, SBF2, CMT4B2	Q86WG5		X	X	X	X				X				
	MTMR14[d]	hJumpy, FLJ22075, hEDTP	Q8NCE2	X												

(continued)

Table 1
(continued)

	Gene	Protein, synonymes	UniProt	1	2	3	4	5	6	7	8	9	10	11	12	13	14	15	16	17	18	19	20	21	22	23	24	25	26	27	28	29	30	31	32	33	34	35
Inositol 4-phosphatases	*INPP4A*	Inositol 4-phosphatase type I	Q96PE3	X																																		
	INPP4B	Inositol 4-phosphatase type II	O15327	X																																		
Class II PTP	*ACP1*	LMPTP, low Mr PTP, LMWPTP, BHPTP	P24666	X																																		
Class III PTPs	*CDC25A*	CDC25A	P30304	X																																		
	CDC25B	CDC25B	P30305	X																																		
	CDC25C	CDC25C	P30307	X																																		

Current data indicate that there are 104 PTP genes in humans, including 100 class I, one class II, and three class III PTP genes

The earlier predicted human OST-PTP (Andersen et al., 2004, FASEB J, 18, 8–30) is not included here. OST-PTP is an active PTP in mouse and rat, but the human OST-PTP cDNA sequence exhibits numerous disablements, indicating that it does not code for a PTP but is rather a pseudogene (Cousin et al., 2004, Biochem Biophys Res Commun, 321, 259–65)

The earlier predicted myotubularin-related phosphatase MTMR15 (Alonso et al., 2004, Cell, 117, 699–711) is not included here. The protein (KIAA1018/FAN1) was identified as a DNA repair nuclease (MacKay et al., 2010, Cell, 142, 65–76)

[a]STEP exists as two major alternatively spliced isoforms, $STEP_{61}$ and $STEP_{46}$. $STEP_{61}$ includes additional 172 amino acids in its amino-terminal region, which contains two transmembrane domains and two proline-rich motifs

[b]The human DUSP13 gene contains two DSP genes: DUSP13A (BEDP; 3 exons), two noncoding exons, followed by DUSP13B (TMDP; 3 exons). In the mouse, the two genes are designated as separate genes

[c]Formerly DUPD1, but does not contain an N-terminal cyclophilin domain, as originally annotated. The mouse ortholog also lacks cyclophilin domain. Thus, the name DUPD1 is a misnomer

[d]hJumpy lacks any of the typical domains characteristic of the myotubularin family, but it does contain the myotubularin-specif WDR motif within the P-loop and does have phosphatidylinositol-3-phosphate and phosphatidylinositol-(3,5)-bisphosphate phosphatase activity (Tosch et al., 2006, Hum Mol Genet, 15, 3098–106)

Domains and motifs are ordered by frequency of occurrence:

1 PTP domain, 2 Pseudo-PTP domain, 3 Fibronectin type III-like domain, 4 Coiled-coil motif, 5 Pleckstrin homology-glucosyltransferase, Rab-like GTPase activator and myotubularin (PH-GRAM) domain, 6 CDC25-homolgy (CH2) domain, 7 Immunoglobulin-like domain, 8 Proline-rich motif, 9 FERM domain, 10 Phosphotyrosine binding domain, 11 Src-homology-2 (SH2) domain, 12 Cadherin-like juxtamembrane sequence, 13 Meprin/A5/μ domain (MAM), 14 PDZ-binding sequence, 15 C2 domain, 16 CAAX sequence/farnesylation site, 17 Kinase-interaction motif (KIM), 18 PDZ domain, 19 Slingshot domain, 20 Carbonic anhydrase-like domain, 21. DENN domain, 22 DSP-like domain, 23 FYVE domain, 24 Heavily glycosylated, 25 Potential myristoylation site, 26 Putative transmembrane segment, 27 RDGS-adhesion recognition motif, 28 BRO1-homolgy domain, 29 C1 domain, 30 Carbohydrate-binding domain, 31 Cysteine-rich zinc-binding domain, 32 Guanylyltransferase domain , 33 Histidine domain, 34 Pleckstrin homology (PH) domain, 35 SEC14/cellular retinaldehyde-binding protein-like domain

structurally related to bacterial arsenate reductases and appears to be more ancient than class I PTPs. Representatives of this family are found in all major phyla, including plants, numerous prokaryotes, and archaea [9]. The class III PTP family comprises the pThr/pTyr-specific CDC25s, which appear to have evolved from bacterial rhodanese-like enzymes [20].

3 The Mechanism of PTP Catalysis

The mechanism of PTP catalysis was elucidated by site-directed mutagenesis and kinetic analyses [24–26], in conjunction with structural information [27–29, 44, 58]. All PTPs share a common catalytic mechanism based on a nucleophilic cysteine with a low pKa that forms a thiophosphate intermediate during catalysis (Fig. 1). The transition state of the reaction is stabilized by an invariant arginine, and hydrolysis is assisted by a catalytic acid/base aspartate. The ordered uni–bi two-step reaction is facilitated by an optimal arrangement of these residues, which are located in a number of highly conserved loops that form the active site. In the first step of the reaction, the catalytic cysteine of the phosphate-binding loop (P-loop) initiates the breaking of the phosphorus–oxygen bond via a nucleophilic attack on the phosphorous atom, while the catalytic aspartate of the WPD-loop acts as a general acid, donating a proton to the OH-leaving group. This step results in the formation of a phosphocysteine intermediate and the release of the dephosphorylated substrate. In the second, rate-limiting step, the thiophosphate intermediate is hydrolyzed, again with the assistance of the catalytic aspartate in the WPD-loop, which now acts as a general base to abstract a proton from a water molecule, thus facilitating hydrolysis of the scissile phosphorous-sulfur bond and producing free phosphate. In classical PTPs, the water molecule used for hydrolysis is positioned by a conserved glutamine located in the Q-loop. The guanidinium group of an invariant arginine of the P-loop coordinates the phosphate group during substrate binding and, more importantly, it stabilizes the transition state of the first step of the enzymatic reaction. While mutational studies have shown that PTPs lacking the aspartate or arginine still have residual phosphatase activity, PTPs with mutations in the catalytic cysteine are completely inactive [30]. Interestingly, despite a highly conserved active site, different PTPs, even within one subfamily, show different catalytic efficiencies toward the general PTP substrate *p*-nitrophenylphosphate (pNPP). For example, the *Yersinia* PTP YopH [31] is approximately tenfold more catalytically efficient than human PTP1B [64], which is in turn approximately tenfold more catalytically efficient than human hematopoietic PTP (HePTP) [32]. Thus, even small changes in the active site microenvironment can significantly affect the efficiency of PTP catalysis. One known determinant of the catalytic rate is the flexibility of the WPD-loop, which cycles between an

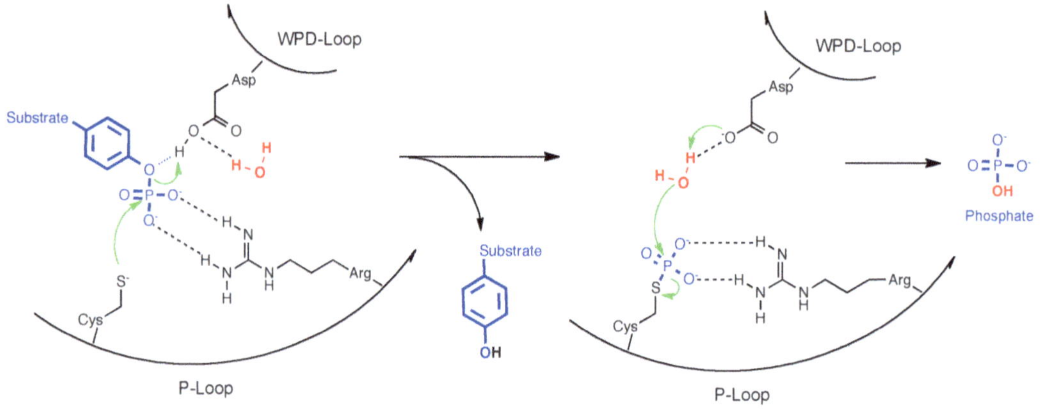

Fig. 1 Common PTP catalytic mechanism

open conformation in the apo structure, and a closed, active conformation when substrate is bound in the active site [33]. Point mutations or specific ligands that decrease the flexibility of this loop result in substantially lower PTP catalytic activity [34, 35].

4 Modular Domain Arrangement of Human PTPs

Alongside the ~280 residue PTP catalytic domain, most human PTPs also contain additional non-catalytic domains or motifs (Table 1), which provide a variety of different functions. For example, MKPs contain CDC25-homology (CH2) domains that mediate binding to their MAP kinase substrates. Similarly, HePTP (*PTPN7*), STEP (*PTPN5*), and PTP-SL (*PTPRR*) contain a 16-residue kinase interaction motif (KIM) that is essential for their ability to bind to MAP kinase substrates. In contrast, binding of a proline-rich motif in LYP (*PTPN22*) to the Src-homology 3 (SH3) domain in C-src kinase (CSK) dictates subcellular localization and restrains LYP substrate access [36]. Other examples of PTPs with domains/motifs that dictate subcellular localization include PTPH1 (*PTPN3*), PTP-MEG1 (*PTPN4*), and PTP-BAS (*PTPN13*), which contain PDZ domains for targeting to the cytoskeleton, as well as PAC-1 (*DUSP2*), MKP-1 (*DUSP1*), MKP-2 (*DUSP4*), MKP-3 (*DUSP6*), and MKP-7 (*DUSP16*), which contain signals for nuclear localization. Interestingly, some non-catalytic domains/motifs play multiple roles in PTP regulation. For example, SHP1 (*PTPN6*) and SHP2 (*PTPN11*) each contain two tandem Src-homology 2 (SH2) domains for targeting to specific tyrosine-phosphorylated proteins upon cell stimulation [37]. However, in unstimulated cells, one of the two SH2 domains folds upon the PTP domain and auto-inhibits SHP1/2 activity by blocking access to the active site [38, 39]. RPTPs have extracellular domains such as immunoglobulin, fibronectin, MAM, and carbonic anhydrase domains that are utilized to interact with the extracellular milieu in a receptor-like fashion. Many RPTPs also contain a second PTP domain that lacks phosphatase activity due to a single or multiple mutations of catalytically essential amino acids. These C-terminal pseudo-PTP domains presumably function as regulatory [40] or substrate-binding [41, 42] domains. Intriguingly, the multidomain structure of PTPs is in stark contrast to the architecture of protein serine/threonine phosphatases (PSPs). The PSP superfamily constitutes only ten apoenzymes, consisting of a small catalytic subunit that binds regulatory or targeting subunits encoded by separate genes [43]. Together, these subunits form a large number of distinct holoenzymes with different biological functions. Presumably, this combinatorial principle can generate more diversity and flexibility among the PSPs, but may lack some of the specificity and tight regulation possible with single-chain multidomain PTPs [9]. In this context it is noteworthy that several members of the atypical DSP subgroup of PTPs consist of only a very small catalytic core structure and do not contain any other domains or motifs [9]. It is possible that these enzymes participate in multisubunit complexes, similarly to PSPs, but no examples have yet been found. To summarize, most PTPs have a modular arrangement of individual extracatalytic domains and motifs that provide substrate specificity, dictate subcellular localization, furnish receptor-like functions, and even alter phosphatase activity.

5 Structure of the PTP Domain

Significant progress has been made in the past two decades toward understanding PTP specificity on a molecular level. Numerous atomic resolution structures of human PTP catalytic domains are now available, with representatives from every human PTP family and subfamily (Fig. 2). PTP1B was the first PTP whose structure was determined to high resolution [44]. With >100 structures solved to date, PTP1B often serves as a reference for PTP structures, the overall fold of which is extremely well conserved, consisting of a central twisted β-sheet surrounded by an arrangement of α-helices (Fig. 3). The active site architecture is formed by several loop regions: (1) the phosphate-binding loop (P-loop), which contains the PTP signature motif with the catalytic cysteine and the invariant arginine; (2) the WPD-loop, which contains the catalytic acid/base aspartate; (3) the Q-loop, which contains a conserved glutamine that coordinates the water molecule necessary for hydrolysis; (4) the pTyr-recognition loop (pTyr-loop), which contains a conserved motif that provides specificity for pTyr in classical PTPs; and (5) the E-loop, which contains multiple conserved residues that appear to coordinate the dynamics of the WPD-loop. Interestingly, PTP

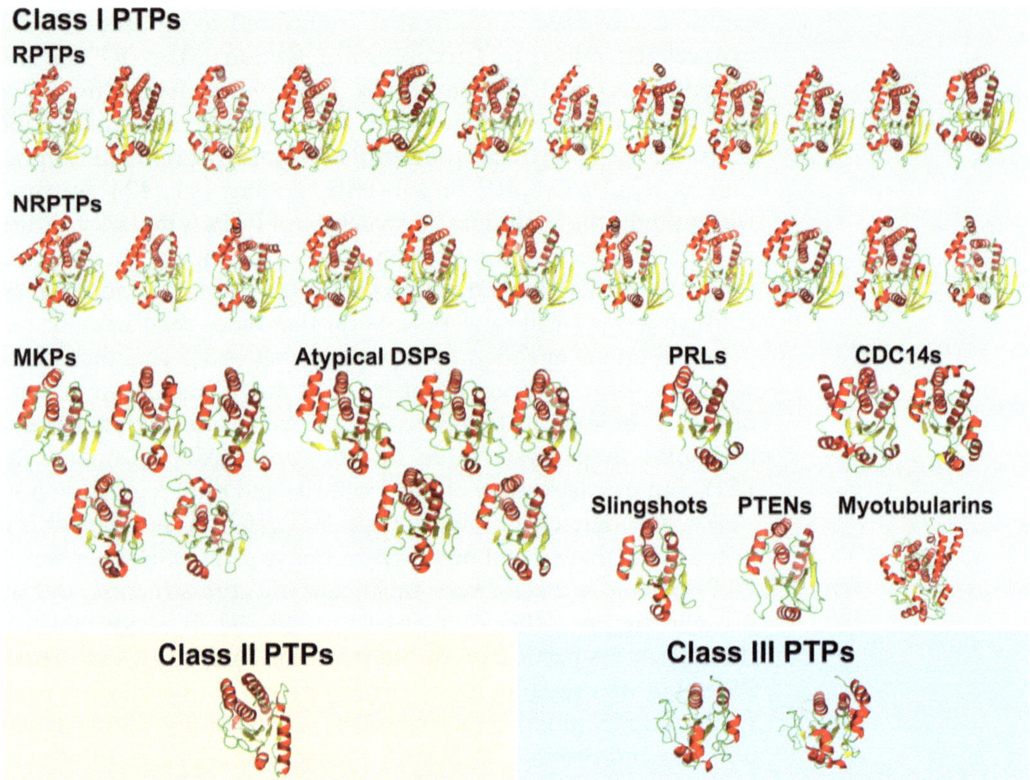

Fig. 2 Structures of human PTP catalytic domains. Cartoon representation of human PTP catalytic domain structures, divided by family/subfamily, and colored by secondary structure (α-helices in *red*, β-strands in *yellow*, loops in *green*)

function has different structural requirements among the various subfamilies. In fact, the P-loop is the only active site feature that is present in all PTPs. For instance, the WPD-loop with the catalytic aspartate, although highly conserved, is not found in the myotubularins and CDC25s, and apparently is not necessary for PTP activity in these enzymes. The pTyr-recognition loop is only present in the classical, pTyr-specific PTPs, which use this loop to form a deep catalytic cleft in order to discriminate between pTyr and the shorter pSer/pThr residues. In contrast, DSPs, which do not possess an equivalent loop, have a much shallower active site, allowing pSer/pThr to fully penetrate the catalytic center.

5.1 The P-Loop

The base of the active site of all PTPs is formed by the P-loop, which contains the conserved PTP signature motif $\mathbf{C}(X)_5\mathbf{R}$. The backbone nitrogen atoms of P-loop residues project into the catalytic pocket, creating a positively charged microenvironment that is further enhanced by the guanidinium group of the conserved P-loop arginine. The positively charged environment of the catalytic pocket has two functions: (1) to provide exceptional affinity for tetrahedral oxyanions such as the phosphate group, and (2) to lower the pKa of the

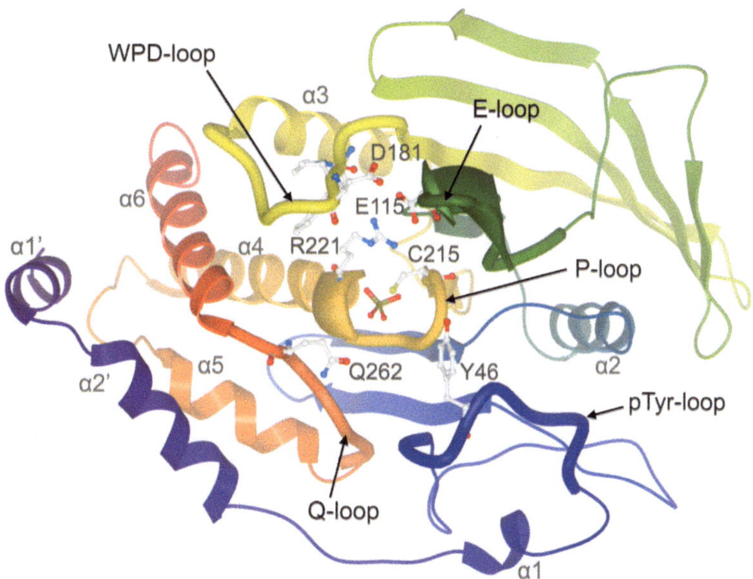

Fig. 3 Ribbon representation of the classical class I PTP catalytic domain, colored by NtoC, with tungstate (shown in ball-and-stick representation) bound into the active site (PTP1B; PDB ID: 2HNQ). Conserved residues important for catalysis are highlighted in ball-and-stick representation: catalytic cysteine (C215) and invariant arginine (R221) of the P-loop; WPD-loop residues, including the catalytic acid/base aspartate (D181); conserved glutamine (Q262) of the Q-loop; tyrosine (Y46) of the pTyr-recognition loop (pTyr-loop); conserved glutamate (E115) of the E-loop

catalytic cysteine by stabilizing the thiolate anion (Cys-S^-). A deprotonated thiol group of the catalytic cysteine is crucial for its role as nucleophile in the catalytic mechanism (*see* Fig. 1). While cysteine residues within proteins usually have pKa values of ~8.5, which would not allow for effective catalysis due to the low nucleophilicity under physiological conditions, the unique environment of the active site confers an unusually low pKa (between 4.5 and 5.5), permitting the enzymatic reaction to be efficient at physiological pH (or even at the pH optimum of 5–6) [45]. On the other hand, due to its low pKa, the catalytic cysteine is also highly susceptible to oxidation [46], nitrosylation [47], and sulfhydration [48], resulting in reversible or irreversible modifications that abrogate its nucleophilic function and thereby inactivate enzyme activity.

5.2 The WPD-Loop

The WPD-loop is present in almost all PTPs, apart from a few exceptions, which include the myotubularins and the CDC25s. This loop is located approximately 30–40 residues upstream of the PTP signature motif in the primary structures of both pTyr-specific PTPs and DSPs. The single class II PTP LMPTP is the exception, with the P-loop located near the N-terminus, and the WPD-loop ~120 amino acids downstream from there. The WPD-loop is so named because in classical PTPs it contains the highly conserved

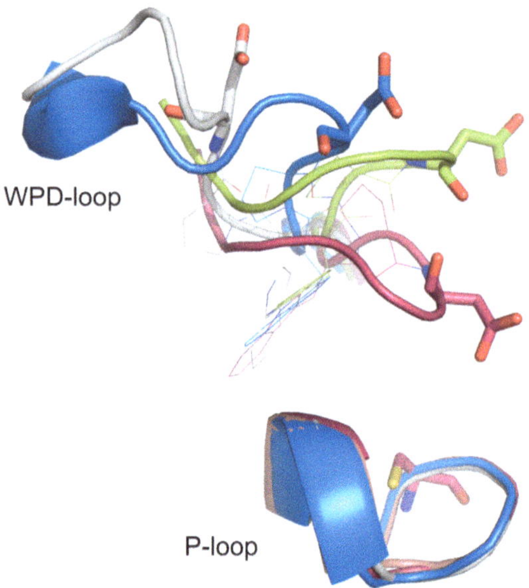

Fig. 4 WPD-loop conformations. PTP1B closed state (magenta; PDB ID: 1SUG), PTP1B open state (*lime*; PDB ID: 2HNP), STEP atypical open state (*blue*, PDB ID: 2BV5), and LYP atypical open state (*white*; PDB ID: 2P6X). Catalytic cysteine and catalytic aspartate residues shown in stick representation, conserved tryptophan and proline shown in line representation

tryptophan–proline–aspartate (WPD) motif. Interestingly, both the tryptophan and proline residues (100 and 97 % conserved in the pTyr-specific PTPs) are not present in the corresponding loops of the DSPs and LMPTP. This is somewhat surprising, as in classical PTPs the tryptophan has been identified as an essential hinge residue important for loop flexibility [49, 50]. On the other hand, the aspartate residue, which functions as the general acid/base during catalysis, is conserved in all PTP subfamilies. The WPD-loop acts as a flexible gate to the active site and has been observed in the "closed" (active) conformation, as well as a variety of "open" (inactive) conformations (Fig. 4) [33]. In the absence of substrate, the WPD-loop likely fluctuates between the open and closed conformations [51]. However, substrate binding can only occur when the loop is in the open state. Upon substrate binding, the WPD-loop closes about the active site so that the catalytic aspartate is positioned to participate in catalysis. Compromised mobility of the WPD-loop results in substantially lower catalytic activity, as shown by studies with mutated tryptophan hinge residue [49, 50], or with allosteric inhibitors that decrease WPD-loop flexibility [34, 35]. PTPs with a different amino acid at the position of the catalytic aspartate, including the RPTPs IA-2 (*PTPRN*; D→A) [52], PTPS31 (*PTPRQ*; D→E) [53], and PTPλ (*PTPRU*; D→E) [54], and the NRPTP HD-PTP (*PTPN23*; D→E) [55], are often inactive or have a very low activity. Exceptions are PTPD1 (*PTPN21*; D→E), which has been shown to effectively dephosphorylate Src

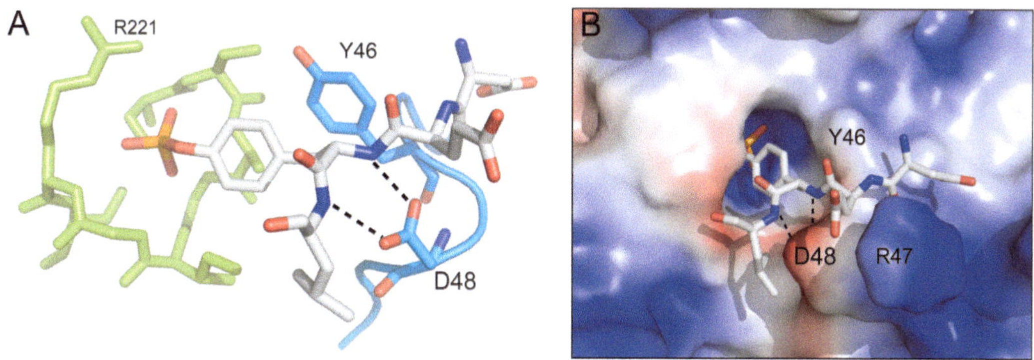

Fig. 5 The pTyr-loop. (**a**) pTyr-loop (*blue*) relative to the P-loop (*lime*), with bound pTyr-peptide (*white*) (PTP1B; PDB ID: 1PTT). The conserved tyrosine (Y46 in PTP1B) defines the depth of the catalytic pocket and facilitates pTyr binding through aromatic π–π interactions. The conserved aspartate or asparagine (D48 in PTP1B) stabilizes substrate binding through bipartite hydrogen bonding interaction with backbone nitrogen atoms of the substrate peptide. **(b)** Same complex as in (**a**) but with PTP1B rendered in surface representation (*blue*, most positive; *red*, most negative). (In addition to the interactions listed in (**a**), R47 of the PTP1B pTyr-loop is labeled, the side chain of which interacts with the glutamate in the −1 position of the substrate peptide, highlighting the fact that PTP1B favors substrates with acidic residues at this position)

at Tyr527 [56], and PTPS31 (*PTPRQ*; D→E), which can dephosphorylate phosphoinositite [53]. A different amino acid at the position of the conserved proline in the WPD motif is only found in IA-2β (*PTPRN2*, P→Y), an RPTP that has no PTP activity [57].

5.3 The pTyr-Recognition Loop (pTyr-Loop)

The pTyr-loop, which sometimes also is referred to as the substrate-binding loop and is present in all classical PTPs, defines the depth of the catalytic pocket, thereby creating selectivity for pTyr over the shorter pSer/pThr (Fig. 5). The loop contains the pTyr-recognition motif KNRY, with the 84 % conserved tyrosine positioned such that its side chain acts a causeway to the catalytic pocket, defining its considerable depth of ~9 Å, and ensuring that only pTyr can reach the bottom of the pocket. In addition to defining the depth of the phosphate-binding pocket, the conserved tyrosine also forms aromatic π–π interactions with the substrate pTyr residue, thereby facilitating substrate binding to the active site [58]. The guanidinium group of the 100 % conserved arginine of the KNRY motif interacts with nearby backbone oxygen atoms, thereby stabilizing the loop conformation. In addition, the surface-exposed arginine side chain may also act as substrate binding site. There are six PTPs in which the tyrosine of the KNRY motif is replaced by another amino acid [20]. Interestingly, five of these six phosphatases, IA-2 (*PTPRN*; Y->H), IA-2β (*PTPRN2*, Y->S), PTPS31 (*PTPRQ*; Y->F), PTPλ (*PTPRU*; Y->Q), and PTPD1 (*PTPN21*; Y->F), also have altered WPD motifs. Thus, the function of these PTPs may have changed during the course of evolution in a concerted fashion, i.e., through mutations in both the WPD- and pTyr-loops. Two residues C-terminal of the conserved

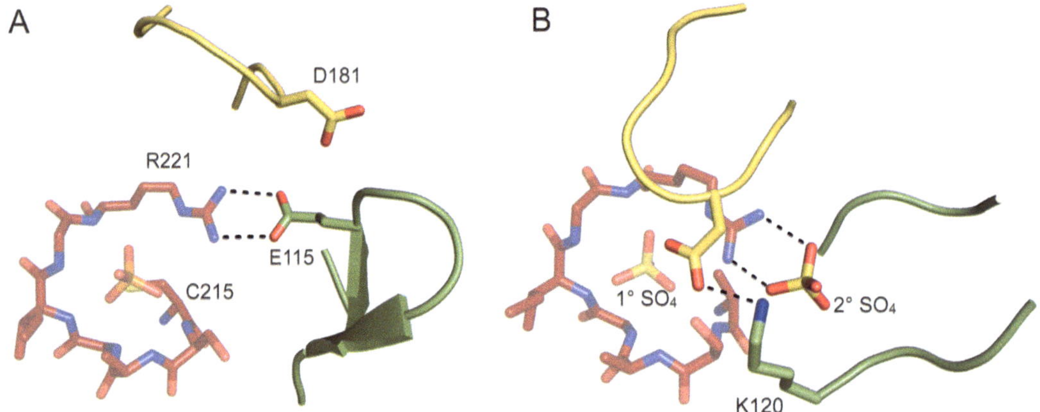

Fig. 6 The E-loop. **(a)** The E-loop (*green*) relative to the P-loop (*red*) and WPD-loop (*yellow*) (PTP1B; PDB ID: 2HNQ). In this structure, the WPD-loop is in the open conformation and the E-loop forms a tight β-hairpin, with the E-loop glutamate (E115, shown in stick representation) neutralizing the charge of the conserved P-loop arginine (R221). **(b)** The E-loop (*green*) relative to the P-loop (*red*) and WPD-loop (*yellow*) (PTP1B; PDB ID: 2B4S). In this structure, the WPD-loop is in the closed conformation and the E-loop does not form a β-hairpin and is partially disordered. The E-loop lysine (K120) and catalytic aspartate of the WPD-loop (both shown in stick representation) form a hydrogen bond, and a sulfate molecule is bound at the active site (1° SO_4) and at a secondary binding site (2° SO_4), which neutralizes the charge of the P-loop arginine

tyrosine of the pTyr-loop KNRY motif is an aspartate or asparagine residue (84 % conserved). This residue forms a bipartite hydrogen bond interaction with the backbone amide nitrogen atoms of the substrate pTyr and adjacent residue, providing additional stabilization to the substrate pTyr in the active site (Fig. 5).

5.4 The E-Loop

In classical PTPs, the E-loop contains a 100 % conserved glutamate residue and an 89 % conserved lysine residue. In >85 % of known classical PTP structures, the E-loop forms a tight β-hairpin, with the conserved glutamate forming a bipartite hydrogen bonding interaction with the side chain of the invariant P-loop arginine, presumably stabilizing the guanidinium group in a position that favors phosphate binding to the P-loop (Fig. 6a). Although the E-loop has also been observed in multiple unique conformations [59–61] or to be completely disordered [32, 62], the glutamate-arginine interaction is found in nearly all of the reported classical PTP structures. The conserved lysine often is observed to hydrogen bond with the catalytic aspartate of the WPD-loop when this loop is in the closed conformation, an interaction that likely stabilizes the WPD-loop in its active, substrate-bound conformation (Fig. 6b). Indeed, mutating this residue to alanine results in reduced catalytic efficiency in HePTP [63]. Interestingly, there appears to be no equivalent function of the E-loop outside of the classical PTP family. Although all PTP subfamilies have a somewhat similarly positioned loop, the conserved glutamate and lysine residues are only present in the pTyr-specific PTPs. No other conserved residues are found at the corresponding positions in DSPs.

5.5 The Q-Loop

The Q-loop is present in all classical PTPs and is so named because it contains a 97 % conserved glutamine residue. (A similarly positioned loop is also present in DSPs, but does not contain a glutamine or equivalently conserved residue.) Kinetic studies [64–66] in conjunction with structural analyses [27–29, 44, 58] revealed that the glutamine side chain, which reaches toward the catalytic center, positions the water molecule that is used for the phosphoester hydrolysis. Interestingly, LMPTP (which does not contain an equivalent loop) or DSPs (which do not have the glutamine) catalyze phosphoryl transfer also to alcohols in addition to water [65]. In contrast, in the classical PTPs, the Q-loop is responsible for maintaining strict hydrolytic activity. Mutations of the glutamine to residues that cannot hydrogen bond with the nucleophilic water result in less restrictive enzymes and confer phosphotransferase activity [65]. Thus, the Q-loop prevents classical PTPs from acting as kinase-like phosphotransferases, which otherwise may phosphorylate indiscriminately. In addition to this function, it was proposed that the water molecule positioned by the Q-loop glutamine might also be important for closure of the WPD-loop [33].

5.6 Structural Comparison Between Class I, Class II, and Class III PTPs

Based on their conserved P-loop region, including the signature motif $\mathbf{C}(X)_5\mathbf{R}$, all PTPs have evolved from a common ancestor gene. While the more ancient class II PTPs are structurally related to bacterial arsenate reductases, the class III PTPs likely stem from bacterial rhodaneses. In fact, both arsenate reductases and rhodaneses also contain the same $\mathbf{C}(X)_5\mathbf{R}$ P-loop structural element, suggesting that these three enzyme classes may have evolved from an ancestral oxyanion-binding protein. The structure of the class II PTP domain shares key similarities with that of the classical class I PTP domain. Like that of the class I PTPs, the class II PTP domain consists of a central twisted β-sheet surrounded by an arrangement of α-helices, with the base of the active site formed by residues of the P-loop. Also like the class I PTPs, the class II PTP domain contains a WPD-loop with an aspartate residue that functions as a catalytic acid/base. Additionally, although not directly corresponding to the pTyr-loop in classical class I PTPs, the class II PTPs contain a similarly positioned loop that defines the depth of the PTP active site [67]. Only a single member, LMPTP (*ACP1*), represents the class II PTP family in humans [23]. The primary transcript of LMPTP has been shown to be alternatively spliced, resulting in excision of either exon 3 (LMPTP isoform A) or exon 4 (LMPTP isoform B) [68], or both exons 3 and 4 (LMPTP isoform C) [69]. The latter splice variant encodes an approximately 15 kDa protein with no detectable in vitro phosphatase activity, whereas LMPTP splice variants lacking either exon 3 or exon 4 alone encode proteins with pTyr-specific phosphatase activity [70]. Exons 3 and 4 encode amino acids 39–75, which includes the loop region that corresponds to the pTyr-loop in the classical class I PTPs, as well as the substrate-binding region that is responsible for the differing substrate specificities of the two isoenzymes [68].

The major difference between the class I and class II PTPs is the sequential order of the typical PTP structural elements. Overall, the positioning of α-helices, β-sheets, and active site loops is reminiscent of class I PTPs. However, crystal structures of LMPTP cannot be superimposed onto those of classical PTPs because of the quite different sequential alignment of these structural elements. For instance, the P-loop containing the PTP signature sequence is located at the extreme N-terminus of LMPTP. In addition, the molecular weight of catalytically active LMPTP is only approximately 18 kDa; in contrast, the classical class I PTP catalytic domains exceed 30 kDa in size. Nevertheless, LMPTP-A and LMPTP-B are both highly active pTyr-specific phosphatases with important and nonredundant functions in human cell physiology [23].

The class III PTP family is represented in humans by three members, CDC25A, CDC25B, and CDC25C, each of which have different splice variants [71]. The CDC25 proteins are typified by their N-terminal and C-terminal regions. The N-terminal region is highly divergent in sequence and contains multiple sites for post-translational modifications, which regulate expression levels [72, 73], protein–protein interactions [74–76], and/or catalytic activity [77, 78]. CDC25 homology with other PTPs is limited entirely to the C-terminal region, which, unsurprisingly, contains the PTP signature sequence. Crystal structures have been determined of the PTP catalytic domain from CDC25A [79] and CDC25B [80]. The class III PTP catalytic domain fold is most similar to that of the rhodanese sulfur transport protein from mitochondria and some bacteria [81]. Similar to the class II PTPs, the class III PTP catalytic domain is considerably smaller than that of the classical class I PTPs, which is attributed to a narrower central β-sheet surrounded by fewer, shorter α-helices. Importantly, and in contrast to the classical class I PTP catalytic domain, the class III PTP catalytic domain does *not* contain equivalents of the pTyr-, WPD-, or Q-loops. As such, the comparatively shallow CDC25 active site permits access of both pTyr- and pThr-containing substrates to the catalytic cysteine.

6 PTP Substrate Recognition and Specificity

Despite the highly conserved structure of the PTP catalytic domain, PTPs have distinct substrate preferences from one another. For example, PTP1B and TCPTP preferentially dephosphorylate receptor tyrosine kinases and related adaptor molecules, whereas the KIM-family PTPs HePTP, STEP, and PTP-SL dephosphorylate specific MAP kinases. While the presence of distinct non-catalytic domains/motifs facilitate specific localization or binding to substrate proteins, PTP substrate specificity is also dictated by differences within the PTP catalytic domain itself.

6.1 Active Site Substrate Specificity Determinants

The most effective active site determinant is the pTyr-loop in the classical PTPs, which dictates strict specificity for pTyr over pSer/pThr in this PTP subfamily (*see* Subheading 5.3). In addition to excluding substrates other than pTyr from reaching the catalytic residues at the bottom of the catalytic pocket, the pTyr-loop also controls specificity among the classical PTPs. In 84 % of classical PTPs, the pTyr-loop contains an aspartate or asparagine residue at the KNRY+2-position. The side chain of this residue forms a bipartite hydrogen bond interaction with backbone residues of the substrate peptide, and thereby facilitates substrate binding at the active site (Fig. 5) [82]. However, in a small subset of PTPs, the aspartate/asparagine is replaced with an alanine, proline, or threonine residue, which are incapable of forming the equivalent interaction, and thereby restrain substrate binding at the active site in these PTPs. For example, the KIM-family PTPs HePTP, STEP, and PTP-SL are unique in that they have a threonine at the corresponding position. Studies with HePTP have shown that the threonine (Thr106 in HePTP) serves as a negative determinant, which reduces phosphatase activity towards pTyr-containing substrates other than MAP kinases [83]. Mutation of Thr106 in HePTP to aspartate or asparagine results in a 10- to 70-fold increase in catalytic efficiency towards several pTyr-peptides, including a peptide corresponding to the activation loop of its direct substrate ERK2. This is in agreement with earlier studies that indicated the importance of the common aspartate/asparagine at this position for the catalytic efficiency in PTP1B [82]. In contrast, dephosphorylation of full-length ERK2 by HePTP is not affected by the Thr106 mutations. In fact, catalytic efficiency of wild-type HePTP and aspartate/asparagine mutants is >300-times greater towards full-length ERK2 than towards the ERK2 peptide, owing to the presence of the ERK2-binding KIM in the N-terminal region of HePTP [83]. Thus, the threonine in the pTyr-loop in HePTP (and probably also in STEP and PTP-SL) acts as a negative determinant that restrains binding of pTyr-containing proteins and, in combination with a non-catalytic motif (the KIM and also the kinase specificity sequence (KIS) [84, 85]), provides specificity for the MAP kinases.

Altered substrate specificity and increased phosphatase activity upon substrate binding has also been reported for the MAP kinase phosphatase-3 (MKP-3), a DSP that dephosphorylates the activation loop pThr and pTyr residues specifically in ERK1/2. When tested against the corresponding ERK2 peptide (DHTGFLpTEpYVATR), MKP-3 catalytic efficiency was very low ($k_{cat}/K_m = 5.0\ M^{-1}\ s^{-1}$), and only pTyr was dephosphorylated [86]. In contrast, MKP3-catalyzed hydrolysis of the intact ERK2 protein was 10^6-fold more efficient ($k_{cat}/K_m = 3.8 \times 10^6\ M^{-1}\ s^{-1}$), and both pTyr and pThr were dephosphorylated [87]. Structural and biochemical studies suggested that binding of ERK to MKP-3 causes a significant change of the active site loop conformations, resulting in much higher catalytic activity and also a

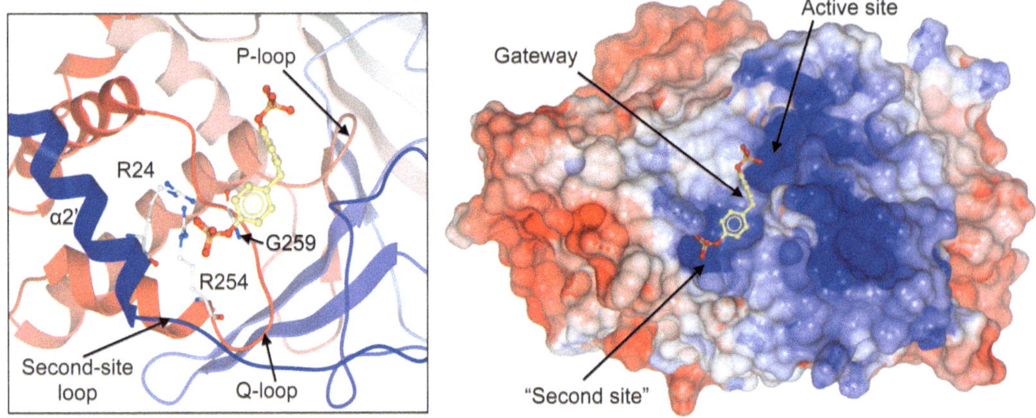

Fig. 7 Binding of a small-molecule phosphate (methylenebis(4,1-phenylene) bis(dihydrogen phosphate); shown in stick representation) to the "second site" in PTP1B (PDB ID: 1AAX; shown as ribbon diagram (*left panel*) and surface representation (*right panel*; *blue*, most positive; *red*, most negative). Second-site residues Arg24 and Arg 254, as well as gateway residue Gly259 are shown in stick representation (*left panel*)

change in substrate phospho-amino acid recognition (reviewed in ref. 30). In light of these results, it is not surprising that numerous published specificity studies utilizing combinatorial peptide libraries fail to identify true substrate sequence specificity among the PTPs.

Finally, sequence motifs within the P-loop also appear to govern substrate specificity. For instance, all 15 myotubularins share a Trp-Asp-Arg (WDR) sequence as part of the PTP signature motif (C(X)$_3$**WDR**). This sequence seems to confer specificity for their substrates phosphatidylinositol-3-phosphate and phosphatidylinositol-(3,5)-bisphosphate. Interestingly, because the myotubularins do not contain a WPD-loop, the aspartate residue of the WDR motif may function as the general acid/base during catalysis; mutating the corresponding Asp422 in MTMR2 to alanine renders the enzyme inactive [88].

6.2 Gateway Residues, Secondary Substrate-Binding Pocket, and Second-Site Loop

Based on a second-site binding pocket first identified in PTP1B [89], the combination of structural information, sequence alignment, and Cα-regiovariation score analysis led to the identification of non-conserved but similarly located putative binding pockets in other PTPs [20]. In PTP1B, this "second site" is important for substrate recognition [90], accommodating a second pTyr residue at the substrate +1 position. Commonly, the second site is defined by residues of helix α2′ and the loop connecting helix α2′ and helix α1 (Fig. 7). Residues of this loop, which is also termed the "second-site loop" [33], are not conserved among the PTPs and, together with loop length and conformation, determine the diverse shape and nature of the PTP second site. In PTP1B, Arg24 and Arg254 are responsible for the positively charged nature of the second-site pocket, facilitating strong binding interactions with the second pTyr residue [89]. Access to the second site is controlled by a so-called gateway residue (Gly259 in PTP1B), which is located in the Q-loop and is a key determinant in substrate recognition and catalysis [91]. The small size of Gly259 in PTP1B allows unhindered access to the second site, whereas in CD45, for instance, a leucine residue (Leu869) blocks the narrow cleft and impedes access. Based on the nature of and access to the second site, Barr and colleagues have proposed to group PTPs into five different categories [33], a classification that should be very useful for the future development of specific PTP inhibitors.

6.3 Tools to Determine PTP Substrate Recognition by X-ray Crystallography

In order to study the complex of a PTP with its phosphorylated substrate using X-ray crystallography, the PTP must be inactivated so that a stable PTP:substrate complex is formed. Inactivation of PTPs is achieved by mutating the residues directly involved in catalysis, including the catalytic cysteine of the P-loop, the catalytic aspartate of the WPD-loop, and the conserved glutamine of the Q-loop. Collectively, these mutants are called substrate-trapping mutants (STMs) [92]. In some cases, effective STMs have been formed by mutating the catalytic cysteine, either to serine [58, 93, 94] or alanine [90, 95]. In other cases, effective STMs have been formed by mutating the catalytic aspartate alone [92, 96] or in combination with the conserved glutamine [64]. STMs have been used not only to crystallize PTP:substrate complexes [58, 61, 90, 93, 94], but also to identify novel PTP targets in intact cells [92, 97, 98]. Thus, STMs represent invaluable tools for investigating PTP substrate recognition both in vitro and in vivo.

Apo PTP crystals can often be readily obtained by including tetrahedral oxyanion, such as phosphate, sulfate or tungstate, in the crystallization drop. This almost always results in binding of the oxyanion at the PTP active site (i.e., P-loop). For this reason, in order to obtain crystals of PTP:substrate complexes, crystallization precipitants containing tetrahedral oxyanions should be *avoided*, as they will compete with substrates for binding the PTP active site. Still, apo PTP crystals containing oxyanion at the active site can actually be depleted of active site-bound oxyanion [63, 99, 100], after which alternative ligands can populate the active site. In one example, crystals of the classical PTP HePTP were obtained in the presence of sulfate and contained sulfate bound at the active site with the WPD-loop in the closed conformation [63]. Transfer of these crystals from the original crystallization drop into drops containing decreasing concentrations of sulfate, as well as increasing concentrations of the oxyanion tartrate, resulted in gradual depletion of active site-bound sulfate (PDB ID: 3O4T), opening of the WPD-loop, and subsequent repopulation of the active site by tartrate (PDB ID: 3O4U). Importantly, in this HePTP crystal form, the WPD-loop faced a solvent channel, which was distal to crystallographic symmetry mates, a critical feature for opening of the WPD-loop without concomitant disintegration of the crystal.

It is expected that, for other PTPs, the WPD-loop must also be distal from symmetry mates in order to allow for movement of the WPD-loop and depletion/repopulation of the active site using this methodology.

7 PTPs in Human Disease

Disturbance of the dynamic balance between protein tyrosine phosphorylation and dephosphorylation of signaling molecules is known to be crucial for the development of many human diseases, ranging from cancer to cardiovascular, immunological, infectious, neurological, and metabolic diseases (reviewed in refs. 13, 19, 21, 101–104). In cancer, this is best exemplified by the common loss of PTEN in many malignancies including breast and prostate cancer [105]. Similarly, SHP1 (*PTPN6*) is frequently lost in myelodysplastic syndrome (MDS) [106] and lymphomas [107]. However, PTPs not only act as tumor suppressors but also as "positive" components of signaling pathways and cell functions, illustrated by a growing number of examples of overexpressed or hyperactive PTPs in cancer cells [103]. For instance, SHP2 (*PTPN11*) is a well-known oncogene with various gain-of-function mutants occurring in several forms of leukemia [108, 109]; overexpression of PRL3 (*PTP4A3*) and MKP1 (*DUSP1*) has been found in metastatic colon cancer [110] and prostate cancer [111], respectively; CDC25, which is a rate-limiting enzyme for cyclin-dependent kinase (CDK)-dependent transition from G1 to S phase and G2 to M phase during cell cycle, is overexpressed in a number of cancers [112]; or PTP1B (*PTPN1*), first discovered as a crucial negative regulator of insulin and leptin signaling [113], was recently found to be a positive regulator of ErbB2 (HER2/neu) induced signals that trigger breast tumorigenesis and metastasis [114–116]. The lymphoid tyrosine phosphatase LYP (*PTPN22*), which is best known for its causative role in numerous autoimmune diseases including type 1 diabetes (T1D) [117], rheumatoid arthritis (RA) [118], systemic lupus erythematosus (SLE) [119], and others (reviewed in refs. 17, 102, 120), was recently found overexpressed in chronic lymphocytic leukemia (CLL) [121], a common malignancy of autoreactive B cells. Interestingly, autoimmunity is related to a single nucleotide polymorphism (SNP) in LYP, whereas inhibition of antigen-induced apoptosis and activation of anti-apoptotic pathways in CLL occurs independently of the SNP. HePTP (*PTPN7*), one of the three KIM-family PTPs and a critical negative regulator of the MAP kinases ERK1/2 and p38 in hematopoietic cells [122], is upregulated in acute myeloid leukemia (AML) and in T cell acute lymphoblastic leukemia (T-ALL) [123, 124]. Additionally, the HePTP gene is often duplicated in bone marrow cells from patients with MDS [125, 126]. Another KIM-family

PTP, the brain-specific STEP (*PTPN5*), which modulates key signaling molecules involved in synaptic plasticity and neuronal function, has been implicated in a number of neuropsychiatric disorders, including Alzheimer's disease, schizophrenia, fragile X syndrome, epileptogenesis, alcohol-induced memory loss, Huntington's disease, drug abuse, stroke/ischemia, and inflammatory pain (reviewed in ref. 104). Genetic polymorphisms of the ubiquitously expressed LMPTP (*ACP1*) is linked to several common diseases, including allergy, asthma, obesity, myocardial hypertrophy, and Alzheimer's disease [23, 127]. Loss of the DSP laforin (*EPM2A*) causes Lafora's epilepsy [128], whereas loss of myotubularin (*MTM1*) leads to an X-linked muscle dystrophy [129]. Mutations in myotubularin-related proteins 2 and 13 (MTMR2 and MTMR13) are associated with the inherited nerve myelination disease Charcot-Marie-Tooth syndrome type 4B [130, 131]. Loss of the hematopoietic-specific transmembrane PTP CD45 (*PTPRC*), which regulates Src kinases required for T and B cell activation, has been linked to severe combined immunodeficiency disease [132]. These are just a few examples of the rapidly growing number of human diseases associated with aberrant PTP function. Not surprisingly, this has begun to elicit growing interest in novel therapeutics that target specific PTPs. In the following we highlight three of the currently most interesting/validated PTP drug targets.

7.1 PTP1B (PTPN1)

In 1999 Elchebly and colleagues reported increased insulin sensitivity and obesity resistance in mice lacking the PTP1B gene [133]. It was this paper that truly ignited the quest for PTP inhibitors. *PTPN1*$^{-/-}$ mice showed increased phosphorylation of the insulin receptor and were resistant to weight gain and insulin insensitivity when fed a high-fat diet. The data suggested that inhibition of PTP1B would alleviate insulin resistance in type 2 diabetes and would improve the effects of insulin on both glucose balance and fatty acid metabolism. Laboratories in both industry and academia heavily invested in the development of PTP1B inhibitors as potential therapeutics in the treatment of type 2 diabetes and obesity. As a result, a large number of PTP1B inhibitors have been developed during the last decade (reviewed in refs. 134–137). However, only few compounds progressed into clinical trials, and so far none have advanced beyond phase II. The reasons for failure and common challenges in PTP inhibitor design are discussed in Subheading 8.

More recently, PTP1B was found as an activator of ErbB2 (HER2/neu)-induced signals that trigger breast tumorigenesis and metastasis [114–116]. Genetic or pharmacological deletion of PTP1B activity in transgenic mice containing an activating mutation in ErbB2 (NDL2 mice, ref. 138) resulted in significant mammary tumor latency and resistance to lung metastasis [114, 115], providing proof-of-concept and validating PTP1B as a novel drug target in breast cancer. The PTP1B gene is located at the chromosomal region

20q13 that is frequently amplified in breast and ovarian cancer and is associated with poor prognosis [139, 140]. PTP1B overexpression was found in 72 % of human breast tumors, correlating with ErbB2 overexpression [141], as well as in ErbB2-transformed human breast epithelial cells and tumors derived from such cells [142]. The underlying molecular mechanism of PTP1B function in tumorigenesis is not fully understood and is somewhat controversial. Studies conducted with breast cancer cell lines suggest that PTP1B function is required for c-Src activation by dephosphorylation of the inhibitory Tyr527, thereby positively affecting ErbB2 signaling [143, 144]. However, in animal studies, no change of Src phosphorylation and activity was observed with overactive ErbB2 receptors [114, 115]. Consistent with these findings, an earlier study demonstrated that Src kinase function is not necessary for ErbB2-mediated tumorigenesis [145]. Instead, loss of PTP1B function led to reduced Ras/MAP kinase pathway and PI3 kinase/AKT pathway activation [114, 115]. Indeed, PTP1B is required for the activation of the small GTPases Ras and Rac, which are typically associated with increased cell proliferation and motility [146, 147]. To bypass Src activation, an alternative mechanism was suggested, in which $p62^{Dok}$, an inhibitor of the Ras/MAPK pathway, is deactivated by PTP1B, and loss of PTP1B function leads to hyperphosphorylation of $p62^{Dok}$ with subsequent inactivation of Ras and its downstream effectors [147, 148].

ErbB2 is overexpressed in 20–30 % of early-stage breast cancers and is associated with poor prognosis [149, 150]. The current therapeutic approach in ErbB2 positive tumors is a combination of chemotherapy and a monoclonal antibody that selectively binds ErbB2 (trastuzumab, ref. 151). However, the development of trastuzumab resistances in the majority of patients has limited the use of the drug [152]. Nonetheless, the therapeutic efficacy of trastuzumab is strong evidence for the critical role of ErbB2 in human breast cancer and clearly justifies more targeted approaches downstream of the ErbB2 receptor, including PTP1B.

7.2 SHP2 (PTPN11)

Src homology 2 domain-containing protein tyrosine phosphatase 2 (SHP2) is a widely expressed tyrosine-specific PTP that appears to have a net positive role in cell activation in response to growth factors, cytokines, and hormones, regulating cell survival, growth, and differentiation [12, 153]. In addition, SHP2 modulates cell adhesion-induced signal transduction and plays a role in cell migration and motility [154]. Germ-line mutations in *PTPN11* were first observed in Noonan syndrome (NS) [155], an autosomal dominant disorder that is associated with craniofacial abnormalities, cardiac defects, short stature, and learning disabilities [156–158]. NS affects 1 in 1,000–2,500 live births [155, 159], 40–50 % of which carry mutations in SHP2. Similar germ-line mutations cause two related genetic disorders, Noonan-like disorder with

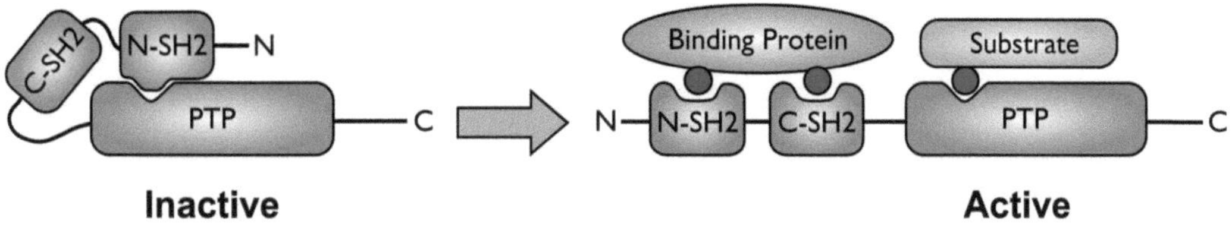

Fig. 8 Activation of SHP2 by binding of specific pTyr-containing proteins

multiple giant cell lesion syndrome and LEOPARD syndrome [160]. Somatic mutations in *PTPN11* occur in ~35 % of cases of juvenile myelomonocytic leukemia (JMML) [161, 162], a rare but aggressive myeloid neoplasm of childhood, clinically characterized by overproduction of monocytic cells that can infiltrate organs including spleen, liver, gastrointestinal tract, and lung. Somatic mutations in *PTPN11* have also been found in solid tumors [162] and other types of leukemia (reviewed in refs. 108, 153, 163, 164).

Under basal conditions, SHP2 activity is inhibited via intramolecular interactions between the N-terminal SH2 domain (N-SH2) and the PTP domain (Fig. 8) [39, 165]. Activation of SHP2 in response to extracellular stimuli involves binding of pTyr residues via the tandem SH2 domains, resulting in a conformational change that allows substrate access to the catalytic site. Most of the *PTPN11* mutations in NS and JMML (as well as other leukemias) affect amino acid residues at the interface between the N-SH2 and PTP domains and directly interfere with the autoinhibitory state of SHP2. The resulting gain-of-function (GOF) effect of the mutant SHP2 proteins promotes sustained activation of ERK MAP kinases in transfected cells and in animal models [166–171]. Hyperactivation of ERK most likely is the underlying cause of the developmental dysfunctions seen in NS patients. Indeed, mutations in other components of the Ras/ERK pathway, such as KRAS, RAF, and SOS1, have been identified in NS patients lacking *PTPN11* mutations [172–175]. Similarly, 75–85 % of JMML cases directly result from GOF-mutations of components of the Ras/ERK signaling cascade (NRAS, KRAS, NF1, SHP2). Interestingly, *PTPN11* mutations in JMML appear to be mutually exclusive of NRAS/KRAS2 or other genetic lesions [176]. Moreover, while *PTPN11* mutations observed at disease presentation were undetectable at remission or in control individuals, the same mutations could be detected at initial diagnosis and relapse [176]. These data suggest that *PTPN11* mutations represent events that directly contribute to leukemogenesis.

In hematopoietic cells, SHP2 appears to be involved in various signaling pathways (Fig. 9), with the positive regulatory role generally ascribed to activation of the ERK pathway upstream of Ras [177, 178]. This function could be mediated by inhibiting the recruitment of Ras-GAP or the DOK proteins, or more indirectly

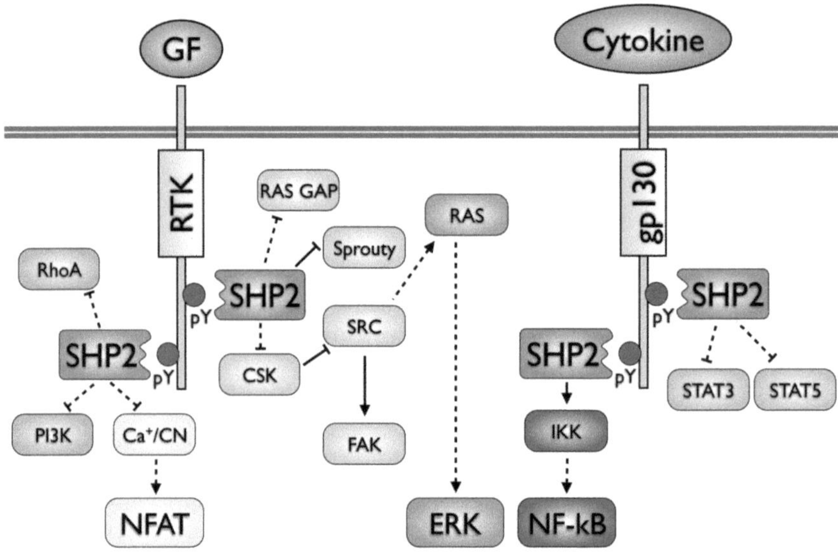

Fig. 9 SHP2 interactions and signaling pathways in hematopoietic cells. Direct interactions are indicated by *solid lines*, indirect interactions by *dashed lines* (adapted from ref. 158)

by inhibiting CSK recruitment to CBP/PAG and thus promoting Src family kinase activity [179]. One of the primary groups of cytosolic adapters which recruit SHP2 are the GAB proteins, which amplify PI3K signaling and likely place SHP2 in the proximity of appropriate targets [179]. Additional identified interaction partners include growth factor/cytokine receptors, SIRPα/SHPS-1, PZR, GRB2, FRS, IRS-1, p85, STAT5/3/1, and Sprouty proteins [180]. However, none of the putative substrates identified to date can fully account for the various signaling effects of SHP2 or its oncogenic potential.

JMML is unresponsive to radiation or existing chemotherapy. The current standard of care relies on allogeneic hematopoietic stem cell transplant, resulting in ~50 % survival rate [181]. However, relapse is the most frequent cause of treatment failure. Several therapeutic strategies targeting the Ras/RAF/ERK signaling pathway have been investigated, including inhibition of farnesyltransferase [181], inhibition of RAF (Sorafenib is currently in early phase testing), and inhibition of MEK [182]. In particular the disappointing results from the MEK inhibition trials suggest that aberrant signaling through ERK is not sufficient to drive leukemogenesis, at least not in JMML. Inhibiting constitutively active SHP2 seems to be more intriguing, given its causal role in disease development and the complexity of interaction networks. Several inhibitors of SHP2 have been reported [183–188]. However, most of these compounds suffer from either poor selectivity for SHP2, lack of efficacy in cell-based assays, or both. Future efforts need to address these shortcomings in order to generate specific SHP2 inhibitors with efficacy in vivo.

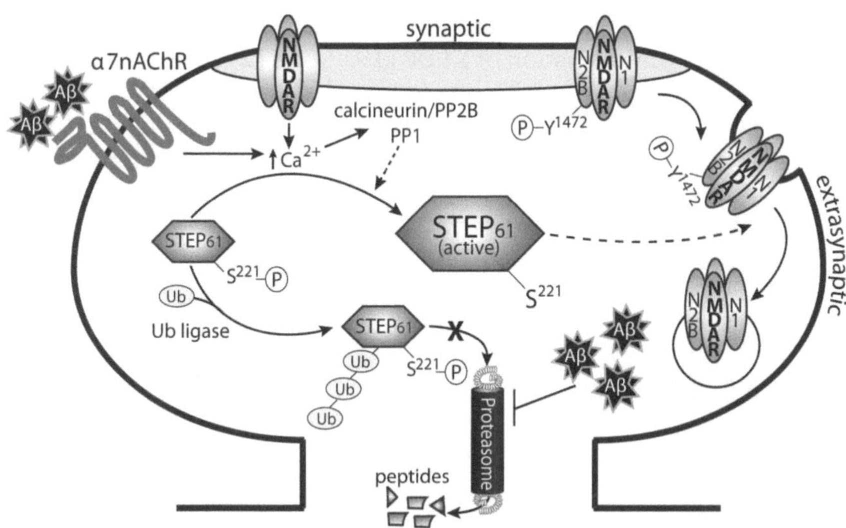

Fig. 10 Model of glutamate receptor internalization through Aβ-mediated activation of STEP. Aβ activates the α7 nicotinic receptor, leading to Ca^{2+} influx and activation of calcineurin [197]. Calcineurin subsequently dephosphorylates and activates STEP. Concomitantly, Aβ also elevates STEP protein levels through inhibition of the ubiquitin proteasome system [203]. STEP dephosphorylates a regulatory tyrosine residue in both the NR2B and GluR2 glutamate receptor subunits, leading to internalization of the receptors [199, 200]. As a result, synaptic function is disrupted (Figure from ref. 131, with permission)

7.3 STEP (PTPN5)

Striatal-enriched protein tyrosine phosphatase (STEP) is a brain-specific PTP that exists as two major alternatively spliced isoforms, $STEP_{61}$ and $STEP_{46}$, which appear to have distinct substrate specificities and functions (reviewed in ref. 131, 189). Membrane-associated $STEP_{61}$ is targeted to the postsynaptic density, extrasynaptic sites, and the endoplasmic reticulum; $STEP_{46}$ is primarily found in the cytoplasm. STEP function is regulated by several mechanisms, including phosphorylation, cleavage, dimerization, and ubiquitination. STEP recently gained attention when Lombroso and colleagues reported that genetic deletion of STEP attenuates the cognitive and cellular deficits observed in 6-months old 3×Tg-AD mice, a triple transgenic Alzheimer's disease (AD) model [190]. Previous studies indicated that STEP levels are elevated in the frontal cortex of AD patients and in three transgenic AD mouse models [191, 192]. STEP was shown to dephosphorylate glutamate receptor subunits, resulting in the internalization of NMDA and AMPA glutamate receptors that control synaptic plasticity and memory function [193, 194]. The classical hypothesis of AD proposes that accumulation of β-amyloid (Aβ) in the brain is responsible for disease-related pathology. Synaptic function is thought to be disrupted through Aβ-induced internalization of NMDA and AMPA receptors [195, 196]. The current data suggest an advanced model in which Aβ-induced internalization of glutamate receptors is mediated through STEP (Fig. 10). More precisely, STEP activity is increased due to (1)

Aβ-mediated activation of calcineurin, resulting in dephosphorylation and activation of STEP [191], and (2) Aβ-inhibition of the proteasome, resulting in decreased degradation of STEP [197]. As a result, STEP activity is increased in AD, leading to loss of NMDA and AMPA receptors from synaptosomal membranes. Collectively, these studies suggest that inhibition of STEP activity may be beneficial in AD treatment, and validate STEP as a novel drug target in AD (and perhaps other neuropsychiatric disorders, reviewed in ref. 131).

8 PTP Inhibitor Development

PTPs have been considered as therapeutic targets for more than a decade, and many promising compounds have been published (reviewed in refs. 19, 136, 198–201). However, past efforts to develop drugs targeting a specific PTP have been plagued by issues related to bioavailability and selectivity. This is due to the fact that the majority of PTP inhibitors carry a pTyr-mimicking group that provides most of the binding energy through interaction with the highly conserved active site residues. In early efforts, high-affinity peptide substrates were converted into competitive PTP inhibitors by changing the pTyr moiety to a nonhydrolyzable pTyr mimetic such as a (fluoro)phosphonate group [202]. However, despite the incredible potency of some of these compounds in vitro, little to no cell-based activity could be achieved, owing to the multicharged nature of the phosphonates and their poor drug-likeness according to Lipinski's rules [203], resulting in the lack of cell membrane permeability of these molecules. Subsequently, small molecules containing pTyr mimetics with only one negative charge (e.g., a carboxylic acid or tetrazole group) have been developed; these compounds are often not as potent as the phosphonates, but usually have better drug-like properties. Prodrug strategies to deliver PTP inhibitors more easily into cells have also been utilized [204, 205]. However, a significant challenge yet to be mastered is posed by the highly conserved catalytic core structure among the members of the PTP family. This makes it difficult to generate inhibitors with selectivity for a particular target, and to generate drugs without serious side effects. Although evidence in the form of many solved 3D structures has now made it clear that there are indeed differences in surface topology and charge distribution in the terrain that surrounds the catalytic pocket [33], examples of truly specific inhibitors are still missing. The fact that the majority of PTP inhibitors owe most of their binding energy to their interactions within the highly conserved phosphate-binding pocket (catalytic pocket) illustrates the challenge in designing potent and at the same time selective inhibitors. Taking advantage of unique amino acid residues and surface features peripheral of the catalytic pocket has resulted in inhibitors with somewhat increased selectivity; however, the trade off is usually expanded size and molecular weight of the inhibitor, making the compound less likely to enter cells or to be absorbed in the gut. Clearly, new approaches are needed to overcome this significant hurdle in the development of novel PTP-based treatment strategies in human disease.

8.1 Open State Binding

Recent co-crystal structures solved by Zhong-Yin Zhang's group and structure-based computational approaches by our laboratory have shown that small-molecule inhibitors can bind a PTP with its WPD-loop in the inactive, open conformation (*see* Subheading 5.2) and can stabilize the loop in this position [187, 206–208]. Although crystal structures with the WPD-loop in open conformation have long been available, it was previously thought that ligand binding to the P-loop always causes the WPD-loop to close, restricting the ligand binding mode to resemble the one of the natural substrate. As shown by recent crystal structures of LYP:inhibitor [206], PTP1B:inhibitor [207], and SHP2:inhibitor [187] complexes, this is not necessarily the case. These findings suggest that it is indeed possible to generate molecules that bind and inhibit PTPs in a way that locks the WPD-loop in its open, inactive state. Such a binding mode has profound implications on the design of inhibitors. Surface properties of the active site are much more diverse among PTPs with the WPD-loop in the open state conformation, suggesting that compounds could be generated that specifically bind the open state of the target PTP of choice. Moreover, the active site pocket in the open state not only is less conserved, but also significantly larger, allowing the design of inhibitors that bind entirely in the pocket. A comparison of PTP1B:inhibitor complexes in closed and open conformations is shown in Fig. 11. In the closed-state binding mode, only the head group of the inhibitor binds into the catalytic pocket (Fig. 11a, b). In the open-state binding mode, the entire inhibitor occupies the very large depression between P-loop and WPD-loop in PTP1B (Fig. 11c, d). A comparison of open state structures of different PTPs reveals significant differences in shape and electrostatic potential of the large pocket (Fig. 11f–j). Why "open state binding" has only been observed in a few cases so far is not entirely clear. It may be more difficult for a small molecule to stabilize a rather flexible loop in the open state. Or co-crystallization may be more successful with compounds that bind the closed state, which in turn may favor closed state binders in the selection of lead compounds. Regardless, given the existence of many crystal structures in the open conformation, structure-based methods such as virtual ligand screening may be employed in the search for hits that specifically bind the open state [208].

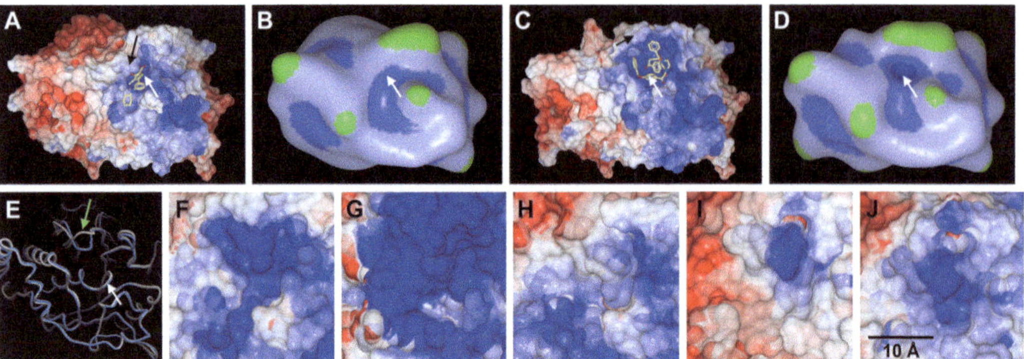

Fig. 11 (**a–d**) Comparison of ligand binding to PTP1B in closed state (**a/b**; PDB ID: 2QBS) and open state (**c/d**; PDB ID: 3EAX). The protein surface in (**a**) and (**c**) is colored by electrostatic potential as calculated and rendered in ICM (*blue*, most positive; *red*, most negative; the colors were capped at ±5 kcal/electron units). Macroshape representations as rendered in ICM (*blue*, deepest depression; *green*, largest protrusion) illustrate the differences between the active site pockets in closed (**b**) and open conformation (**d**). *White arrows* indicate the position of the P-loop, *black arrows* indicate the WPD-loop in closed state (**a**) and open state (**c**). **(e)** Comparison of PTP1B with WPD-loop in closed (*blue*; PDB ID 2QBS) and open (*grey*; PDB ID 3EAX) conformation. The *white arrow* indicates the P-loop, the *green arrow* indicates the WPD-loop. **(f–j)** Comparison of open state conformation in PTP1B (**f**; PDB ID 3EB1), TCPTP (**g**; PDB ID 1L8K), LYP (**h**; PDB ID 2P6X), LAR (**i**; PDB ID 1LAR), and RPTPγ (**j**; PDB ID 2H4V). The protein surface is colored by electrostatic potential as calculated and rendered in ICM (*blue*, most positive; *red*, most negative; the colors were capped at ±5 kcal/electron units)

8.2 Allosteric Inhibition

Another solution to the inherent selectivity problem could be to target allosteric sites at the PTP protein surface. So far allosteric inhibitors have only been reported for PTP1B [34]. In their paper, Wiesmann and colleagues determined the crystal structure of the complex between PTP1B and 3-(3,5-dibromo-4-hydroxybenzoyl)-2-ethyl-*N*-(4-sulfamoylphenyl) benzofuran-6-sulfonamide ("compound 2," Fig. 12a). They found that compound 2 binds to a site located ~20 Å away from the catalytic site at the "back" of the phosphatase. They showed that compound 2 acts as an allosteric inhibitor ($IC_{50} = 25$ μM) by blocking the mobility of the WPD-loop and thereby preventing the formation of the active, closed form of the enzyme. Interestingly, the residues that form the allosteric site in PTP1B are not conserved, and the surface properties of this site differ even between closely related PTPs (Fig. 12b). In agreement with this notion, compound 2, which also enhanced insulin signaling in cells, exhibited good selectivity for PTP1B. In a follow-up study, Kamerlin and colleagues used molecular dynamics simulations to study WPD-loop mobility in the presence of 3-(3,5-dibromo-4-hydroxybenzoyl)-2-ethyl-*N,N-dimethylbenzofuran* -6-sulfonamide ("compound 1" in ref. 34, $IC_{50} = 350$ μM) [35]. Their data suggest that the reduced flexibility of the WPD-loop may be the result of the allosteric inhibitor making the α3-helix more rigid and blocking a potential contraction of the

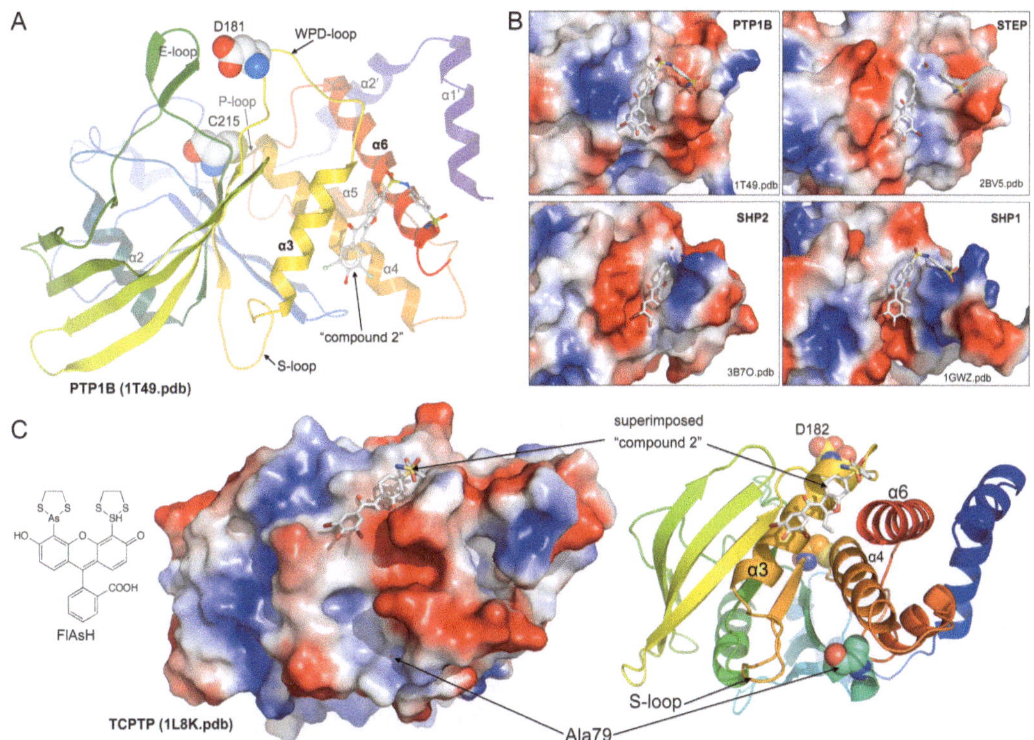

Fig. 12 Allosteric inhibition of PTPs. (**a**) Crystal structure of PTP1B complexed with an allosteric inhibitor (compound 2, shown in ball-and-stick representation); catalytic Cys215 and Asp181 are shown as spheres (PDB ID: 1T49, ref. 34). (**b**) Comparison of the allosteric site in PTP1B (as in (**a**)) with corresponding sites in STEP (PDB ID: 2BV5), SHP2 (PDB ID: 3B7O), and SHP1 (PDB ID: 1GWZ). Structures are superimposed; proteins are shown in surface representation (*blue*, most positive; *red*, most negative); compound 2 is shown as reference in all structures. (**c**) Scanning-insertional mutagenesis using FlAsH and a FlAsH-binding peptide (TetraCys: CCPGCC) inserted at position Ala79 identifies a potential allosteric site in TCPTP. To illustrate the location of the identified allosteric site, the crystal structure of TCPTP (PDB ID: 1L8K) is shown as surface and ribbon representation; Ala79, Cys216, and D182 are shown as *spheres*. The PTP allosteric inhibitor (compound 2) is superimposed for orientation

helix that is necessary for WPD-loop flexibility. Additionally, their data shows that the reduced flexibility in the WPD-loop is accompanied by a reduced flexibility in the S-loop, which is the loop that follows the α3-helix and precedes a β-sheet that connects to the P-loop. Thus, allosteric inhibition of PTP1B, and perhaps of other PTPs, may not be restricted to binding of compounds to the site identified by Wiesmann and colleagues, but may also be mediated through binding of compounds to additional surface areas that involve the α3-helix and the S-loop. In fact, Zhang and Bishop used scanning-insertional mutagenesis to engineer mutants of TCPTP in the search for potential allosteric sites [209]. Specifically, they employed a small molecule fluorescein arsenical hairpin binder (FlAsH) and a FlAsH-binding peptide (TetraCys: CCPGCC) inserted at various loop regions in the catalytic domain (Fig. 12c).

They then tested the phosphatase activity in the presence or absence of FlAsH. They found that TCPTP catalytic efficiency was decreased by more than threefold in the presence of FlAsH when TetraCys was inserted at the Ala79 position. Interestingly, Ala79 is located near the S-loop in the TCPTP 3D structure, suggesting that the allosteric effect of FlAsH-binding could be due to reduced flexibility of the WPD-loop via stabilization of the S-loop/α3-helix. Collectively, these studies provide proof of principle for an allosteric approach in PTP inhibitor development and justify future studies that specifically search for ligands that bind corresponding sites in PTP1B, TCPTP, or other PTPs.

8.3 Possible Alternative Approaches

To date, numerous screening and medicinal chemistry campaigns have failed to produce highly specific and efficacious PTP inhibitors with drug-like properties. However, PTP specificity in vivo is indisputable, raising the question whether the commonly used assay systems [210] are capable of identifying suitable compounds that can be developed into PTP-based small-molecule therapeutics. Given that binding of intact protein substrates appears to determine PTP substrate specificity and even changes the nature of the PTP active site (*see* Subheading 6.1), it seems reasonable to develop assay systems that employ intact protein substrates and PTP proteins that, in addition to the minimal catalytic domain, also include all relevant substrate binding domains/motifs. A limiting factor for such an approach will be the availability of suitable recombinant proteins in quantities necessary for high-throughput screening (HTS). While recombinant PTP catalytic domains can be easily expressed and isolated from bacteria in high yields and purity, larger protein entities that include additional domains/motifs are not as easily accessible or may not be attainable at all. A second limiting factor will be the implementation of a reliable PTP assay system. K_m values of intact substrate proteins are usually in the nanomolar range (e.g., 610 nM for ERK2/pTpY with HePTP). Using the protein substrates at K_m concentration in the screening assay will produce free phosphate at levels below the detection limit of the commonly used malachite green-based reagents [210]. Thus, enzyme-coupled or pTyr antibody-based assays with much greater sensitivity will need to be adopted and implemented for HTS. Besides active site inhibitors with possibly greater selectivity for the PTP of interest, an assay that uses intact protein substrates with suitable PTPs is expected to also identify compounds that interfere with substrate binding distal from the active site. Because the distal substrate binding sites, such as the "second site" (*see* Subheading 6.2), appear not nearly as conserved as the active site-proximal substrate interface, such inhibitors are expected to be more selective for the PTP of interest.

A major drawback in PTP activity-based HTS assays is that the nucleophilic cysteine (i.e., thiolate) is extremely susceptible to inactivation through alkylation [211–213], oxidation [46], nitrosylation [47], and sulfhydration [48]. For instance, trace amounts of Cu^{2+} (widely used as a catalyst in synthetic chemistry) can effectively abrogate PTP activity via oxidation that is only partially impeded/reversed by the use of reducing agents such as dithiothreitol (DTT) [214]. In fact, from our own experience in utilizing libraries of commercial compounds for HTS in the search for PTP inhibitors, we find that the majority of hit compounds, when repurchased as powders from commercial sources, contain impurities that are responsible for the observed PTP inhibition. Thus, significantly improved quality of compound libraries would greatly enhance the quality of the data from PTP activity-based HTS assays. Alternatively, assays could be applied that do not rely on measuring PTP activity, but instead measure ligand binding to proteins. One such technology suitable for HTS is differential scanning fluorimetry (DSF; aka Thermofluor or fluorescence thermal shift assay). DSF is a rapid and inexpensive screening method to identify small-molecule ligands that bind and stabilize globular proteins [215]. Applicable to 384- and even 1536-well formats, DSF uses relatively little protein and provides a fluorescence readout measurement of protein melting temperatures, which correlate with ligand binding. Confirmed binders can subsequently be repurchased, repurified, and tested in PTP activity assays. By blocking the active site during the DSF measurement (e.g., with orthovanadate), it should also be possible to search specifically for ligands that do not bind to the active site and possibly interfere with protein substrate binding or act as allosteric inhibitors.

Another advantage of using an assay system for HTS that does not rely on measuring PTP activity is the possibility to utilize PTPs in an inactive, e.g., oxidized state, and specifically screen for molecules that bind the protein in this state. Under physiological conditions, optimal tyrosine phosphorylation responses are also controlled by reactive oxygen species (ROS), which transiently inactivate PTPs by oxidizing the active site cysteine [216, 217]. ROS can oxidize reactive thiol groups to sulfenic acid (–SOH, oxidation state +1) [218], a transient modification that is reversed by thiol-containing reductants such as glutathione [219]. However, under stronger oxidizing conditions, sulfenic acid can be further oxidized to sulfinic acid (–SO_2H, oxidation state +2) and sulfonic acid (– SO_3H, oxidation state +4). Both of these oxidation states cannot be reversed with reducing agents. Given that reversibility is essential for normal cell signaling, different mechanisms are in place that prevent further oxidation of the active site cysteine in PTPs. In many PTPs, an additional cysteine residue is located in close proximity to the active site cysteine. For a number of these PTPs, including CDC25, PTEN, LMPTP, MKP-3, DUSP12, SHP1, SHP2, and LYP, it was shown that disulfide bond formation between the two cysteine residues prevents irreversible

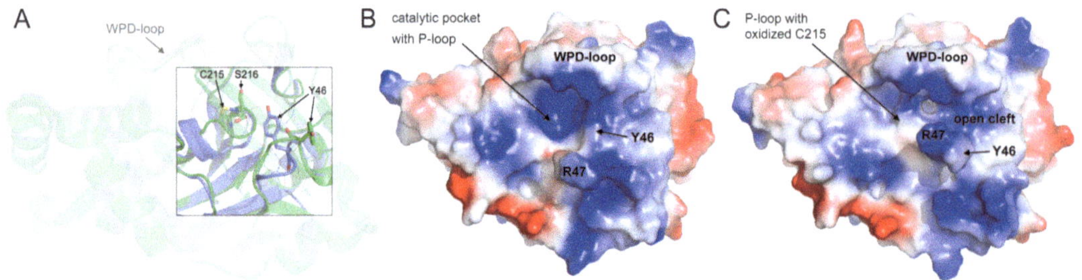

Fig. 13 Comparison of reduced and oxidized states of PTP1B. (**a**) Ribbon diagram of PTP1B in reduced state (*blue*; PDB ID: 2HNP) and PTP1B in oxidized, sulphenyl-amide state (*green*, PDB ID: 1OEM). P-loop Cys215 and Ser216 residues, which form the sulphenyl-amide five-membered ring, and pTyr-loop Tyr46 residues are shown in stick representation. (**b/c**) Surface representation (*blue*, most positive; *red*, most negative) of PTP1B in the reduced state (**b**) and oxidized state (**c**)

oxidation (reviewed in ref. 220). Active site cysteine residues in PTPs that do not contain an additional cysteine near the active site can be protected by formation of a cyclic sulphenyl-amide, in which the cysteine sulfur atom is covalently linked to the backbone nitrogen atom of an adjacent residue. This was first observed in PTP1B [221, 222], and later also in the membrane distal pseudo-PTP domain (D2) in RPTPα [223]. Interestingly, sulphenyl-amide formation is accompanied by a major rearrangement of the active site P- and pTyr-loops (Fig. 13). In fact, Tonks and colleagues recently reported the generation of antibodies that specifically stabilize PTP1B in the oxidized, sulphenyl-amide state and thereby inhibit PTP1B function [224]. Expression of these conformation-sensing antibodies enhanced insulin-induced tyrosine phosphorylation of the insulin receptor and its substrate IRS-1 and increased insulin-induced phosphorylation of PKB/AKT. Importantly, the antibodies were specific for PTP1B and did not recognize oxidized TCPTP, a closely related PTP. These data suggest that stabilization of the oxidized, inactive form of PTP1B may (1) be beneficial in the treatment of type 2 diabetes, and (2) be attained very specifically. Judging from the crystal structure of the sulphenyl-amide form of PTP1B, the surface properties of the active site not only appear dramatically different from the reduced form, but also seem to be amenable to high affinity small-molecule interaction (Fig. 13). Importantly, due to the conformational change of the P-loop, the highly charged active site pocket in the reduced, active state (Fig. 13b) does not exist as such in the oxidized state (Fig. 13c). As a consequence of these major changes, active site PTP inhibitors are not expected to bind to the oxidized state and vice versa. Furthermore, the conformational change of the pTyr-loop, which leads to a major rearrangement of the conserved tyrosine residue (Tyr46 in PTP1B), opens up a cleft that could be exploited for small-molecule binding (Fig. 13c). Collectively, these data provide rationale for developing a novel screening strategy, which seeks to identify compounds that specifically bind PTP1B in the oxidized state. Non-activity-based HTS assays such as DSF could be employed for this purpose. Such assays will also benefit from a mutant form of PTP1B (PTP1B-CASA), in which the catalytic cysteine and adjacent serine residues are mutated to alanine. Due to the loss of two critical hydrogen bonding interactions, PTP1B-CASA adopts a stable P-loop conformation that is identical to PTP1B in the oxidized, sulphenyl-amide state [224]. Targeting the oxidized state would circumvent the common difficulties related to the highly charged PTP active site inhibitors, because high-affinity binding would no longer depend on pTyr-mimicking groups, which usually account for low membrane permeability and nonselective inhibition of PTPs. If such efforts yield potent and efficacious compounds, and if future studies show that under physiological conditions sulphenyl-amide formation is a general feature of the PTP family, PTP inhibitor development could be up for a major paradigm shift.

Acknowledgments

This work was supported by NIH grants R03MH095532, R03MH084230, and R21CA132121 (to L.T.). We thank Robert Parker for his help in preparing Table 1.

References

1. Hunter T (2000) Signaling—2000 and beyond. Cell 100:113–127
2. Pawson T, Scott JD (2005) Protein phosphorylation in signaling—50 years and counting. Trends Biochem Sci 30:286–290
3. Luan S (2003) Protein phosphatases in plants. Annu Rev Plant Biol 54:63–92
4. Deutscher J, Saier MHJ (2005) Ser/Thr/Tyr protein phosphorylation in bacteria—for long time neglected, now well established. J Mol Microbiol Biotechnol 9:125–131
5. Feher Z, Szirak K (1999) Signal transduction in fungi—the role of protein phosphorylation. Acta Microbiol Immunol Hung 46:269–271
6. Moorhead GB, De Wever V, Templeton G et al (2009) Evolution of protein phosphatases in plants and animals. Biochem J 417:401–409
7. Olsen JV, Blagoev B, Gnad F et al (2006) Global, in vivo, and site-specific phosphorylation dynamics in signaling networks. Cell 127:635–648
8. Hunter T, Sefton BM (1980) Transforming gene product of Rous sarcoma virus phosphorylates tyrosine. Proc Natl Acad Sci U S A 77:1311–1315
9. Hunter T (1998) The role of tyrosine phosphorylation in cell growth and disease. Harvey Lect 94:81–119
10. Larsen M, Tremblay ML, Yamada KM (2003) Phosphatases in cell-matrix adhesion and migration. Nat Rev Mol Cell Biol 4:700–711
11. Alonso A, Sasin J, Bottini N et al (2004) Protein tyrosine phosphatases in the human genome. Cell 117:699–711
12. Mustelin T, Vang T, Bottini N (2005) Protein tyrosine phosphatases and the immune response. Nat Rev Immunol 5:43–57
13. Tonks NK (2006) Protein tyrosine phosphatases: from genes, to function, to disease. Nat Rev Mol Cell Biol 7:833–846
14. Halle M, Tremblay ML, Meng TC (2007) Protein tyrosine phosphatases: emerging regulators of apoptosis. Cell Cycle 6:2773–2781
15. Pao LI, Badour K, Siminovitch KA et al (2007) Nonreceptor protein-tyrosine phosphatases in immune cell signaling. Annu Rev Immunol 25:473–523

16. Hunter T (2009) Tyrosine phosphorylation: thirty years and counting. Curr Opin Cell Biol 21:140–146
17. Rhee I, Veillette A (2012) Protein tyrosine phosphatases in lymphocyte activation and autoimmunity. Nat Immunol 13:439–447
18. Cohen P (2002) Protein kinases—the major drug targets of the twenty-first century? Nat Rev Drug Discov 1:309–315
19. Tautz L, Pellecchia M, Mustelin T (2006) Targeting the PTPome in human disease. Expert Opin Ther Targets 10:157–177
20. Andersen JN, Mortensen OH, Peters GH et al (2001) Structural and evolutionary relationships among protein tyrosine phosphatase domains. Mol Cell Biol 21:7117–7136
21. Pulido R, Hooft van Huijsduijnen R (2008) Protein tyrosine phosphatases: dual-specificity phosphatases in health and disease. FEBS J 275:848–866
22. Patterson KI, Brummer T, O'Brien PM et al (2009) Dual-specificity phosphatases: critical regulators with diverse cellular targets. Biochem J 418:475–489
23. Bottini N, Bottini E, Gloria-Bottini F et al (2002) Low-molecular-weight protein tyrosine phosphatase and human disease: in search of biochemical mechanisms. Arch Immunol Ther Exp (Warsz) 50:95–104
24. Zhang ZY, Wang Y, Dixon JE (1994) Dissecting the catalytic mechanism of protein-tyrosine phosphatases. Proc Natl Acad Sci U S A 91:1624–1627
25. Zhang ZY (1995) Kinetic and mechanistic characterization of a mammalian protein-tyrosine phosphatase, PTP1. J Biol Chem 270:11199–11204
26. Denu JM, Stuckey JA, Saper MA et al (1996) Form and function in protein dephosphorylation. Cell 87:361–364
27. Denu JM, Lohse DL, Vijayalakshmi J et al (1996) Visualization of intermediate and transition-state structures in protein-tyrosine phosphatase catalysis. Proc Natl Acad Sci U S A 93:2493–2498
28. Stuckey JA, Schubert HL, Fauman EB et al (1994) Crystal structure of Yersinia protein tyrosine phosphatase at 2.5 A and the complex with tungstate. Nature 370:571–575
29. Hengge AC, Denu JM, Dixon JE (1996) Transition-state structures for the native dual-specific phosphatase VHR and D92N and S131A mutants. Contributions to the driving force for catalysis. Biochemistry 35: 7084–7092
30. Zhang ZY (2002) Protein tyrosine phosphatases, structure and function, substrate specificity, and inhibitor development. Annu Rev Pharmacol Toxicol 42:209–234
31. Zhang ZY, Maclean D, McNamara DJ et al (1994) Protein tyrosine phosphatase substrate specificity: size and phosphotyrosine positioning requirements in peptide substrates. Biochemistry 33:2285–2290
32. Mustelin T, Tautz L, Page R (2005) Structure of the hematopoietic tyrosine phosphatase (HePTP) catalytic domain: structure of a KIM phosphatase with phosphate bound at the active site. J Mol Biol 354:150–163
33. Barr AJ, Ugochukwu E, Lee WH et al (2009) Large-scale structural analysis of the classical human protein tyrosine phosphatome. Cell 136:352–363
34. Wiesmann C, Barr KJ, Kung J et al (2004) Allosteric inhibition of protein tyrosine phosphatase 1B. Nat Struct Mol Biol 11:730–737
35. Kamerlin SC, Rucker R, Boresch S (2007) A molecular dynamics study of WPD-loop flexibility in PTP1B. Biochem Biophys Res Commun 356:1011–1016
36. Vang T, Liu WH, Delacroix L et al (2012) LYP inhibits T-cell activation when dissociated from CSK. Nat Chem Biol 8:437–446
37. Lorenz U (2009) SHP-1 and SHP-2 in T cells: two phosphatases functioning at many levels. Immunol Rev 228:342–359
38. Pei D, Lorenz U, Klingmuller U et al (1994) Intramolecular regulation of protein tyrosine phosphatase SH-PTP1: a new function for Src homology 2 domains. Biochemistry 33: 15483–15493
39. Hof P, Pluskey S, Dhe-Paganon S et al (1998) Crystal structure of the tyrosine phosphatase SHP-2. Cell 92:441–450
40. Johnson P, Ostergaard HL, Wasden C et al (1992) Mutational analysis of CD45. A leukocyte-specific protein tyrosine phosphatase. J Biol Chem 267:8035–8041
41. Streuli M, Krueger NX, Thai T et al (1990) Distinct functional roles of the two intracellular phosphatase like domains of the receptor-linked protein tyrosine phosphatases LCA and LAR. EMBO J 9:2399–2407
42. Kashio N, Matsumoto W, Parker S et al (1998) The second domain of the CD45 protein tyrosine phosphatase is critical for interleukin-2 secretion and substrate recruitment of TCR-zeta in vivo. J Biol Chem 273: 33856–33863
43. Shi Y (2009) Serine/threonine phosphatases: mechanism through structure. Cell 139: 468–484
44. Barford D, Flint AJ, Tonks NK (1994) Crystal structure of human protein tyrosine phosphatase 1B. Science 263:1397–1404
45. Zhang ZY, Dixon JE (1993) Active site labeling of the Yersinia protein tyrosine phosphatase: the determination of the pKa of the

active site cysteine and the function of the conserved histidine 402. Biochemistry 32: 9340–9345
46. Tonks NK (2005) Redox redux: revisiting PTPs and the control of cell signaling. Cell 121:667–670
47. Chen YY, Chu HM, Pan KT et al (2008) Cysteine S-nitrosylation protects protein-tyrosine phosphatase 1B against oxidation-induced permanent inactivation. J Biol Chem 283:35265–35272
48. Krishnan N, Fu C, Pappin DJ et al (2011) H2S-Induced sulfhydration of the phosphatase PTP1B and its role in the endoplasmic reticulum stress response. Sci Signal 4:ra86
49. Keng YF, Wu L, Zhang ZY (1999) Probing the function of the conserved tryptophan in the flexible loop of the Yersinia protein-tyrosine phosphatase. Eur J Biochem 259:809–814
50. Hoff RH, Hengge AC, Wu L et al (2000) Effects on general acid catalysis from mutations of the invariant tryptophan and arginine residues in the protein tyrosine phosphatase from Yersinia. Biochemistry 39:46–54
51. Kurkcuoglu Z, Bakan A, Kocaman D et al (2012) Coupling between catalytic loop motions and enzyme global dynamics. PLoS Comput Biol 8:e1002705
52. Cui L, Yu WP, DeAizpurua HJ et al (1996) Cloning and characterization of islet cell antigen-related protein-tyrosine phosphatase (PTP), a novel receptor-like PTP and autoantigen in insulin-dependent diabetes. J Biol Chem 271:24817–24823
53. Seifert RA, Coats SA, Oganesian A et al (2003) PTPRQ is a novel phosphatidylinositol phosphatase that can be expressed as a cytoplasmic protein or as a subcellularly localized receptor-like protein. Exp Cell Res 287: 374–386
54. Cheng J, Wu K, Armanini M et al (1997) A novel protein-tyrosine phosphatase related to the homotypically adhering kappa and mu receptors. J Biol Chem 272:7264–7277
55. Gingras MC, Zhang YL, Kharitidi D et al (2009) HD-PTP is a catalytically inactive tyrosine phosphatase due to a conserved divergence in its phosphatase domain. PLoS One 4:e5105
56. Cardone L, Carlucci A, Affaitati A et al (2004) Mitochondrial AKAP121 binds and targets protein tyrosine phosphatase D1, a novel positive regulator of src signaling. Mol Cell Biol 24:4613–4626
57. Rabin DU, Pleasic SM, Shapiro JA et al (1994) Islet cell antigen 512 is a diabetes-specific islet autoantigen related to protein tyrosine phosphatases. J Immunol 152: 3183–3188
58. Jia Z, Barford D, Flint AJ et al (1995) Structural basis for phosphotyrosine peptide recognition by protein tyrosine phosphatase 1B. Science 268:1754–1758
59. Eswaran J, von Kries JP, Marsden B et al (2006) Crystal structures and inhibitor identification for PTPN5, PTPRR and PTPN7: a family of human MAPK-specific protein tyrosine phosphatases. Biochem J 395:483–491
60. Iversen LF, Moller KB, Pedersen AK et al (2002) Structure determination of T cell protein-tyrosine phosphatase. J Biol Chem 277:19982–19990
61. Yang J, Cheng Z, Niu T et al (2000) Structural basis for substrate specificity of protein-tyrosine phosphatase SHP-1. J Biol Chem 275:4066–4071
62. Asante-Appiah E, Patel S, Desponts C et al (2006) Conformation-assisted inhibition of protein-tyrosine phosphatase-1B elicits inhibitor selectivity over T-cell protein-tyrosine phosphatase. J Biol Chem 281:8010–8015
63. Critton DA, Tautz L, Page R (2011) Visualizing active-site dynamics in single crystals of HePTP: opening of the WPD loop involves coordinated movement of the E loop. J Mol Biol 405:619–629
64. Xie L, Zhang YL, Zhang ZY (2002) Design and characterization of an improved protein tyrosine phosphatase substrate-trapping mutant. Biochemistry 41:4032–4039
65. Zhao Y, Wu L, Noh SJ et al (1998) Altering the nucleophile specificity of a protein-tyrosine phosphatase-catalyzed reaction. Probing the function of the invariant glutamine residues. J Biol Chem 273:5484–5492
66. Pedersen AK, Guo XL, Moller KB et al (2004) Residue 182 influences the second step of protein-tyrosine phosphatase-mediated catalysis. Biochem J 378:421–433
67. Zabell AP, Schroff ADJ, Bain BE et al (2006) Crystal structure of the human B-form low molecular weight phosphotyrosyl phosphatase at 1.6-A resolution. J Biol Chem 281: 6520–6527
68. Bryson GL, Massa H, Trask BJ et al (1995) Gene structure, sequence, and chromosomal localization of the human red cell-type low-molecular-weight acid phosphotyrosyl phosphatase gene, ACP1. Genomics 30:133–140
69. Tailor P, Gilman J, Williams S et al (1999) A novel isoform of the low molecular weight phosphotyrosine phosphatase, LMPTP-C, arising from alternative mRNA splicing. Eur J Biochem 262:277–282
70. Ramponi G, Manao G, Camici G et al (1989) The 18 kDa cytosolic acid phosphatase from bovine live has phosphotyrosine phosphatase activity on the autophosphorylated epidermal

growth factor receptor. FEBS Lett 250: 469–473
71. Rudolph J (2007) Cdc25 phosphatases: structure, specificity, and mechanism. Biochemistry 46:3595–3604
72. Mailand N, Falck J, Lukas C et al (2000) Rapid destruction of human Cdc25A in response to DNA damage. Science 288: 1425–1429
73. Donzelli M, Squatrito M, Ganoth D et al (2002) Dual mode of degradation of Cdc25 A phosphatase. EMBO J 21:4875–4884
74. Conklin DS, Galaktionov K, Beach D (1995) 14-3-3 proteins associate with cdc25 phosphatases. Proc Natl Acad Sci U S A 92:7892–7896
75. Forrest A, Gabrielli B (2001) Cdc25B activity is regulated by 14-3-3. Oncogene 20: 4393–4401
76. Giles N, Forrest A, Gabrielli B (2003) 14-3-3 acts as an intramolecular bridge to regulate cdc25B localization and activity. J Biol Chem 278:28580–28587
77. Hoffmann I, Clarke PR, Marcote MJ et al (1993) Phosphorylation and activation of human cdc25-C by cdc2–cyclin B and its involvement in the self-amplification of MPF at mitosis. EMBO J 12:53–63
78. Hoffmann I, Draetta G, Karsenti E (1994) Activation of the phosphatase activity of human cdc25A by a cdk2-cyclin E dependent phosphorylation at the G1/S transition. EMBO J 13:4302–4310
79. Fauman EB, Cogswell JP, Lovejoy B et al (1998) Crystal structure of the catalytic domain of the human cell cycle control phosphatase, Cdc25A. Cell 93:617–625
80. Reynolds RA, Yem AW, Wolfe CL et al (1999) Crystal structure of the catalytic subunit of Cdc25B required for G2/M phase transition of the cell cycle. J Mol Biol 293:559–568
81. Hofmann K, Bucher P, Kajava AV (1998) A model of Cdc25 phosphatase catalytic domain and Cdk-interaction surface based on the presence of a rhodanese homology domain. J Mol Biol 282:195–208
82. Sarmiento M, Zhao Y, Gordon SJ et al (1998) Molecular basis for substrate specificity of protein-tyrosine phosphatase 1B. J Biol Chem 273:26368–26374
83. Huang Z, Zhou B, Zhang ZY (2004) Molecular determinants of substrate recognition in hematopoietic protein-tyrosine phosphatase. J Biol Chem 279:52150–52159
84. Francis DM, Rozycki B, Tortajada A et al (2011) Resting and Active States of the ERK2:HePTP Complex. J Am Chem Soc 133:17138–17141
85. Francis DM, Rozycki B, Koveal D et al (2011) structural basis of p38a regulation by hematopoietic tyrosine phosphatase. Nat Chem Biol 7:916–924
86. Wiland AM, Denu JM, Mourey RJ et al (1996) Purification and kinetic characterization of the mitogen-activated protein kinase phosphatase rVH6. J Biol Chem 271: 33486–33492
87. Zhao Y, Zhang ZY (2001) The mechanism of dephosphorylation of extracellular signal-regulated kinase 2 by mitogen-activated protein kinase phosphatase 3. J Biol Chem 276:32382–32391
88. Begley MJ, Taylor GS, Kim SA et al (2003) Crystal structure of a phosphoinositide phosphatase, MTMR2: insights into myotubular myopathy and Charcot-Marie-Tooth syndrome. Mol Cell 12:1391–1402
89. Puius YA, Zhao Y, Sullivan M et al (1997) Identification of a second aryl phosphate-binding site in protein-tyrosine phosphatase 1B: a paradigm for inhibitor design. Proc Natl Acad Sci U S A 94:13420–13425
90. Salmeen A, Andersen JN, Myers MP et al (2000) Molecular basis for the dephosphorylation of the activation segment of the insulin receptor by protein tyrosine phosphatase 1B. Mol Cell 6:1401–1412
91. Peters GH, Iversen LF, Branner S et al (2000) Residue 259 is a key determinant of substrate specificity of protein-tyrosine phosphatases 1B and alpha. J Biol Chem 275: 18201–18209
92. Flint AJ, Tiganis T, Barford D et al (1997) Development of "substrate-trapping" mutants to identify physiological substrates of protein tyrosine phosphatases. Proc Natl Acad Sci U S A 94:1680–1685
93. Sarmiento M, Puius YA, Vetter SW et al (2000) Structural basis of plasticity in protein tyrosine phosphatase 1B substrate recognition. Biochemistry 39:8171–8179
94. Song H, Hanlon N, Brown NR et al (2001) Phosphoprotein-protein interactions revealed by the crystal structure of kinase-associated phosphatase in complex with phosphoCDK2. Mol Cell 7:615–626
95. Wang S, Tabernero L, Zhang M et al (2000) Crystal structures of a low-molecular weight protein tyrosine phosphatase from Saccharomyces cerevisiae and its complex with the substrate p-nitrophenyl phosphate. Biochemistry 39:1903–1914
96. Garton AJ, Flint AJ, Tonks NK (1996) Identification of p130(cas) as a substrate for the cytosolic protein tyrosine phosphatase PTP-PEST. Mol Cell Biol 16:6408–6418
97. Blanchetot C, Chagnon M, Dube N et al (2005) Substrate-trapping techniques in the identification of cellular PTP targets. Methods 35:44–53

98. Agazie YM, Hayman MJ (2003) Development of an efficient "substrate-trapping" mutant of Src homology phosphotyrosine phosphatase 2 and identification of the epidermal growth factor receptor, Gab1, and three other proteins as target substrates. J Biol Chem 278:13952–13958
99. Sohn J, Parks JM, Buhrman G et al (2005) Experimental validation of the docking orientation of Cdc25 with its Cdk2-CycA protein substrate. Biochemistry 44:16563–16573
100. Buhrman G, Parker B, Sohn J et al (2005) Structural mechanism of oxidative regulation of the phosphatase Cdc25B via an intramolecular disulfide bond. Biochemistry 44:5307–5316
101. Ostman A, Hellberg C, Bohmer FD (2006) Protein-tyrosine phosphatases and cancer. Nat Rev Cancer 6:307–320
102. Vang T, Miletic AV, Arimura Y et al (2008) Protein tyrosine phosphatases in autoimmunity. Annu Rev Immunol 26:29–55
103. Julien SG, Dube N, Hardy S et al (2011) Inside the human cancer tyrosine phosphatome. Nat Rev Cancer 11:35–49
104. Goebel-Goody SM, Baum M, Paspalas CD et al (2012) Therapeutic implications for striatal-enriched protein tyrosine phosphatase (STEP) in neuropsychiatric disorders. Pharmacol Rev 64:65–87
105. Keniry M, Parsons R (2008) The role of PTEN signaling perturbations in cancer and in targeted therapy. Oncogene 27: 5477–5485
106. Mena-Duran AV, Togo SH, Bazhenova L et al (2005) SHP1 expression in bone marrow biopsies of myelodysplastic syndrome patients: a new prognostic factor. Br J Haematol 129: 791–794
107. Zhang Q, Wang HY, Marzec M et al (2005) STAT3- and DNA methyltransferase 1-mediated epigenetic silencing of SHP-1 tyrosine phosphatase tumor suppressor gene in malignant T lymphocytes. Proc Natl Acad Sci U S A 102:6948–6953
108. Mohi MG, Neel BG (2007) The role of Shp2 (PTPN11) in cancer. Curr Opin Genet Dev 17:23–30
109. Xu R (2007) Shp2, a novel oncogenic tyrosine phosphatase and potential therapeutic target for human leukemia. Cell Res 17:295–297
110. Saha S, Bardelli A, Buckhaults P et al (2001) A phosphatase associated with metastasis of colorectal cancer. Science 294:1343–1346
111. Loda M, Capodieci P, Mishra R et al (1996) Expression of mitogen-activated protein kinase phosphatase-1 in the early phases of human epithelial carcinogenesis. Am J Pathol 149:1553–1564
112. Jiang ZX, Zhang ZY (2008) Targeting PTPs with small molecule inhibitors in cancer treatment. Cancer Metastasis Rev 27:263–272
113. Bourdeau A, Dube N, Tremblay ML (2005) Cytoplasmic protein tyrosine phosphatases, regulation and function: the roles of PTP1B and TC-PTP. Curr Opin Cell Biol 17:203–209
114. Julien SG, Dube N, Read M et al (2007) Protein tyrosine phosphatase 1B deficiency or inhibition delays ErbB2-induced mammary tumorigenesis and protects from lung metastasis. Nat Genet 39:338–346
115. Bentires-Alj M, Neel BG (2007) Protein-tyrosine phosphatase 1B is required for HER2/Neu-induced breast cancer. Cancer Res 67:2420–2424
116. Tonks NK, Muthuswamy SK (2007) A brake becomes an accelerator: PTP1B—a new therapeutic target for breast cancer. Cancer Cell 11:214–216
117. Bottini N, Musumeci L, Alonso A et al (2004) A functional variant of lymphoid tyrosine phosphatase is associated with type I diabetes. Nat Genet 36:337–338
118. Begovich AB, Carlton VE, Honigberg LA et al (2004) A missense single-nucleotide polymorphism in a gene encoding a protein tyrosine phosphatase (PTPN22) is associated with rheumatoid arthritis. Am J Hum Genet 75:330–337
119. Kyogoku C, Langefeld CD, Ortmann WA et al (2004) Genetic association of the R620W polymorphism of protein tyrosine phosphatase PTPN22 with human SLE. Am J Hum Genet 75:504–507
120. Stanford SM, Mustelin TM, Bottini N (2010) Lymphoid tyrosine phosphatase and autoimmunity: human genetics rediscovers tyrosine phosphatases. Semin Immunopathol 32: 127–136
121. Negro R, Gobessi S, Longo PG et al (2012) Overexpression of the autoimmunity-associated phosphatase PTPN22 promotes survival of antigen-stimulated CLL cells by selectively activating AKT. Blood 119:6278–6287
122. Saxena M, Williams S, Brockdorff J et al (1999) Inhibition of T cell signaling by mitogen-activated protein kinase-targeted hematopoietic tyrosine phosphatase (HePTP). J Biol Chem 274:11693–11700
123. Zanke B, Squire J, Griesser H et al (1994) A hematopoietic protein tyrosine phosphatase (HePTP) gene that is amplified and overexpressed in myeloid malignancies maps to chromosome 1q32.1. Leukemia 8:236–244
124. Sergienko E, Xu J, Liu WH et al (2012) Inhibition of hematopoietic protein tyrosine phosphatase augments and prolongs ERK1/2 and p38 activation. ACS Chem Biol 7: 367–377

125. Fonatsch C, Haase D, Freund M et al (1991) Partial trisomy 1q. A nonrandom primary chromosomal abnormality in myelodysplastic syndromes? Cancer Genet Cytogenet 56: 243–253
126. Mamaev N, Mamaeva SE, Pavlova VA et al (1988) Combined trisomy 1q and monosomy 17p due to translocation t(1;17) in a patient with myelodysplastic syndrome. Cancer Genet Cytogenet 35:21–25
127. Souza AC, Azoubel S, Queiroz KC et al (2009) From immune response to cancer: a spot on the low molecular weight protein tyrosine phosphatase. Cell Mol Life Sci 66:1140–1153
128. Minassian BA, Lee JR, Herbrick JA et al (1998) Mutations in a gene encoding a novel protein tyrosine phosphatase cause progressive myoclonus epilepsy. Nat Genet 20: 171–174
129. Laporte J, Hu LJ, Kretz C et al (1996) A gene mutated in X-linked myotubular myopathy defines a new putative tyrosine phosphatase family conserved in yeast. Nat Genet 13: 175–182
130. Bolino A, Muglia M, Conforti FL et al (2000) Charcot-Marie-Tooth type 4B is caused by mutations in the gene encoding myotubularin-related protein-2. Nat Genet 25:17–19
131. Azzedine H, Bolino A, Taieb T et al (2003) Mutations in MTMR13, a new pseudophosphatase homologue of MTMR2 and Sbfl, in two families with an autosomal recessive demyelinating form of Charcot-Marie-Tooth disease associated with early-onset glaucoma. Am J Hum Genet 72:1141–1153
132. Kung C, Pingel JT, Heikinheimo M et al (2000) Mutations in the tyrosine phosphatase CD45 gene in a child with severe combined immunodeficiency disease. Nat Med 6: 343–345
133. Elchebly M, Payette P, Michaliszyn E et al (1999) Increased insulin sensitivity and obesity resistance in mice lacking the protein tyrosine phosphatase-1B gene. Science 283: 1544–1548
134. Zhang S, Zhang ZY (2007) PTP1B as a drug target: recent developments in PTP1B inhibitor discovery. Drug Discov Today 12: 373–381
135. Lee S, Wang Q (2007) Recent development of small molecular specific inhibitor of protein tyrosine phosphatase 1B. Med Res Rev 27:553–573
136. Barr AJ (2010) Protein tyrosine phosphatases as drug targets: strategies and challenges of inhibitor development. Future Med Chem 2:1563–1576
137. Popov D (2011) Novel protein tyrosine phosphatase 1B inhibitors: interaction requirements for improved intracellular efficacy in type 2 diabetes mellitus and obesity control. Biochem Biophys Res Commun 410: 377–381
138. Siegel PM, Ryan ED, Cardiff RD et al (1999) Elevated expression of activated forms of Neu/ErbB-2 and ErbB-3 are involved in the induction of mammary tumors in transgenic mice: implications for human breast cancer. EMBO J 18:2149–2164
139. Tanner MM, Grenman S, Koul A et al (2000) Frequent amplification of chromosomal region 20q12-q13 in ovarian cancer. Clin Cancer Res 6:1833–1839
140. Ginestier C, Cervera N, Finetti P et al (2006) Prognosis and gene expression profiling of 20q13-amplified breast cancers. Clin Cancer Res 12:4533–4544
141. Wiener JR, Kerns BJ, Harvey EL et al (1994) Overexpression of the protein tyrosine phosphatase PTP1B in human breast cancer: association with p185c-erbB-2 protein expression. J Natl Cancer Inst 86:372–378
142. Zhai YF, Beittenmiller H, Wang B et al (1993) Increased expression of specific protein tyrosine phosphatases in human breast epithelial cells neoplastically transformed by the neu oncogene. Cancer Res 53:2272–2278
143. Bjorge JD, Pang A, Fujita DJ (2000) Identification of protein-tyrosine phosphatase 1B as the major tyrosine phosphatase activity capable of dephosphorylating and activating c-Src in several human breast cancer cell lines. J Biol Chem 275:41439–41446
144. Arias-Romero LE, Saha S, Villamar-Cruz O et al (2009) Activation of Src by protein tyrosine phosphatase 1B Is required for ErbB2 transformation of human breast epithelial cells. Cancer Res 69:4582–4588
145. Kaminski R, Zagozdzon R, Fu Y et al (2006) Role of SRC kinases in Neu-induced tumorigenesis: challenging the paradigm using Csk homologous kinase transgenic mice. Cancer Res 66:5757–5762
146. Dadke S, Chernoff J (2003) Protein-tyrosine phosphatase 1B mediates the effects of insulin on the actin cytoskeleton in immortalized fibroblasts. J Biol Chem 278:40607–40611
147. Dube N, Cheng A, Tremblay ML (2004) The role of protein tyrosine phosphatase 1B in Ras signaling. Proc Natl Acad Sci U S A 101: 1834–1839
148. Kashige N, Carpino N, Kobayashi R (2000) Tyrosine phosphorylation of p62dok by p210bcr-abl inhibits RasGAP activity. Proc Natl Acad Sci U S A 97:2093–2098
149. Slamon DJ, Clark GM, Wong SG et al (1987) Human breast cancer: correlation of relapse and survival with amplification of the HER-2/neu oncogene. Science 235:177–182

150. Bange J, Zwick E, Ullrich A (2001) Molecular targets for breast cancer therapy and prevention. Nat Med 7:548–552
151. Carter P, Presta L, Gorman CM et al (1992) Humanization of an anti-p185HER2 antibody for human cancer therapy. Proc Natl Acad Sci U S A 89:4285–4289
152. Nahta R, Esteva FJ (2007) Trastuzumab: triumphs and tribulations. Oncogene 26: 3637–3643
153. Grossmann KS, Rosario M, Birchmeier C et al (2010) The tyrosine phosphatase Shp2 in development and cancer. Adv Cancer Res 106:53–89
154. Liu X, Qu CK (2011) Protein tyrosine phosphatase SHP-2 (PTPN11) in hematopoiesis and leukemogenesis. J Signal Transduct 2011:195–239
155. Tartaglia M, Mehler EL, Goldberg R et al (2001) Mutations in PTPN11, encoding the protein tyrosine phosphatase SHP-2, cause Noonan syndrome. Nat Genet 29:465–468
156. Noonan JA (1968) Hypertelorism with turner phenotype. a new syndrome with associated congenital heart disease. Am J Dis Child 116: 373–380
157. Allanson JE (1987) Noonan syndrome. J Med Genet 24:9–13
158. Marino B, Digilio MC, Toscano A et al (1999) Congenital heart diseases in children with Noonan syndrome: an expanded cardiac spectrum with high prevalence of atrioventricular canal. J Pediatr 135:703–706
159. Opitz JM (1985) The Noonan syndrome. Am J Med Genet 21:515–518
160. Tartaglia M, Gelb BD (2005) Noonan syndrome and related disorders: genetics and pathogenesis. Annu Rev Genomics Hum Genet 6:45–68
161. Tartaglia M, Niemeyer CM, Fragale A et al (2003) Somatic mutations in PTPN11 in juvenile myelomonocytic leukemia, myelodysplastic syndromes and acute myeloid leukemia. Nat Genet 34:148–150
162. Bentires-Alj M, Paez JG, David FS et al (2004) Activating mutations of the noonan syndrome-associated SHP2/PTPN11 gene in human solid tumors and adult acute myelogenous leukemia. Cancer Res 64: 8816–8820
163. Chan RJ, Feng GS (2007) PTPN11 is the first identified proto-oncogene that encodes a tyrosine phosphatase. Blood 109:862–867
164. Chan G, Kalaitzidis D, Neel BG (2008) The tyrosine phosphatase Shp2 (PTPN11) in cancer. Cancer Metastasis Rev 27:179–192
165. Barford D, Neel BG (1998) Revealing mechanisms for SH2 domain mediated regulation of the protein tyrosine phosphatase SHP-2. Structure 6:249–254
166. Araki T, Mohi MG, Ismat FA et al (2004) Mouse model of Noonan syndrome reveals cell type- and gene dosage-dependent effects of Ptpn11 mutation. Nat Med 10:849–857
167. Chan RJ, Leedy MB, Munugalavadla V et al (2005) Human somatic PTPN11 mutations induce hematopoietic-cell hypersensitivity to granulocyte-macrophage colony-stimulating factor. Blood 105:3737–3742
168. Fragale A, Tartaglia M, Wu J et al (2004) Noonan syndrome-associated SHP2/PTPN11 mutants cause EGF-dependent prolonged GAB1 binding and sustained ERK2/MAPK1 activation. Hum Mutat 23:267–277
169. Krenz M, Gulick J, Osinska HE et al (2008) Role of ERK1/2 signaling in congenital valve malformations in Noonan syndrome. Proc Natl Acad Sci U S A 105:18930–18935
170. Mohi MG, Williams IR, Dearolf CR et al (2005) Prognostic, therapeutic, and mechanistic implications of a mouse model of leukemia evoked by Shp2 (PTPN11) mutations. Cancer Cell 7:179–191
171. Yu WM, Daino H, Chen J et al (2006) Effects of a leukemia-associated gain-of-function mutation of SHP-2 phosphatase on interleukin-3 signaling. J Biol Chem 281:5426–5434
172. Carta C, Pantaleoni F, Bocchinfuso G et al (2006) Germline missense mutations affecting KRAS Isoform B are associated with a severe Noonan syndrome phenotype. Am J Hum Genet 79:129–135
173. Schubbert S, Zenker M, Rowe SL et al (2006) Germline KRAS mutations cause Noonan syndrome. Nat Genet 38:331–336
174. Roberts AE, Araki T, Swanson KD et al (2007) Germline gain-of-function mutations in SOS1 cause Noonan syndrome. Nat Genet 39:70–74
175. Tartaglia M, Pennacchio LA, Zhao C et al (2007) Gain-of-function SOS1 mutations cause a distinctive form of Noonan syndrome. Nat Genet 39:75–79
176. Tartaglia M, Martinelli S, Cazzaniga G et al (2004) Genetic evidence for lineage-related and differentiation stage-related contribution of somatic PTPN11 mutations to leukemogenesis in childhood acute leukemia. Blood 104:307–313
177. Shi ZQ, Yu DH, Park M et al (2000) Molecular mechanism for the Shp-2 tyrosine phosphatase function in promoting growth factor stimulation of Erk activity. Mol Cell Biol 20:1526–1536
178. Nguyen TV, Ke Y, Zhang EE et al (2006) Conditional deletion of Shp2 tyrosine phosphatase in thymocytes suppresses both pre-TCR and TCR signals. J Immunol 177:5990–5996
179. Matozaki T, Murata Y, Saito Y et al (2009) Protein tyrosine phosphatase SHP-2: a proto-

oncogene product that promotes Ras activation. Cancer Sci 100:1786–1793
180. Neel BG, Gu H, Pao L (2003) The 'Shp'ing news: SH2 domain-containing tyrosine phosphatases in cell signaling. Trends Biochem Sci 28:284–293
181. Loh ML (2011) Recent advances in the pathogenesis and treatment of juvenile myelomonocytic leukaemia. Br J Haematol 152:677–687
182. Le DT, Kong N, Zhu Y et al (2004) Somatic inactivation of Nf1 in hematopoietic cells results in a progressive myeloproliferative disorder. Blood 103:4243–4250
183. Chen L, Sung SS, Yip ML et al (2006) Discovery of a novel shp2 protein tyrosine phosphatase inhibitor. Mol Pharmacol 70:562–570
184. Noren-Muller A, Reis-Correa IJ, Prinz H et al (2006) Discovery of protein phosphatase inhibitor classes by biology-oriented synthesis. Proc Natl Acad Sci U S A 103: 10606–10611
185. Hellmuth K, Grosskopf S, Lum CT et al (2008) Specific inhibitors of the protein tyrosine phosphatase Shp2 identified by high-throughput docking. Proc Natl Acad Sci U S A 105:7275–7280
186. Lawrence HR, Pireddu R, Chen L et al (2008) Inhibitors of Src homology-2 domain containing protein tyrosine phosphatase-2 (Shp2) based on oxindole scaffolds. J Med Chem 51:4948–4956
187. Zhang X, He Y, Liu S et al (2010) Salicylic acid based small molecule inhibitor for the oncogenic Src homology-2 domain containing protein tyrosine phosphatase-2 (SHP2). J Med Chem 53:2482–2493
188. Liu S, Yu Z, Yu X et al (2011) SHP2 is a target of the immunosuppressant tautomycetin. Chem Biol 18:101–110
189. Baum ML, Kurup P, Xu J et al (2010) A STEP forward in neural function and degeneration. Commun Integr Biol 3:419–422
190. Zhang Y, Kurup P, Xu J et al (2010) Genetic reduction of striatal-enriched tyrosine phosphatase (STEP) reverses cognitive and cellular deficits in an Alzheimer's disease mouse model. Proc Natl Acad Sci U S A 107: 19014–19019
191. Snyder EM, Nong Y, Almeida CG et al (2005) Regulation of NMDA receptor trafficking by amyloid-beta. Nat Neurosci 8:1051–1058
192. Chin J, Palop JJ, Puolivali J et al (2005) Fyn kinase induces synaptic and cognitive impairments in a transgenic mouse model of Alzheimer's disease. J Neurosci 25: 9694–9703
193. Zhang Y, Venkitaramani DV, Gladding CM et al (2008) The tyrosine phosphatase STEP mediates AMPA receptor endocytosis after metabotropic glutamate receptor stimulation. J Neurosci 28:10561–10566
194. Zhang Y, Kurup P, Xu J et al (2011) Reduced levels of the tyrosine phosphatase STEP block beta amyloid-mediated GluA1/GluA2 receptor internalization. J Neurochem 119: 664–672
195. Hardy JA, Higgins GA (1992) Alzheimer's disease: the amyloid cascade hypothesis. Science 256:184–185
196. Selkoe DJ (1991) The molecular pathology of Alzheimer's disease. Neuron 6:487–498
197. Kurup P, Zhang Y, Xu J et al (2010) Abeta-mediated NMDA receptor endocytosis in Alzheimer's disease involves ubiquitination of the tyrosine phosphatase STEP61. J Neurosci 30:5948–5957
198. Bialy L, Waldmann H (2005) Inhibitors of protein tyrosine phosphatases: next-generation drugs? Angew Chem Int Ed Engl 44:3814–3839
199. Vintonyak VV, Antonchick AP, Rauh D et al (2009) The therapeutic potential of phosphatase inhibitors. Curr Opin Chem Biol 13: 272–283
200. Sobhia ME, Paul S, Shinde R et al (2012) Protein tyrosine phosphatase inhibitors: a patent review (2002–2011). Expert Opin Ther Pat 22:125–153
201. He R, Zeng LF, He Y et al (2012) Small molecule tools for functional interrogation of protein tyrosine phosphatases. FEBS J 280: 731–750
202. Burke TRJ, Kole HK, Roller PP (1994) Potent inhibition of insulin receptor dephosphorylation by a hexamer peptide containing the phosphotyrosyl mimetic F2Pmp. Biochem Biophys Res Commun 204:129–134
203. Lipinski CA, Lombardo F, Dominy BW et al (2001) Experimental and computational approaches to estimate solubility and permeability in drug discovery and development settings. Adv Drug Deliv Rev 46:3–26
204. Andersen HS, Olsen OH, Iversen LF et al (2002) Discovery and SAR of a novel selective and orally bioavailable nonpeptide classical competitive inhibitor class of protein-tyrosine phosphatase 1B. J Med Chem 45:4443–4459
205. Erbe DV, Klaman LD, Wilson DP et al (2009) Prodrug delivery of novel PTP1B inhibitors to enhance insulin signalling. Diabetes Obes Metab 11:579–588
206. Yu X, Sun JP, He Y et al (2007) Structure, inhibitor, and regulatory mechanism of Lyp, a lymphoid-specific tyrosine phosphatase implicated in autoimmune diseases. Proc Natl Acad Sci U S A 104:19767–19772
207. Liu S, Zeng LF, Wu L et al (2008) Targeting inactive enzyme conformation: aryl diketoacid

derivatives as a new class of PTP1B inhibitors. J Am Chem Soc 130:17075–17084

208. Wu S, Bottini M, Rickert RC et al (2009) In silico screening for PTPN22 inhibitors: active hits from an inactive phosphatase conformation. ChemMedChem 4:440–444
209. Zhang XY, Bishop AC (2007) Site-specific incorporation of allosteric-inhibition sites in a protein tyrosine phosphatase. J Am Chem Soc 129:3812–3813
210. Tautz L, Mustelin T (2007) Strategies for developing protein tyrosine phosphatase inhibitors. Methods 42:250–260
211. Pot DA, Dixon JE (1992) Active site labeling of a receptor-like protein tyrosine phosphatase. J Biol Chem 267:140–143
212. Zhang ZY, Davis JP, Van Etten RL (1992) Covalent modification and active site-directed inactivation of a low molecular weight phosphotyrosyl protein phosphatase. Biochemistry 31:1701–1711
213. Liu S, Zhou B, Yang H et al (2008) Aryl vinyl sulfonates and sulfones as active site-directed and mechanism-based probes for protein tyrosine phosphatases. J Am Chem Soc 130: 8251–8260
214. Kim JH, Cho H, Ryu SE et al (2000) Effects of metal ions on the activity of protein tyrosine phosphatase VHR: highly potent and reversible oxidative inactivation by Cu^{2+} ion. Arch Biochem Biophys 382:72–80
215. Niesen FH, Berglund H, Vedadi M (2007) The use of differential scanning fluorimetry to detect ligand interactions that promote protein stability. Nat Protoc 2:2212–2221
216. Rhee SG, Chang TS, Bae YS et al (2003) Cellular regulation by hydrogen peroxide. J Am Soc Nephrol 14:S211–S215
217. Miki H, Funato Y (2012) Regulation of intracellular signalling through cysteine oxidation by reactive oxygen species. J Biochem 151:255–261
218. Roos G, Messens J (2011) Protein sulfenic acid formation: from cellular damage to redox regulation. Free Radic Biol Med 51: 314–326
219. Poole LB, Karplus PA, Claiborne A (2004) Protein sulfenic acids in redox signaling. Annu Rev Pharmacol Toxicol 44:325–347
220. Ostman A, Frijhoff J, Sandin A et al (2011) Regulation of protein tyrosine phosphatases by reversible oxidation. J Biochem 150: 345–356
221. Salmeen A, Andersen JN, Myers MP et al (2003) Redox regulation of protein tyrosine phosphatase 1B involves a sulphenyl-amide intermediate. Nature 423:769–773
222. van Montfort RL, Congreve M, Tisi D et al (2003) Oxidation state of the active-site cysteine in protein tyrosine phosphatase 1B. Nature 423:773–777
223. Yang J, Groen A, Lemeer S et al (2007) Reversible oxidation of the membrane distal domain of receptor PTPalpha is mediated by a cyclic sulfenamide. Biochemistry 46: 709–719
224. Haque A, Andersen JN, Salmeen A et al (2011) Conformation-sensing antibodies stabilize the oxidized form of PTP1B and inhibit its phosphatase activity. Cell 147:185–198

Chapter 14

High-Throughput Screening for Protein Tyrosine Phosphatase Activity Modulators

Lutz Tautz and Eduard A. Sergienko

Abstract

Reversible phosphorylation of proteins, principally on serine, threonine, or tyrosine residues, is central to the regulation of most aspects of eukaryotic cell function. Dysregulation of protein kinases and protein phosphatases is linked to numerous human diseases. Consequently, many efforts have been made to target these enzymes with small molecules in order to develop new therapeutic agents. While protein kinase inhibitors have been successfully brought to the market, the development of specific protein phosphatase inhibitors is still in its infancy. The largest and most diverse protein phosphatase superfamily in humans is comprised by the protein tyrosine phosphatases, a group of over 100 enzymes. Here, we describe high-throughput screening methods to search for protein tyrosine phosphatase activity modulators. We illustrate the implementation of relatively simple phosphatase assays, using generic absorbance- or fluorescence-based substrates, in 384- or 1536-well microtiter plates. We discuss steps to optimize HTS assay quality and performance, and describe several PTP screening methods on the basis of previously performed successful HTS campaigns. Finally, we discuss how to confirm, follow up, and prioritize hit compounds, and point out a number of common pitfalls that are encountered in this process.

Key words High-throughput screening, HTS, Protein tyrosine phosphatase, PTP, Inhibitors, pNPP, OMFP, DiFMUP, Biomol green

1 Introduction

Tyrosine phosphorylation [1] is a fundamental mechanism for numerous important aspects of eukaryotic cell physiology, including the regulation of cell-to-cell communication and signal transduction, cell growth and proliferation, cell cycle control, differentiation, malignant transformation, cell morphology, regulation of the cytoskeleton, neurotransmission, adhesion, gene regulation and transcription, intracellular vesicle transport, endocytosis, exocytosis, angiogenesis, embryogenesis, and development [2–10]. Not surprisingly, both protein tyrosine kinases (PTKs) [2, 11] and protein tyrosine phosphatases (PTPs) [12–15] have been implicated in many human diseases. Despite increasing efforts to generate PTP inhibitors [12, 16–20], compared to several approved

José Luis Millán (ed.), *Phosphatase Modulators*, Methods in Molecular Biology, vol. 1053,
DOI 10.1007/978-1-62703-562-0_14, © Springer Science+Business Media, LLC 2013

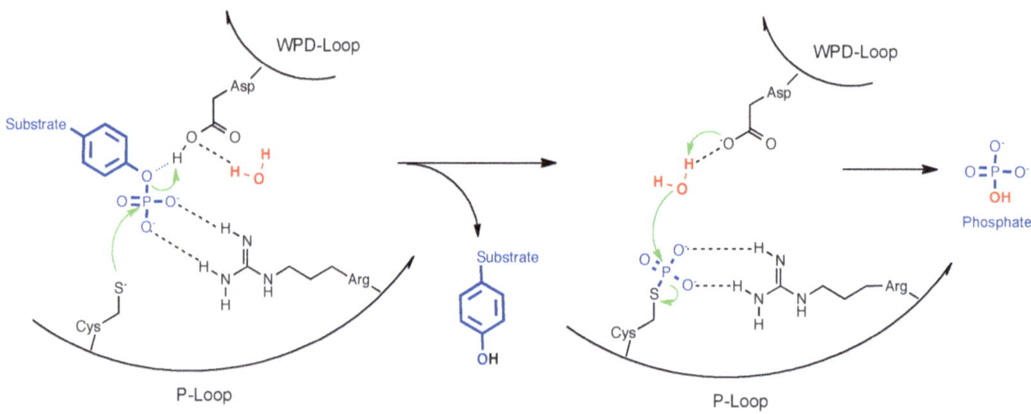

Fig. 1 Common PTP catalytic mechanism. The PTP catalysis is based on a nucleophilic cysteine with a low pK_a that forms a thiophosphate intermediate during the first step of the reaction, the transition state of which is stabilized by an invariant arginine. Cysteine and arginine are located in the conserved phosphate-binding loop (P-loop), which contains the PTP signature motif $C(X)_5R$. Hydrolysis is further assisted by the conserved WPD-loop aspartate, which acts as a general acid in the first step of the reaction, donating a proton to the OH-leaving group. In the second, rate-limiting step, the thiophosphate intermediate is hydrolyzed, and the aspartate acts now as a general base to abstract a proton from a water molecule, thus facilitating hydrolysis of the scissile phosphorous–sulfur bond and producing free phosphate

drugs (e.g., imatinib, dasatinib, sunitinib, nilotinib, ruxolitinib), and many more in clinical trials, that target specific PTKs, the development of novel therapeutics that target specific PTPs is still in its infancy. Major hurdles to overcome are related to insufficient bioavailability and selectivity of compounds, owing to the highly charged and highly conserved PTP active site (*see* also Chapter 13 in this book). All PTPs share a common catalytic mechanism based on a nucleophilic cysteine, which has an unusually low pK_a and forms a thiophosphate intermediate during the first step of the catalysis. In the second step of the ordered uni–bi reaction, the phosphoenzyme intermediate is hydrolyzed, producing free phosphate (Fig. 1).

High throughput screening (HTS) of chemical libraries is commonly used to identify hit compounds, mostly small molecules that bind to and modulate the function of biomolecules like enzymes, receptors, ion channels, or nucleic acids [21, 22]. With the advent of robotic liquid handling and detection systems, HTS became readily available in the early 1990s. Since then, continuous improvements of robotic devices, paired with miniaturization of assays, i.e., going from 96- to 384- to 1,536-well standard assay microtiter plates, has allowed HTS to become an extremely powerful tool in early drug discovery, able to test hundreds of thousands compounds per day. While HTS assays have been developed for all existing drug target classes, enzymes are particularly well suited for HTS, because robust enzymatic assays with recombinant proteins

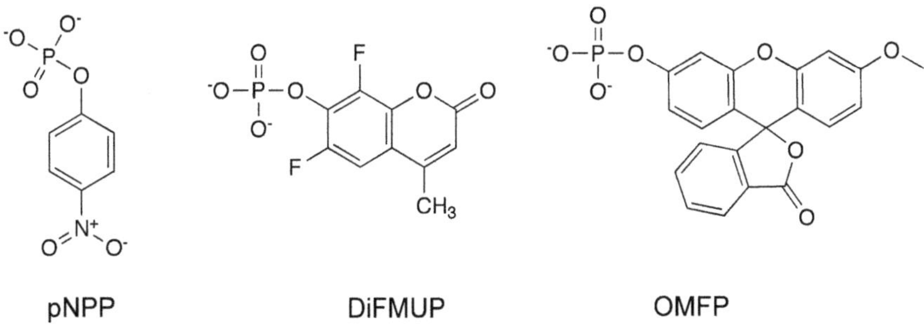

Fig. 2 Common generic PTP substrates. *p*-Nitrophenylphosphate (pNPP), 6,8-difluoro-4-methylumbelliferyl phosphate (DiFMUP), and *O*-methyl-fluorescein phosphate (OMFP)

and straightforward readouts can usually be set up very easily. In that sense, PTPs are no exception, with several relatively simple in vitro assays to choose from [23, 24]. These assays mainly vary by the utilized phosphatase substrate and the means of detecting the progress of the phosphatase reaction.

Widely used generic substrates include *p*-nitrophenylphosphate (pNPP), 6,8-difluoro-4-methylumbelliferyl phosphate (DiFMUP), and *O*-methyl-fluorescein phosphate (OMFP) (Fig. 2). To determine the rate of the PTP reaction, the concentration of the reaction products can be monitored by utilizing absorbance- or fluorescence-based detection, respectively. Hydrolysis of pNPP generates *p*-nitrophenol and free inorganic phosphate. The amount of *p*-nitrophenol can be directly determined via its absorbance at 405 nm, whereas an acidic molybdate–malachite green-based reagent (e.g., Biomol Green or PiColorLock) can be used to measure released phosphate, which forms a green colored complex with the reagent that absorbs light between 590 and 650 nm. Either of the assays have some advantages over the other. The malachite green-based assays are about five times more sensitive and therefore requiring lower enzyme concentration compared to measuring *p*-nitrophenol. However, they have a relatively narrow linear range of detection, ranging from about 1 to 50 μM free phosphate. In addition, they require an extra dispense and a 20–30 min incubation step in order for the colored complex to develop. Dispense of the detection reagent effectively terminates the reaction. In contrast, the absorbance of *p*-nitrophenol is linear over a much wider range of concentrations, from about 5 to 500 μM, but many colored compounds can interfere with the reading at 405 nm, resulting in potential false negatives. Both the malachite green and *p*-nitrophenol assays are typically run as end point assays. Adding the Biomol Green reagent, which contains 1 M HCl, effectively quenches the enzymatic reaction. Reading the absorbance of *p*-nitrophenol requires a pH > 7 in order to shift the equilibrium toward the deprotonated *p*-nitrophenolate anion, which is the actual colored species. Because the majority of the PTPs are most efficient at pH 5.5–6.0, a buffer system with a corresponding pH range is commonly used for HTS, requiring the addition of NaOH for *p*-nitrophenol absorbance reading, which also quenches the reaction.

More recently, fluorogenic phosphatase substrates, such as DiFMUP or OMFP, have been developed. These molecules have a low fluorescence in the phosphorylated state, but are strong fluorophores when dephosphorylated under typical PTP assay conditions. Because of the strong fluorescence intensity of the reaction product (and low fluorescence of all other assay components), PTP assays utilizing these substrates are usually 1–2 orders of magnitude more sensitive than either of the absorbance assays, requiring significantly less enzyme (typically low nanomolar or even subnanomolar concentrations). For recombinant enzymes that are difficult to produce in large quantities, this is a not to be underestimated advantage. Moreover, the signal to noise ratio is excellent for these substrates, and the fluorescence emission of the dephosphorylated products can be measured over a wide range of concentrations (~10 nM to >100 μM, depending on the used plate reader and plate density). Another advantage of fluorogenic substrates is the possibility to run the PTP assay in continuous mode (also called kinetic mode), which allows a much better control over the assay, the determination of more accurate values for the initial velocity rates, and also minimizes compound optical interference.

Fluorogenic substrates are characterized with minor drawbacks. DiFMUP relies on the near-UV/blue spectral range, thus similarly to pNPP may encounter compound spectral interference. Increasing the concentration of fluorescent product generated in the reaction (achieved by increase of either enzyme concentration or reaction incubation time) could lessen the compound's interference. OMFP-based assays generating methylfluorescein are much more resistant to compound optical interference. On the other hand, OMFP has limited aqueous solubility and requires initial dissolution in DMSO at mid-millimolar concentration, thus resulting in extra DMSO added to the reaction mixtures. The latter is usually manageable, because observed K_m values of OMFP for most PTPs are in the mid- to low-micromolar range, keeping the final DMSO concentration <5 % when used at K_m concentration in the HTS assay.

2 Materials

2.1 Reagents and Consumables

- *p*-nitrophenylphosphate (pNPP; Sigma).
- 6,8-difluoro-4-methylumbelliferyl phosphate (DiFMUP; Life Technologies).

- *O*-methyl-fluorescein phosphate (OMFP; Sigma).
- BioMol Green reagent (Biomol).
- Bis–Tris (2-(bis(2-hydroxyethyl)amino)-2-(hydroxymethyl) propane-1,3-diol).
- Sodium orthovanadate (Na_3VO_4; Sigma).
- Tween-20.
- Polyethylene glycol-8000 (PEG-8000).
- Dithiothreitol (DTT).
- Microtiter plates: 384- or 1,536-well plates, black solid (for fluorescence readings) and clear bottom (for absorbance readings).
- Aluminum adhesive seals.

2.2 Equipment

- Echo 550 acoustic dispenser (Labcyte) for dispensing DMSO compound solutions at volumes as low as 2.5 nL.
- Reagent dispenser, e.g., Multidrop Combi (Thermo Scientific), Bravo Automated Liquid Handling Platform (Agilent Technologies).
- Small tube metal/plastic tip dispensing cassettes for Multidrop Combi (Thermo Scientific).
- Microplate reader/imager, e.g., PHERAstar (BMG Labtech), Envision (Perkin Elmer), ViewLux (Perkin Elmer).
- Multichannel pipettes: 8- or 12-channel pipettes.

2.3 Stock and Working Solutions

- Use Milli-Q water for preparing buffers and dissolving reagents.
- Make up buffer without additive (Tween-20 or PEG) in large scale and store at 4 °C.
- Prepare reaction buffer at volumes as needed by adding fresh additive (Tween-20 or PEG) on the day of use.
- Prepare enzyme buffer by adding fresh DTT on the day of use and keep on ice (*see* **Note 1**).
- Prepare 1 M Na_3VO_4 stock solution in water or buffer, make aliquots as necessary, and store at −20 °C.
- Prepare 1 M DTT stock solution in water or buffer, make aliquots as necessary, and store at −20 °C.
- Prepare 10 mM OMFP stock solution in DMSO by sonicating for 10 min until solution is clear. Make aliquots as necessary and store at −20 °C. Prepare working solutions by diluting in water or buffer on the day of use and keep on ice in the dark until use.
- Prepare 10 mM DiFMUP stock solution in water or buffer, make aliquots as necessary, and store at −80 °C. Prepare work-

ing solutions by diluting in water or buffer on the day of use and keep on ice in the dark until use.

- Prepare pNPP working solutions fresh from powder/tablets on the day of use and keep on ice until use.
- Make aliquots of enzyme stock solution as necessary and store at −80 °C. Prepare enzyme working solutions in enzyme buffer just before use and keep on ice.

2.4 Recommended Software Programs

- GraphPad Prism (GraphPad Software, Inc.).

3 Methods

3.1 Setting Up and Optimizing Phosphatase Assays for HTS

This stage represents a series of iterative tests aimed at determination and optimization of a set of assay parameters, including optimal buffer composition and pH, optimal enzyme concentration, Michaelis–Menten constant (K_m), DMSO tolerance, and assay statistics such as signal to background ratio (S/B), signal to noise ratio (S/N), and Z′-factor [25]. An appropriate positive control should also be selected and tested at this stage. Furthermore, depending on the available equipment for liquid handling and absorbance/fluorescence measurements, and the number of assay plates to be tested, optimal timing that ensures equal assay conditions and treatment for each plate during the screening needs to be determined. The reaction time represents the variable that can easily be adjusted, usually somewhere between 20 min and 2 h.

Definitions of statistical assay parameters:

S/B = (mean signal)/(mean background).

S/N = (mean signal − mean background)/(SD of background).

Z′ = 1 − [(3 × SD of signal − 3 × SD of background)/(mean signal − mean background)].

SD, standard deviation; background, no-enzyme control.

1. Determine an approximate suitable enzyme concentration through testing the catalytic activity of a serial dilution of enzyme at a fixed substrate concentration, e.g., 5 mM pNPP or 100 μM OMFP or DiFMUP. Use the assay format (384- or 1,536-well) and reaction time that will be employed for HTS. A suitable enzyme concentration should not deplete more than 10–15 % of the substrate during the course of reaction, in order to keep the enzymatic reaction within the linear range, i.e., at initial velocity conditions with a constant rate. This will ensure that the assay is not influenced by factors such as product inhibition, substrate limitation, and inactivation of the enzyme due to instability. Also, make sure that the assay is run within the capacity range of the instrument. If in doubt,

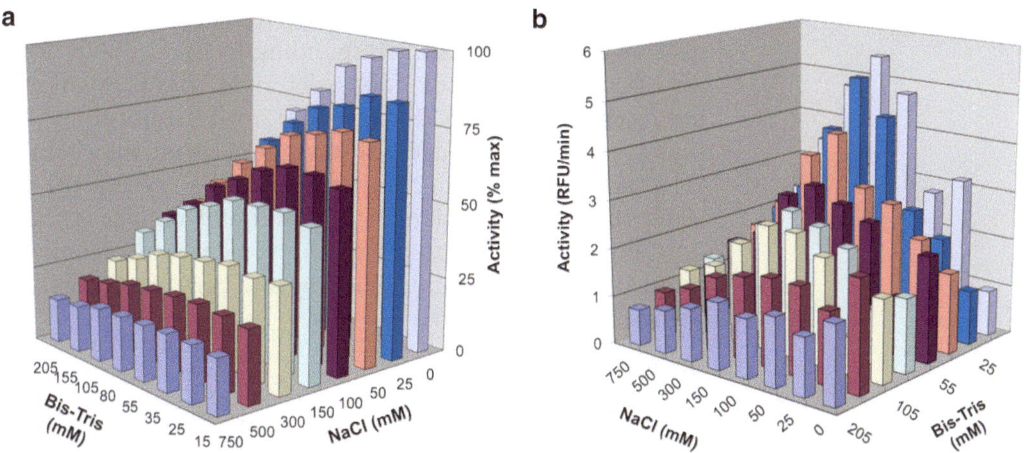

Fig. 3 (**a**) Buffer optimization for a pNPP colorimetric assay in 384-well format using *p*-nitrophenol detection. Activity of HePTP at a concentration of 50 nM was tested in the presence of 1.3 mM pNPP, 1 mM DTT, 0.005 % Tween-20, and various concentrations of Bis–Tris, pH 6.0, and NaCl, over a reaction time of 1 h. The optimal buffer under HTS conditions was 20 mM Bis–Tris, pH. 6.0, 1 mM DTT, and 0.005 % Tween-20. The K_m value for pNPP was 0.4 mM. The reaction demonstrated linearity of the progress curves over a period of 2 h. Enzymatic activity was proportional to enzyme concentration. Assay performance was confirmed using the general PTP inhibitor orthovanadate, the IC_{50} value of which was 150 μM. (**b**) Buffer optimization for an OMFP fluorescent assay in 384-well format. Activity of HePTP at a concentration of 6.25 nM was tested in the presence of 0.5 mM OMFP, 1 mM DTT, 0.005 % Tween-20, and various concentrations of Bis–Tris (pH 6.0) and NaCl. Fluorescence intensity was measured in kinetic mode using fluorescein optics. The optimal buffer condition was 20 mM Bis–Tris, pH 6.0, 150 mM NaCl, 1 mM DTT, and 0.005 % Tween-20. The K_m value for OMFP in the optimized buffer was 0.3 mM. Figure reprinted with permission from ref. [27]. Copyright 2012 American Chemical Society

determine the linear range of detection by measuring and plotting the signal over a wide range of product concentrations. (The dephosphorylated products of pNPP, DiFMUP, and OMFP are commercially available.) To measure background values, i.e., PTP-independent hydrolysis of the substrate, include a no-enzyme control; aim for an S/B of >5.

2. Optimize buffer conditions through testing buffers at different pH values at different concentrations and different ionic strengths, the latter being adjusted by the addition of sodium chloride (NaCl). Examples for a buffer optimization matrix are shown in Fig. 3. PTPs have peak activities usually between pH 5.5 and 6.5. Therefore, the buffer we recommend is Bis–Tris, which has a pK_a value of 6.5 (at 25 °C) and is useful for a pH range of 5.8–7.2. Always add a detergent such as 0.005 % Tween-20 in order to stabilize the protein, prevent it from sticking to the wall of the microtiter well, and reduce the likelihood of promiscuous, aggregate-based inhibition [26]. Bovine serum albumin (BSA) or globulin proteins may be used as detergent substitutes to stabilize the enzyme in order to ensure linearity of the catalytic reaction.

3. To identify competitive inhibitors, the substrate should be used at a concentration corresponding to its K_m value. Using the optimized buffer conditions, determine the K_m by measuring Michaelis–Menten kinetics. This involves measuring PTP activity at serially diluted substrate concentrations (between 0.1- and 10-times K_m concentration) and analyzing the data using the Michaelis–Menten equation and nonlinear regression. For ease of calculations and convenience, we recommend using a dedicated scientific graphing and curve fitting software such as GraphPad Prism.

   Michaelis–Menten equation:

   $$v_0 = (V_{max}[S]) / (K_m + [S])$$

   v_0, initial rate (initial velocity); V_{max}, maximum rate; $[S]$, substrate concentration; K_m, Michaelis–Menten constant.

4. Test the effect of various DMSO concentrations on the activity of the enzyme and the background control values. In our experience with PTPs, we found that assays can be run reliably at up to 5 % (v/v) DMSO.
5. Using the final, optimized assay conditions, reevaluate the enzyme concentration to be utilized for HTS. To do so, record progression curves at different enzyme concentrations and analyze the data using linear regression. Linearity can be assumed for correlation coefficients (r^2) >0.99.
6. Validate the assay by testing a positive control, e.g., a known inhibitor of the PTP of interest. If no specific inhibitor is available, use orthovanadate, which is a general, transition state analogue PTP inhibitor.
7. Evaluate well-to-well assay performance by testing entire plates containing negative (no inhibition) and positive (100 % inhibition) controls, and calculating assay statistics, including S/B, S/N and Z′-factor. Desired quality control statistics are: S/B > 5, S/N > 10, and Z′ > 0.6. Repeat the experiment on at least one more day to evaluate day-to-day performance.
8. Finally, determine an appropriate compound concentration for HTS. Start with for example 20 μM and perform a pilot HTS with a subset of compounds of a larger library (e.g., 5,000–10,000 compounds). Aim for a hit rate between 0.1 and 0.5 % at a set threshold of 50 % inhibition. If the hit rate is higher than 0.5 %, lower the compound concentration; if the hit rate is lower than 0.1 %, increase the compound concentration. Once you settle on a concentration value, test at least one duplicate set of compound plates on different days to confirm reproducibility of hits.

3.2 HTS Protocol for Using pNPP and p-Nitrophenol Detection

This protocol was used to screen for inhibitors of HePTP (*PTPN7*), a potential drug target in acute leukemias [27]. The assay was developed in 384-well clear plates.

Final assay conditions:

- Bis–Tris: 20 mM, pH 6.0.
- Tween-20: 0.005 %.
- DTT: 1 mM.
- $HePTP_{44-339}$: 2.7 nM.
- pNPP: 0.4 mM.
- Final reaction volume: 20 μL.
- Test compound concentration: 20 μM.
- Final DMSO concentration: 2 %.

Plate map:

- Positive (Low) control in column 1: 9 mM sodium orthovanadate, enzyme and substrate.
- Negative (High) control in column 2: DMSO, enzyme and substrate.
- Test compound in columns 3–24: test compounds, enzyme and substrate.

Procedure:

1. Add 4 μL of 45 mM sodium orthovanadate to column 1.
2. Add 4 μL of 10 % DMSO aqueous solution to column 2.
3. Add 4 μL of 5× compounds solutions in 10 % DMSO to columns 3–24.
4. Add 8 μL of 2.5× pNPP solution in water to all wells.
5. Start reaction by adding 8 μL of 2.5× HePTP solution in 2.5× assay buffer to all wells.
6. Briefly spin down the plates (15 s at 223 × *g*).
7. Incubate reaction at room temperature for 1 h.
8. Terminate the reaction by adding 40 μL of 0.1 M NaOH.
9. Measure the absorbance at 405 nm using an appropriate plate reader.

3.3 HTS Protocol for Using pNPP and Free Phosphate Detection with the Biomol Green Reagent

This approach is illustrated on the example of HTS for VHR (*DUSP3*) inhibitors [28]. This assay was developed in 1,536-well black clear-bottom plates. Results are published at PubChem (http://pubchem.ncbi.nlm.nih.gov), Assay ID 1654.

Final assay conditions:

- Bis–Tris: 20 mM, pH 6.0.
- Tween-20: 0.005 %.

- DTT: 1 mM.
- VHR: 40 nM.
- pNPP: 1.474 mM.
- Final reaction volume: 3 μL.
- Test compound concentration: 13.3 μM.
- Final DMSO concentration: 0.67 %.

Plate map:

- Positive (Low) control in columns 1 and 2: DMSO, substrate only.
- Negative (High) control in columns 3 and 4: DMSO, enzyme and substrate.
- Test compound in columns 5–48: test compounds, enzyme and substrate.

Procedure:

1. Add 1.5 μL of assay buffer to columns 1–2 (*see* **Note 2**).
2. Add 1.5 μL of 1.5 mM pNPP in assay buffer to all wells.
3. Dispense 20 nL of 2 mM compound solutions in 100 % DMSO to columns 5–48, and 20 nL DMSO to columns 1–4.
4. Add 1.5 μL of 40 nM VHR in assay buffer to columns 3–48.
5. Briefly spin down the plates (15 s at 223 × *g*).
6. Incubate for 1 h at room temperature.
7. Add 3 μL Biomol Green reagent to all wells.
8. Briefly spin down the plates (15 s at 223 × *g*).
9. Incubate for 30 min.
10. Read plate at 620–650 nm in absorbance mode (*see* **Note 3**).
11. Determine the ratio of inhibition in comparison to the negative control.

Compounds that demonstrated activity of ≥50 % inhibition were defined as hits in this assay and cherry-picked for rescreening in triplicate.

3.4 HTS Protocol for Using OMFP

This assay was developed for identifying inhibitors of STEP (*PTPN5*), a drug target in Alzheimer's disease and other neuropsychiatric disorders [29]. The assay was adapted to 1536-well plate format, using black solid flat bottom plates. Results are published at PubChem (http://pubchem.ncbi.nlm.nih.gov), Assay ID 588621.

Final assay conditions:

- Bis–Tris: 50 mM, pH 6.0.
- Tween-20: 0.005 %.

- DTT: 2.5 mM.
- STEP (catalytic domain): 0.5 nM.
- OMFP: 25 μM.
- Final reaction volume: 4 μL.
- Test compound concentration: 20 μM.
- Final DMSO concentration: 1 %.

Plate map:

- Positive (Low) control in columns 1 and 2: DMSO, substrate only.
- Negative (High) control in columns 3 and 4: DMSO, enzyme and substrate.
- Test compound in columns 5–48: test compounds, enzyme and substrate.

Procedure:

1. Transfer 40 nL from a plate containing 2 mM test compounds into assay plate columns 5–44. Transfer 40 nL DMSO to columns 1–4 for control wells.
2. Spin plates at 223 × *g* for 1 min.
3. Add 2 μL/well of control buffer (no-enzyme control) to columns 1 and 2 for the positive control wells.
4. Add 2 μL/well of enzyme solution to columns 3–48 for both the negative control and test compound wells.
5. Add 2 μL/well of substrate solution to columns 1–48 (all wells).
6. Spin plates at 223 × *g* for 1 min.
7. Incubate plates in the dark at room temperature for 20 min.
8. Detect signals on Perkin Elmer ViewLux (*see* **Note 4**).

Compounds that demonstrated activity of ≥40 % inhibition were defined as hits in this assay.

3.5 HTS Protocol for Using DiFMUP

This approach is illustrated on the example of LYP (*PTPN22*), a potential drug target in autoimmunity [30–32]. This assay was developed in 384-well black plates. Results are published at PubChem (http://pubchem.ncbi.nlm.nih.gov), Assay ID 640.

Final assay conditions:

- Bis–Tris: 150 mM, pH 6.0, ionic strength adjusted to 150 mM.
- PEG: 0.33 %.
- DTT: 1.67 mM.
- GST-LYP (catalytic domain): 2.5 nM.
- DiFMUP: 50 μM.

- Final reaction volume: 40 μL.
- Test compound concentration: 2 μM.
- Final DMSO concentration: 0.6 %.

Plate map:

- Positive (Low) control in columns 23 and 24: 8 mM sodium orthovanadate, enzyme and substrate.
- Positive (Low) control ("background") in column 1: DMSO, substrate only.
- Negative (High) control in column 2: DMSO, enzyme and substrate.
- Test compound in columns 3–22, test compounds, enzyme and substrate.

Procedure:

1. Add 20 μL assay buffer to columns 1 and 2.
2. Add 20 μL of vanadate working solution to columns 23 and 24.
3. Add 60 nL of 2 mM compound in 100 % DMSO to columns 3–22.
4. Add 60 nL of 100 % DMSO to columns 1 and 2.
5. Add 20 μL of enzyme buffer to column 1.
6. Add 20 μL of LYP working solution to columns 2–24.
7. Start reaction by adding 20 μL of DiFMUP working solution to each well.
8. Incubate plates in the dark for 40 min at room temperature.
9. Shake plate for 30 s and measure fluorescence using a plate reader with excitation at 360 nm, emission at 465 nm.

All data were normalized on a per-plate basis. Inhibition for each compound was calculated using the following formula: % inhibition = 100(1 – [test compound – background]/[negative control – background]).

Hits were defined using a hit threshold of 24 % inhibition, corresponding to the average plus three standard deviations.

3.6 Hit Confirmation, Prioritization, and Follow-up Studies

Primary hits are usually cherry-picked for rescreening in triplicate. In order to prioritize compounds for which dry powders will be obtained for retesting, we also recommend assaying serial compound dilutions and discarding hits from further consideration that yield dose–response curves with Hill slopes far from 1. In particular, compounds with very steep IC_{50} curves (Hill slope >4) should be avoided, because unspecific effects are likely at play. Once inhibitors have been confirmed from repurchased powders, compounds with reasonable IC_{50} values (e.g., ≤20 μM) should be tested in counterscreens against a panel of additional PTPs, in order to weed out compounds that non-selectively inhibit PTPs. In addition, publicly available screening data (e.g., from PubChem) can be mined in order to detect and sort out "frequent hitters". Typically, the vast majority of hits will be discarded at this stage, because only few compounds will show selectivity for the PTP of interest. However, depending on the novelty, properties, and tractability of particular chemical scaffolds, hit compounds may be kept for further consideration, even though they are not selective. Importantly, at this stage, we highly recommend repurifying the commercial powders using preparative HPLC and retesting the compounds. In our experience, inhibition of PTPs is often not due to the actual compound, but to impurities in commercial substances. Because the PTP catalytic cysteine residue is not only prone to oxidation but also susceptible to alkylation, e.g., via Michael addition or nucleophilic substitution, compounds should also be tested for time-dependent inhibition by measuring activity progress curves at different compound concentrations (Fig. 4a). Linearity of these curves indicates that there is no time-dependent inhibition of the enzyme. Close ABC (analog-by-catalog) analogs of hit compounds should be obtained and tested to establish a sound structure–activity relationship (SAR). Michaelis–Menten kinetic studies should be performed to determine inhibition constants and the mode of inhibition [23]; an Eadie–Hofstee plot for a typical hit compound exhibiting a mixed inhibition pattern is shown in Fig. 4b.

If a three-dimensional structure of the PTP is available, in silico docking can be utilized to model and dock the inhibitor into the active site (Fig. 5). The docking can then be verified by mutagenesis studies, in which specific amino acids that the docking predicted to interact with the inhibitor are mutated to alanine. Inhibitory activity of the compound is then measured with the mutant protein and is expected to be lower than with the wild-type protein (Fig. 4c). Finally, biophysical binding studies, such as isothermal titration calorimetry (ITC) [33] or ThermoFluor [34] may further aid in prioritizing compounds for cell-based assays and high-resolution structural analysis.

4 Notes

1. The presence of reducing agents such as DTT ensures that the PTP is in fully active state, i.e., the catalytic cysteine is in the reduced thiol/thiolate form. In addition, reducing agents also prevent potentially oxidizing compounds from unspecifically inhibiting PTPs through oxidation of the catalytic cysteine residue. On the other hand, some compounds are known to generate hydrogen peroxide in the presence of DTT, leading to inactivation of PTPs [35, 36].

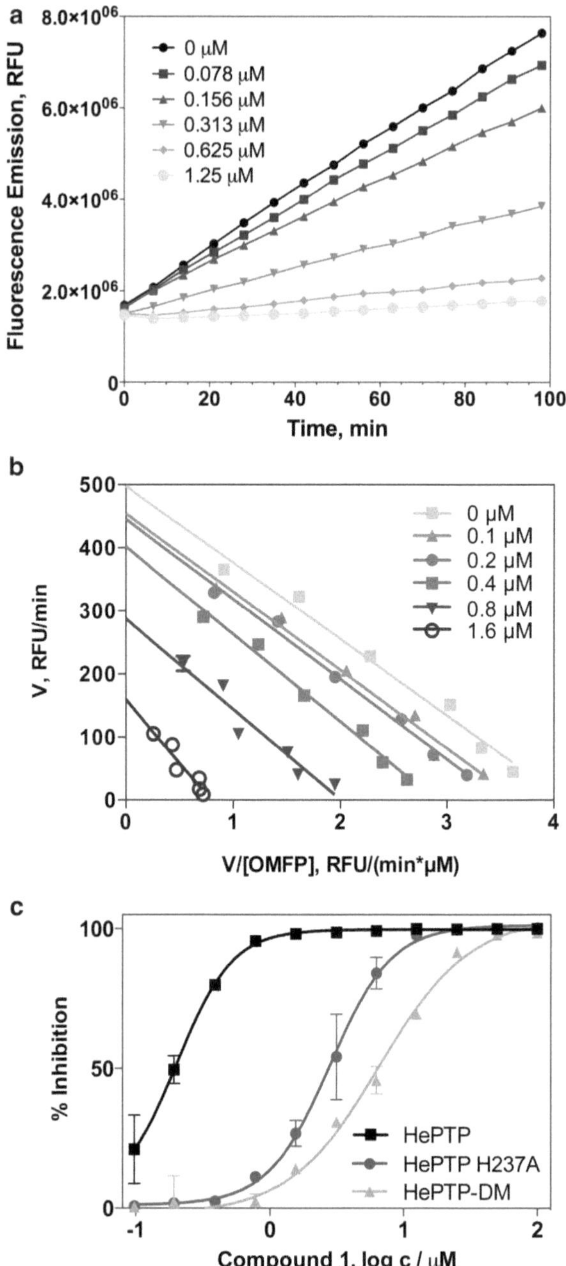

Fig. 4 Mechanism of action (MOA) and inhibition studies of ML119 (compound **1**) with HePTP and HePTP mutants. **(a)** Progress curves of HePTP (6.25 nM) activity in the presence of different doses of compound **1** (0, 0.078, 0.156, 0.313, 0.625, 1.25 μM) and 0.3 mM OMFP in 20 mM Bis–Tris, pH 6.0, 150 mM NaCl, 1 mM DTT, and 0.005 % Tween-20 in 20 μL total assay volume in black 384-well microtiter plates. No time-dependent inhibition was observed as demonstrated by the linear progress curves of the HePTP phosphatase reaction. **(b)** Eadie–Hofstee plot of the Michaelis–Menten kinetic study with compound **1**. The HePTP-catalyzed hydrolysis of OMFP was assayed at room temperature in a 60 μL 96-well format reaction system in 50 mM Bis–Tris, pH 6.0 assay buffer containing 1.7 mM DTT, 0.005 % Tween-20, and 5 % DMSO. Recombinant HePTP (5 nM) was preincubated with various fixed concentrations of inhibitor (0, 0.1, 0.2, 0.4, 0.8, 1.6 μM) for 10 min. The reaction was initiated by addition of various concentrations of substrate (0, 12.5, 25, 50, 100, 200, 400 μM) to the

2. Important: This assay is not compatible with buffers containing carboxylic acid groups, such as sodium citrate or *N*-(2-acetamido)-iminoacetic acid (ADA), because the Biomol Green complex formation is compromised in such buffers.
3. Perkin Elmer Viewlux Settings:
   - Ex1:2 = 630 DF10.
   - Em1:564/250(absorbance).
   - Em2:Clear.
   - Light energy = 100,000.
   - measurement time = 5 s.
   - 1× binning.
4. Perkin Elmer Viewlux Settings:
   - Light Energy: 10,000.
   - Measurement: Time 1 s.
   - Excitation Filter: 480/20 (FITC).
   - Emission Filter: 540/25 (FITC).
   - Mirror: FITC dichroic.

Fig. 4 (continued) reaction mixture. The initial rate was determined using an excitation wavelength of 485 nm and measuring the emission of the fluorescent reaction product at 528 nm. The nonenzymatic hydrolysis of the substrate was corrected by measuring the control without addition of enzyme. Eadie–Hofstee plots were generated using the program GraphPad Prism. The inhibition pattern and inhibition constant were determined as described previously [37]. **(c)** Dose–response curves for compound **1** with wild-type HePTP, HePTP K105A/T106A (HePTP-DM), or HePTP H237A. Compound **1** was 31-fold less active against HePTP-DM (IC_{50} = 6.6 μM), and 13-fold less active against the HePTP H237A (IC_{50} = 2.8 μM) as compared to wild-type HePTP (IC_{50} = 0.21 μM). As predicted by the in silico docking (*see* Fig. 5), these results confirmed that Lys105 and Thr106 as well as His237 are important for the binding of compound **1** to HePTP. Compound **1** was twofold serially diluted in DMSO before being added to the reactions for a 10-point dose–response curve. Reactions contained 50 mM Bis–Tris pH 6.0, 1.7 mM DTT, 0.005 % Tween 20, and 5 % DMSO in 60 μL total assay volume in black 96-well microtiter plates. OMFP concentrations (corresponding to the K_m value of each enzyme) were 117 μM for HePTP (2.75 nM), 144 μM for HePTP-DM (5 nM), and 222 μM for HePTP H237A (5 nM). Fluorescence intensity was measured in kinetic mode over 10 min to determine the slopes of progress curves, using an excitation wavelength of 485 nm, and an emission wavelength of 528 nm. IC_{50} values were determined using nonlinear regression (sigmoidal dose–response with variable slope) and the program GraphPad Prism. Figure reprinted with permission from ref. [27]. Copyright 2012 American Chemical Society

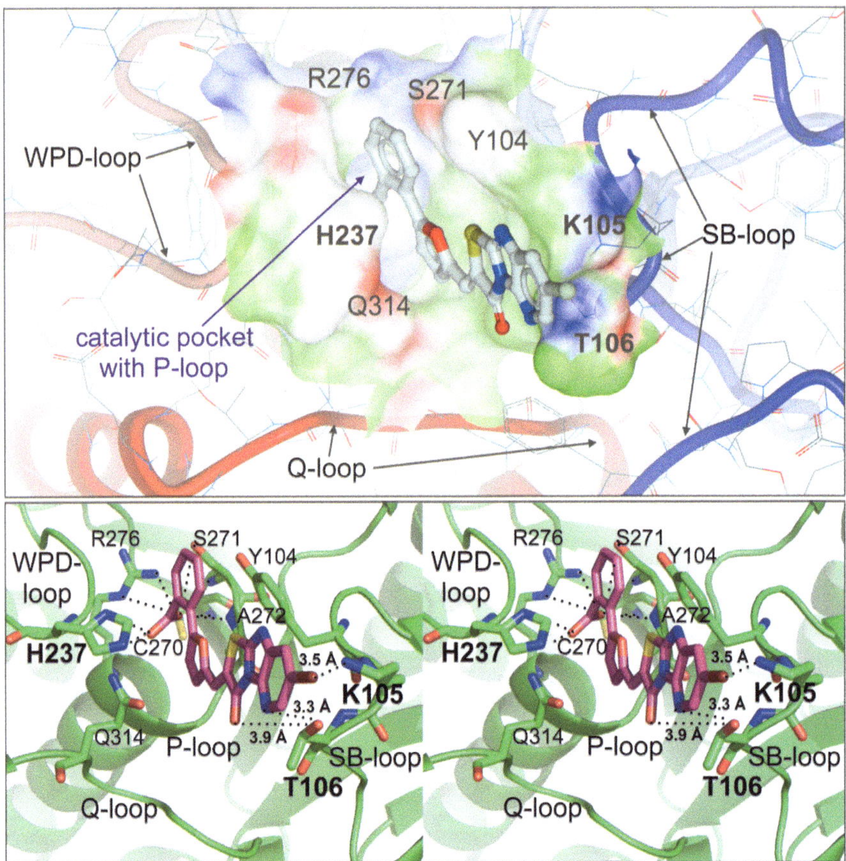

Fig. 5 In silico docking of ML119 (compound **1**) into the HePTP active site. *Upper panel*: Ribbon and surface representation of the HePTP active site (crystal structure, PDB ID: 3D44) with docked inhibitor (stick representation). Surface color code: *white*, neutral; *green*, hydrophobic; *red*, hydrogen bond acceptor potential; *blue*, hydrogen bond donor potential. *Lower panel*: *Stereo ribbon* diagram of the docking pose used in the *upper panel*. Residues that interact with compound **1** (*magenta*) and the catalytic cysteine (C270 in HePTP) are shown in stick representation. Flexible ligand docking calculations were performed with the ICM docking algorithm as implemented in the ICM-Pro program (v3.7-1g, Molsoft, LLC.). The coordinates of the crystal structure were converted into ICM objects, charges were assigned, orientations of side chain amides were corrected, and hydrogen atoms added and their positions optimized by energy minimization using MMFF force field. Mutations in the crystal structures were corrected to wild-type amino acid sequences, and the side chains were optimized using the Optimize Side Chains tool as implemented in ICM-Pro. The binding pocket was defined in an 8 Å radius around the phosphotyrosine of the peptide ligand. Figure reprinted with permission from ref. [27]. Copyright 2012 American Chemical Society

Acknowledgments

This work was supported by NIH Roadmap Initiative grants U54HG003916 and U54HG005033 (to CPCCG), and NIH grants R03MH095532, R03MH084230, and R21CA132121 (to L.T.).

References

1. Hunter T, Sefton BM (1980) Transforming gene product of Rous sarcoma virus phosphorylates tyrosine. Proc Natl Acad Sci U S A 77:1311–1315
2. Hunter T (1998) The role of tyrosine phosphorylation in cell growth and disease. Harvey Lect 94:81–119
3. Hunter T (2000) Signaling–2000 and beyond. Cell 100:113–127
4. Larsen M, Tremblay ML, Yamada KM (2003) Phosphatases in cell-matrix adhesion and migration. Nat Rev Mol Cell Biol 4:700–711
5. Alonso A, Sasin J, Bottini N et al (2004) Protein tyrosine phosphatases in the human genome. Cell 117:699–711
6. Mustelin T, Vang T, Bottini N (2005) Protein tyrosine phosphatases and the immune response. Nat Rev Immunol 5:43–57
7. Halle M, Tremblay ML, Meng TC (2007) Protein tyrosine phosphatases: emerging regulators of apoptosis. Cell Cycle 6:2773–2781
8. Pao LI, Badour K, Siminovitch KA et al (2007) Nonreceptor protein-tyrosine phosphatases in immune cell signaling. Annu Rev Immunol 25:473–523
9. Hunter T (2009) Tyrosine phosphorylation: thirty years and counting. Curr Opin Cell Biol 21:140–146
10. Rhee I, Veillette A (2012) Protein tyrosine phosphatases in lymphocyte activation and autoimmunity. Nat Immunol 13:439–447
11. Cohen P (2002) Protein kinases–the major drug targets of the twenty-first century? Nat Rev Drug Discov 1:309–315
12. Tautz L, Pellecchia M, Mustelin T (2006) Targeting the PTPome in human disease. Expert Opin Ther Targets 10:157–177
13. Tonks NK (2006) Protein tyrosine phosphatases: from genes, to function, to disease. Nat Rev Mol Cell Biol 7:833–846
14. Vang T, Miletic AV, Arimura Y et al (2008) Protein tyrosine phosphatases in autoimmunity. Annu Rev Immunol 26:29–55
15. Julien SG, Dube N, Hardy S et al (2011) Inside the human cancer tyrosine phosphatome. Nat Rev Cancer 11:35–49
16. Bialy L, Waldmann H (2005) Inhibitors of protein tyrosine phosphatases: next-generation drugs? Angew Chem Int Ed Engl 44:3814–3839
17. Vintonyak VV, Antonchick AP, Rauh D et al (2009) The therapeutic potential of phosphatase inhibitors. Curr Opin Chem Biol 13:272–283
18. Barr AJ (2010) Protein tyrosine phosphatases as drug targets: strategies and challenges of inhibitor development. Future Med Chem 2:1563–1576
19. Sobhia ME, Paul S, Shinde R et al (2012) Protein tyrosine phosphatase inhibitors: a patent review (2002–2011). Expert Opin Ther Pat 22:125–153
20. He R, Zeng LF, He Y et al (2013) Small molecule tools for functional interrogation of protein tyrosine phosphatases. FEBS J 280:731–750
21. Mayr LM, Bojanic D (2009) Novel trends in high-throughput screening. Curr Opin Pharmacol 9:580–588
22. Kool J, Lingeman H, Niessen W et al (2010) High throughput screening methodologies classified for major drug target classes according to target signaling pathways. Comb Chem High Throughput Screen 13:548–561
23. Tautz L, Mustelin T (2007) Strategies for developing protein tyrosine phosphatase inhibitors. Methods 42:250–260
24. Montalibet J, Skorey KI, Kennedy BP (2005) Protein tyrosine phosphatase: enzymatic assays. Methods 35:2–8
25. Zhang JH, Chung TD, Oldenburg KR (1999) A simple statistical parameter for use in evaluation and validation of high throughput screening assays. J Biomol Screen 4:67–73
26. Feng BY, Shoichet BK (2006) A detergent-based assay for the detection of promiscuous inhibitors. Nat Protoc 1:550–553
27. Sergienko E, Xu J, Liu WH et al (2012) Inhibition of hematopoietic protein tyrosine phosphatase augments and prolongs ERK1/2 and p38 activation. ACS Chem Biol 7:367–377
28. Bobkova EV, Liu WH, Colayco S et al (2011) Inhibition of the hematopoietic protein tyrosine phosphatase by phenoxyacetic acids. ACS Med Chem Lett 2:113–118
29. Goebel-Goody SM, Baum M, Paspalas CD et al (2012) Therapeutic implications for striatal-enriched protein tyrosine phosphatase (STEP) in neuropsychiatric disorders. Pharmacol Rev 64:65–87
30. Vang T, Congia M, Macis MD et al (2005) Autoimmune-associated lymphoid tyrosine phosphatase is a gain-of-function variant. Nat Genet 37:1317–1319
31. Xie Y, Liu Y, Gong G et al (2008) Discovery of a novel submicromolar inhibitor of the lymphoid specific tyrosine phosphatase. Bioorg Med Chem Lett 18:2840–2844
32. Vang T, Liu WH, Delacroix L et al (2012) LYP inhibits T-cell activation when dissociated from CSK. Nat Chem Biol 8:437–446
33. Freyer MW, Lewis EA (2008) Isothermal titration calorimetry: experimental design, data

analysis, and probing macromolecule/ligand binding and kinetic interactions. Methods Cell Biol 84:79–113

34. Cummings MD, Farnum MA, Nelen MI (2006) Universal screening methods and applications of ThermoFluor. J Biomol Screen 11:854–863

35. Bova MP, Mattson MN, Vasile S et al (2004) The oxidative mechanism of action of ortho-quinone inhibitors of protein-tyrosine phosphatase alpha is mediated by hydrogen peroxide. Arch Biochem Biophys 429: 30–41

36. Johnston PA, Foster CA, Tierno MB et al (2009) Cdc25B dual-specificity phosphatase inhibitors identified in a high-throughput screen of the NIH compound library. Assay Drug Dev Technol 7:250–265

37. Tautz L, Bruckner S, Sareth S et al (2005) Inhibition of Yersinia tyrosine phosphatase by furanyl salicylate compounds. J Biol Chem 280:9400–9408

Chapter 15

Evaluating Effects of Tyrosine Phosphatase Inhibitors on T Cell Receptor Signaling

Souad Rahmouni, Laurence Delacroix, Wallace H. Liu, and Lutz Tautz

Abstract

The importance of tyrosine phosphorylation in normal cell physiology is well established, highlighted by the many human diseases that stem from abnormalities in protein tyrosine kinase (PTK) and protein tyrosine phosphatase (PTP) function. Contrary to earlier assumptions, it is now clear that both PTKs and PTPs are highly specific, non-redundant, and tightly regulated enzymes. Hematopoietic cells express particularly high numbers of PTKs and PTPs, and aberrant function of these proteins have been linked to many hematopoietic disorders. While PTK inhibitors are among FDA approved drugs for the treatment of leukemia and other cancers, efforts to develop therapeutics that target specific PTPs are still in its infancy. Here, we describe methods on how to evaluate effects of PTP inhibitors on T cell receptor signaling. Moreover, we provide a comprehensive strategy for compound prioritization, applicable to any drug discovery project involving T cells. We present a testing funnel that starts with relatively high-throughput luciferase reporter assays, followed by immunoblot, calcium flux, flow cytometry, and proliferation assays, continues with cytokine bead arrays, and finishes with specificity assays that involve RNA interference. We provide protocols for experiments in the Jurkat T cell line, but more importantly give detailed instructions, paired with numerous tips, on how to prepare and work with primary human T cells.

Key words T cells, PBMC, NFAT/AP-1 assay, Phosphotyrosine blot, Calcium flux, Cytokine bead array (CBA), ^{3}H-thymidine incorporation, siRNA, CD25, CD69, TCR, PTP, Small molecule inhibitors

1 Introduction

A dynamic balance between tyrosine phosphorylation and dephosphorylation of signaling molecules is crucial for maintaining the homeostasis of the immune system [1]. This balance is emphasized by the fact that immune cells express more genes encoding protein tyrosine kinases (PTKs) and protein tyrosine phosphatases (PTPs) than any other cell type (with the possible exception of neurons) [1, 2]. In fact, several PTPs are exclusively expressed in hematopoietic cells, including CD45 (*PTPRC*), SHP1 (*PTPN6*), HePTP (*PTPN7*), PTP-HSCF (*PTPN18*), and LYP (*PTPN22*). T cells express some 40 PTPs, acting as both activators or inhibitors of

José Luis Millán (ed.), *Phosphatase Modulators*, Methods in Molecular Biology, vol. 1053,
DOI 10.1007/978-1-62703-562-0_15, © Springer Science+Business Media, LLC 2013

signaling [1]. Acute changes in tyrosine phosphorylation regulate antigen-receptor-mediated T cell activation, resulting in the expression of various cytokines that are secreted into the extracellular matrix, where they initiate a physiological immune response [3, 4].

Disturbances of PTP function have been implicated in many human diseases, ranging from cancer to cardiovascular, immunological, infectious, neurological, and metabolic diseases [5–11]. Hence, there is growing interest in the development of new therapeutics that target these enzymes. However, generating small-molecule PTP inhibitors that are not only potent, but also specific and bioavailable, is a challenging task that still imposes considerable difficulties. Nonetheless, a number of promising examples have been published over the last decade (reviewed in refs. 9, 12–14). Recent work by our laboratories, for instance, has resulted in an efficacious inhibitor of LYP (*PTPN22*) [15], which is implicated in various autoimmune diseases (reviewed in ref. 6) and recently was found overexpressed in B cell leukemia [16]. Most of the methods we describe in this chapter have been utilized to evaluate our LYP inhibitor [15], and we recommend this study for further reading.

The development of novel PTP inhibitors often starts with a high-throughput screening (HTS) of large chemical libraries, utilizing relatively simple phosphatase assays, which allow the screening of hundreds of thousands of drug-like molecules [17]. HTS hits are then confirmed in dose–response assays, and tested for their selectivity for the PTP of interest in counterscreens. Structure–activity relation (SAR) as well as biophysical binding studies are used to prioritize the most promising hits, and typically result in several compound series, based on a number of distinct chemical scaffolds. At that point, evaluating compounds not only for their efficacy in cells, but also for potential unwanted cytotoxic effects, becomes important for further prioritization. A first screening system to test many compounds in living cells commonly employs a human cell line that can be easily propagated and constitutes a pure population of cells. For T cells, T cell antigen receptor (TCR)-induced assays can be conducted on the Jurkat leukemic T cell line, an extensively studied model system that has been instrumental for the development of our current TCR signaling paradigm [18]. Once PTP inhibitors have been shown to be non-toxic and effective in modulating TCR signaling in Jurkat T cells, more sophisticated experiments that use freshly isolated human peripheral blood mononuclear cells (PBMC) or purified primary human T cells will be justified. Since such assays are time and cost intensive, need additional skill sets and experience, and also require access to freshly drawn blood from donors, only the most promising compounds are tested in primary human cells. Nonetheless, testing of PTP inhibitors in PBMC and primary T cells is important because of the common shortcomings related to cell lines in general

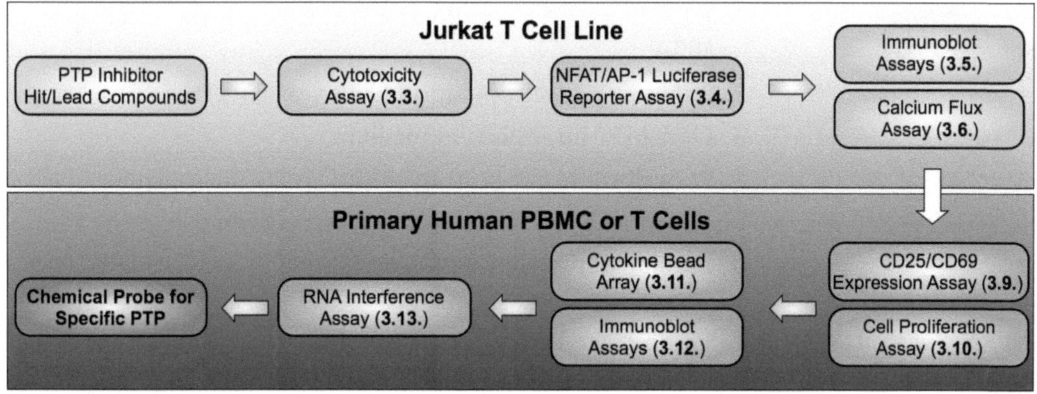

Fig. 1 Testing funnel of PTP inhibitors in T cells

(i.e., altered cell biology due to immortalization), and Jurkat T cells in particular (i.e., lack of certain signaling molecules, including the phosphatases SHIP and PTEN). However, recent technological advances in flow cytometry (e.g., cell sorting and multiplexing) and greatly improved transfection efficiency in mammalian cells have made primary T cell experiments more feasible for drug discovery. Based on the methods presented in this chapter, we recommend a workflow for inhibitor testing as shown in Fig. 1. We also would like to note that most of the described methods are generally applicable for testing the effects of small molecules on TCR-induced signaling, and are not restricted to PTP inhibitors only.

2 Materials

2.1 Common Reagents for All Experiments

1. RPMI 1640 culture medium (1×) and heat-inactivated fetal bovine serum (FBS).
2. Ficoll-Paque PLUS (GE Healthcare).
3. L-glutamine, penicillin–streptomycin, and sterile phosphate-buffered saline (PBS).
4. Dimethyl sulfoxide (DMSO).
5. 0.4 % Trypan Blue (Gibco); 2 % paraformaldehyde (*see* **Note 1**).
6. OKT3 (mouse anti-CD3 clone, 1 mg/mL stock).

2.2 Cell Viability/Cytotoxicity Assay

1. 96-well flat-bottom culture grade microplates.
2. Cell proliferation Kit (MTT; Roche): colorimetric assay for non-radioactive quantification of cell viability and proliferation.
3. ELISA reader at 550–600 nm with a reference wavelength of >650 nm.

2.3 Luciferase Reporter Assay

1. BTX Electro Square Porator T820 bench-top electroporator (BTX).
2. 2 mm Gap Electroporation Cuvettes (BTX).
3. Opti-MEM transfection medium.
4. 96-well white flat-bottom culture grade microplates.
5. Dual-Luciferase Reporter Assay Kit (Promega).
6. Luminometer (Veritas; Turner).

2.4 Immunoblot Assays

1. Vertical acrylamide electrophoresis unit with power supply (Bio-Rad).
2. Electroblotting unit, fully submerged (Bio-Rad).
3. PVDF or nitrocellulose membrane (0.45 μm pore size).
4. Whatman #1 filter paper.
5. Methanol.
6. Precast 4–20 % SDS-PAGE gradient gels.
7. RIPA lysis buffer; 1 % NP-40, 0.1 % SDS, 50 mM Tris–HCl pH 7.4, 150 mM NaCl, 0.5 % sodium deoxycholate, 1 mM EDTA.
8. Coomassie (Bradford) Protein Assay Kit.
9. SDS-sample buffer 10× stock solution (10 mL); 4 % glycerol, 4 % SDS, 100 mM Tris pH 6.8, 0.002 % bromophenol blue, 100 mM dithiothreitol.
10. SDS running buffer 10× stock solution (2 L); 60.4 g Tris base, 288 g glycine, 20 g SDS. Make up the solution close to the desired volume using Milli-Q water. Stir for 10 min. Adjust the pH to be between 8.1 and 8.5. Adjust to the final desired volume.
11. Complete EDTA-free protease inhibitor cocktail (Roche).
12. PhosphoSTOP phosphatase inhibitor cocktail (Roche).
13. TBST buffer 10× stock solution (1 L); 24.2 g Tris base, 80 g NaCl, add 800 mL of cold Milli-Q water. Adjust pH to 7.6 by adding ~15 mL conc. HCl, add 10 mL Tween-20. Stir until Tween is completely dissolved and adjust the final volume to 1 L using cold Milli-Q water.
14. Blocking buffers; 3 % BSA in 1× TBST; 5 % fat free milk in 1× TBST.
15. Molecular weight marker: Mix equal volumes of MagicMark (Life Technologies) and SeeBlue Plus2 Prestained Protein Standards (Life Technologies). This will allow visual control of migration of proteins on the gel (SeeBlue) and visualize the marker (MagicMark) together with proteins of interest after ECL.

16. Optional: phospho-specific antibody and pan-antibody to the signaling molecule that is a direct or indirect target of the PTP of interest.
17. Anti-phosphotyrosine antibody (4G10 clone; Upstate); anti-actin or anti-GAPDH antibodies (Sigma-Aldrich); HRP-conjugated anti-mouse and anti-rabbit secondary antibodies. Dynabeads Human T-Activator CD3/CD28 (Life Technologies). StemSep Human T Cell Enrichment Kit (Stemcell Technology).
18. Chemiluminescent detection kit.

2.5 Ca^{2+} Flux Assay

1. 1 mM Fluo-4 AM in DMSO (Life Technologies).
2. 1 mg/mL Ionomycin in DMSO (Calbiochem).
3. FACSCanto II flow cytometer with Blue Argon laser (BD Biosciences) (or other flow cytometers equipped with Blue Argon laser).
4. 12 mm × 75 mm polystyrene tubes.

2.6 CD25 and CD69 Expression on T Cell Surface of Stimulated PBMC Cultures

1. 48-well flat-bottom sterile tissue culture plates for cells in suspension.
2. Purified anti-CD3 antibody (clone UCHT1).
3. Purified anti-CD28 antibody (clone CD28.2).
4. Fluorescein Isothiocyanate (FITC)-conjugated anti-CD69 (clone L78).
5. Phycoerythrin (PE)-conjugated-anti-CD25 (clone M-A251).
6. Allophycocyanin (APC)-conjugated anti-CD3 (clone UCHT1).
7. Flow cytometer equipped with two lasers (488 and 633 nm) (*see* **Note 2**).

2.7 ^{3}H Thymidine Incorporation Assay in Stimulated PBMC Cultures

1. 96-well round-bottom sterile tissue culture plates.
2. 1 mCi/mL tritiated thymidine (^{3}H-TdR; PerkinElmer); specific activity: 20 Ci (740 GBq)/mmol. Prepare ^{3}H-TdR working solution by adding 100 μL ^{3}H-TdR stock solution to 5.9 mL sterile RPMI.
3. Omnifilter-96 Cell Harvester (PerkinElmer).
4. Scintillation analyzer (Top count NXT microplate scintillation and luminescence counter; Perkin Elmer).
5. MultiScreen harvest 3H, FC plates (96-well plates; Millipore).
6. Scintillation liquid buffer (MicroScint-O; PerkinElmer).
7. White adhesive bottom seal for 96-well backing tape (BackSeal-96/384; Perkin Elmer).
8. Adhesive seal for 96-well plate (MicroAmp Optical adhesive film; Life Technologies).

2.8 siRNA Transfection Assay

1. Amaxa Nucleofector (Lonza).
2. Amaxa Human T Cell Nucleofector Kit for unstimulated human T cells (Lonza).
3. StemSep Human T Cell Enrichment Kit (Stemcell Technologies).

3 Methods

3.1 Maintaining Jurkat T Cells

The complete medium to maintain the Jurkat T cell line is RPMI 1640 supplemented with 10 % FBS, 2 mM glutamine and 1× penicillin–streptomycin.

In culture, Jurkat T cells are characterized by non-adherent growth at a rate of 30–40 % per day.

1. To thaw cells, incubate frozen cell vials at 37 °C, shaking the tube occasionally, until solution is thawed.
2. Transfer cells to 5 mL of sterile FBS, spin at 200 × *g* for 5 min, and remove supernatant.
3. Resuspend pellet in 10 mL complete medium.
4. Seed cells in a cell culture flask and incubate at 37 °C and 5 % CO_2 for 2 days.
5. Add 30–40 % complete medium to cells every day to maintain a concentration of $0.4–1.3 \times 10^6$ cells/mL.

If cells need to be expanded, simply add 30–40 % medium directly into the flasks. To mix, either swirl the flask gently or pipet mix. If cells need to be maintained at a lower volume, remove the appropriate volume of cells, and then add 30–40 % medium to the remaining population. Do not leave Jurkat T cells unfed for more than 1 day, or they will begin to starve and die due to their rapid growth and quick consumption of resources (medium will turn yellow). Make sure that cells are healthy (i.e., having a rounded shape and no large vacuoles). In the event of a contamination, it is best to throw away the cells and start with a fresh batch of cells.

In order to stock Jurkat T cells, freeze them after ~2 weeks of growth. For freezing, take out $1–20 \times 10^6$ cells, spin at 300 × *g* for 7 min, aspirate supernatant, resuspend pellet in 0.5 mL FBS, and slowly add 0.5 mL 20 % DMSO in FBS solution, while swirling the tube continuously. Transfer the cell solution into a cryotube, keep on ice for a few minutes, and then keep at −80 °C over night. The next day, store the cryotube in liquid nitrogen.

3.2 Treatment of Jurkat T Cells with PTP Inhibitors

Prepare 20 mM inhibitor stock solutions in DMSO, which will allow testing of inhibitors in cells at up to 40 μM at the non-toxic concentration of 0.2 % DMSO. Store stock solutions as small aliquots as needed at −20 °C and avoid repeated thawing and freezing. For inhibitor treatment, cells are resuspended in additive-free medium in order to prevent depletion of inhibitor through nonspecific binding to serum proteins.

1. Prepare 500× concentrated inhibitor working solutions in DMSO for each inhibitor concentration to be tested. This will ensure equal amounts of DMSO in inhibitor dose–response assays (A 500× working solution results in 0.2 % DMSO final concentration).
2. Harvest cells in 50 mL conicals by centrifuging at 300 × *g* for 6 min.
3. Resuspend cells with additive-free RPMI and transfer all cells into one conical. Wash down the sides of all conicals with additive-free RPMI and add wash solution to the cells. Adjust volume in final conical with additive-free RPMI to 50 mL. Spin at 300 × *g* for 6 min. Discard supernatant and wash again with 50 mL additive-free RPMI.
4. Resuspend cells in additive-free RPMI.
5. Aliquot the appropriate amount of cells per reaction into 1.5 mL microtubes and incubate at 37 °C for 5 min.
6. Add vehicle (DMSO) or 500× inhibitor working solutions to cells and mix. Incubate cells at 37 °C for 30–45 min.

3.3 Cytotoxicity Assay

Several commercial kits are available for conveniently and reliably measuring potential cytotoxic effects of inhibitors. We commonly use the MTT assay kit, which is a quantitative colorimetric assay that shows linearity over a broad range of cell densities [19]. In this assay, MTT (3-(4,5-dimethylthiazol-2-yl)-2,5-diphenyltetrazolium bromide) is reduced by living cells, resulting in the formation of purple formazan crystals, which then are dissolved in an organic solvent for measuring its absorbance between 550 and 600 nm. Test compounds at a concentration of 50–100× IC_{50} value. Include a positive control (e.g., 10 μM staurosporine). Perform the MTT assay following the kit instructions provided by the vendor. Each condition should be measured in triplicate.

1. Seed 2×10^5 inhibitor- or vehicle-treated cells in 100 μL per well in 96-well plates. Incubate for 48 h at 37 °C and 5 % CO_2.
2. Add 10 μL of MTT solution 1 and incubate for 4 h in humidified atmosphere (37 °C, 5 % CO_2).
3. Add 100 μL of MTT solution 2 and incubate overnight in humidified atmosphere (37 °C, 5 % CO_2).
4. Proceed with absorbance measurement using a spectrophotometer or plate reader. Measure the formazan absorbance at

550–600 nm (according to the filters available in the plate reader). The reference wavelength should be >650 nm.

5. Subtract reference absorbance from test absorbance values. Calculate the percentage of viable cells relative to the vehicle control.
6. Discard compounds with significant cytotoxicity from further evaluation.

3.4 Luciferase Reporter Assay in Jurkat T Cells

Reporter assays are easily conducted on Jurkat T cells, which can be transfected with good efficiency. In addition, reporter assays are relatively high throughput and can be utilized as a first screening assay to prioritize compounds for lower throughput immunoblot assays. To measure TCR-induced activation, a dual-luciferase reporter assay kit is employed. Expression of a firefly luciferase is controlled by the proximal interleukin-2 (IL-2) promoter, containing binding sites for nuclear factor of activated T-cells (NFAT) and activator protein 1 (AP-1) [20]. A *Renilla* luciferase under the control of a SV40-promoter represents the baseline transcriptional activity and is used as a control for nonspecific effects on the cellular transcriptional machinery. This assay captures changes in downstream TCR signaling, i.e., the activation of the IL-2 transcription factors NFAT and AP-1. Inhibitors of negative regulatory PTPs are expected to increase TCR signaling, hence to augment the firefly luciferase reaction. Inhibitors of positive regulatory PTPs are expected to have the opposite effect. We recommend testing inhibitors at a single concentration first (e.g., choose a concentration between 5× and 10× IC_{50}), and then confirm active compounds in a dose–response experiment.

Day 1: Transfection

1. Aliquot and harvest cells (~20×10^6 cells per inhibitor) by centrifuging at $300 \times g$ for 6 min. Save ~1 mL of conditioned RPMI in one conical per transfection.
2. Wash cells twice by resuspending them in ~50 mL Opti-MEM and centrifuging at $300 \times g$ for 6 min.
3. Resuspend cells in Opti-MEM to about 50×10^6 cells/mL. Per transfection, 0.4 mL = 20×10^6 cells are used.
4. Add 10 μg NFAT-AP-luc + 1 μg RLO per 20×10^6 cells.
5. Keep cells, the saved conditioned RPMI, and Opti-MEM on ice.
6. Electroporation:
   (a) Set electroporator to "low-voltage" mode, 230 V, 65 pulse length, 1 pulse.
   (b) Aliquot 0.4 mL cells to a clean cuvette. Place cuvette in holder.

   (c) Press pulse button.
   (d) Pipet transfected cells to conical of saved conditioned RPMI.
   (e) Wash the cuvette with 1 mL Opti-MEM and transfer contents to saved RPMI.
   (f) Repeat for each transfection.
7. Transfer cells to 10 mL cold complete medium per transfection in a cell culture flask.
8. Incubate cells at 37 °C (5 % CO_2) for 14–24 h before harvesting for assay.

Day 2: Inhibitor treatment and stimulation of cells

1. Harvest cells by centrifuging at 300 × *g* for 6 min.
2. Resuspend and transfer all cells into one conical. Wash down the sides of all conicals with additive-free RPMI and add wash solution to the cells. Adjust volume in final conical with additive-free RPMI to 50 mL. Spin at 300 × *g* for 6 min. Discard supernatant and wash again in 50 mL with additive-free RPMI.
3. Resuspend cells in additive-free RPMI to a concentration of 20×10^6 cells/mL.
4. Aliquot 700 μL cell suspension per condition (vehicle + inhibitors) into 1.5 mL microtubes and incubate at 37 °C for 5 min.
5. Add vehicle or inhibitors to cells and mix. Incubate cells at 37 °C for 30–45 min.
6. Meanwhile, prepare OKT3 solution by diluting the OKT3 stock 3:200 in RPMI. For each condition (vehicle + inhibitors), add 20 μL RPMI with OKT3 into one well, and 20 μL RPMI without OKT3 (unstimulated samples) into a second well of a sterile 48-well plate.
7. Expand cell volume from 700 μL to 1.4 mL with pre-warmed complete medium. Add 600 μL of each cell suspension to both wells containing either OKT3 or just RPMI vehicle. (Final OKT3 concentration in the stimulated samples will be 0.5 μg/mL.)
8. Incubate plate for 6 h in incubator (37 °C, 5 % CO_2).
9. Harvest cells from each well into a microtube. Spin at 300 × *g* for 5 min.
10. Prepare 1× Passive Lysis Buffer with RPMI or Milli-Q water.
11. Discard supernatant from each reaction and resuspend cells in 100 μL 1× PLB. Incubate at room temperature for 10 min, and then freeze at −20 °C.

Day 3: Luciferase assay

1. Prepare LAR II reagent (firefly luciferase substrate) and Stop & Glo reagent (*Renilla* luciferase substrate) according to the kit instructions in 15 mL conicals. For each reagent, 50 μL is needed per reaction in addition to ~1.2 mL for priming the luminometer and dead volume (*see* **Note 3**).
2. Plate 20 μL cells (~1.2×10^6 cells) per reaction/well in triplicate into a white 96-well plate.
3. Set up luminometer:
   (a) Injectors 1 and 2: add 50 μL.
   (b) Integration time = 1 s.
   (c) Delay time before first read = 10 s.
   (d) Delay time between each read = 2 s.
4. Flush three times with Milli-Q water, ethanol, Milli-Q water, air.
5. Prime with reagents.
6. Run plate (the luminometer dispenses the LAR II reagent first, measures firefly luciferase activity, and then dispenses the Stop & Glo reagent and measures *Renilla* luciferase activity).
7. Flush injectors.

Calculate the level of NFAT/AP-1 activation by determining the ratio between firefly and Renilla luciferase activity for each sample. Discard inhibitors giving Renilla luciferase readings deviating >20 % from the DMSO control samples, because this suggests that the compounds have nonspecific effects. Also exclude inhibitors from further evaluation that yield higher luciferase levels in the unstimulated versus TCR-stimulated sample. Select inhibitors giving firefly–Renilla luciferase ratios >2-fold higher (or lower) than the corresponding ratios for the vehicle control samples. Higher ratios are expected for inhibitors of negative regulatory PTPs, whereas lower ratios are expected if the inhibited PTP is a positive regulator of TCR signaling. Confirm effects of selected inhibitors by retesting in a dose–response format. Select inhibitors with a confirmed dose-dependent effect on NFAT/AP-1 activation for testing in the immunoblot assays.

3.5 Immunoblot Assays in Jurkat T Cells

The luciferase reporter assay captures the effect of test compounds on downstream signaling events (i.e., activation of NFAT and AP-1). However, such effects could also be due to off-target activity and not (solely) inhibition of the PTP of interest. If the direct substrate(s) of the target PTP is known, and phospho-specific antibodies for this/these substrate(s) are available, immunoblot assays can be employed that directly probe the phosphorylation levels of the substrate(s). If no direct substrate of the target PTP has been identified yet, phosphorylation levels of signaling molecules downstream of the PTP of interest can be measured. For instance, if the target PTP is involved in MAP kinase activation, specific antibodies to the activatory phosphorylation sites in ERK1/2, JNK, or p38 can be utilized to test the effect of inhibitors. If no phosphospecific antibodies to the substrate phosphorylation site are available, immunoprecipitation of the substrate protein using an available pan-antibody, followed by an anti-phosphotyrosine (4G10) western blot, can be performed (*see* **Note 4**). Moreover, specificity of inhibitors can be evaluated by determining the overall change in tyrosine phosphorylation by performing anti-phosphotyrosine (4G10) western blots on cell lysates from activated versus non-activated cells, treated with vehicle or with inhibitor. Specific PTP inhibitors are expected to augment the phosphorylation level of a single or only few proteins/bands. If the inhibitor causes a dramatic change in the global tyrosine phosphorylation, as is the case when cells are treated with the pan-PTP inhibitor pervanadate (*see* **Note 5**), specificity of the compound for the PTP of interest is questionable.

We recommend performing two different immunoblot experiments: First, a dose response experiment to determine an optimal dose of inhibitor, i.e., the lowest concentration that yields maximal response; and second, a time course experiment that captures the effects of inhibitors (used at optimal dose) at various time points after TCR stimulation. Efficacious PTP inhibitors not only are expected to augment phosphorylation levels of substrates, but also to induce a sustained effect on these molecules, compared to the vehicle control. For instance, treatment of Jurkat T cells with the chemical probe ML119 [21] resulted in augmented and sustained phosphorylation of the MAP kinases ERK1/2 and p38 [22]. This effect was due to direct inhibition of HePTP (*PTPN7*), a classical PTP that specifically dephosphorylates the phosphotyrosine residue in the activation loop of ERK1/2 and p38.

3.5.1 Dose–Response Experiment

For six conditions (vehicle + five inhibitor concentrations):

1. Distribute 5×10^6 cells into each of two sets of six 1.5 mL microtubes in 100 μL 37 °C-prewarmed additive-free RPMI (one set for stimulated samples, one set for unstimulated samples).
2. Add vehicle or inhibitor and incubate cells for 30–45 min at 37 °C with gentle shaking using Thermomixer Comfort set at 37 °C and 400 rpm.
3. Transfer tubes on ice.
4. Add 10 μg of purified anti-CD3 antibody to the set of tubes that will be stimulated and mix by pipetting up and down.
5. Incubate all samples on ice for 20 min. (This step allows an equal binding of anti-CD3 without initiating the signaling cascade.)

6. Add 500 μL of 37 °C-prewarmed complete medium to all tubes and incubate at 37 °C for 5 min. To stop the activation reaction, add 500 μL of ice cold PBS to all tubes and immediately centrifuge at 2,000 × *g* for 5 min to pellet cells.
7. Discard the supernatant and immediately add 300 μL of RIPA lysis buffer complemented with protease and PhosphoSTOP phosphatase inhibitors.
8. Mix well by pipetting up and down and by vortexing.
9. Incubate on ice for 20 min and vortex every 5 min during the lysis procedure.
10. Centrifuge at 15,500 × *g* for 15 min at 4 °C.
11. Collect the supernatants and measure the protein concentration in each sample using the Bradford assay kit (5×10^6 Jurkat cells have a total amount of ~250 μg protein).
12. Add the appropriate volume of 4× SDS-PAGE sample buffer (to get 1× final concentration of the sample buffer) and boil the samples at 95 °C for 5 min.
13. Use lysis buffer if needed to adjust for equal loading among the samples according to the protein concentrations determined in the Bradford assay.
14. Separate proteins in each sample via SDS-PAGE using 1 mm or 1.5 mm thick 4–20 % precast gels (*see* **Note 6**). Load a minimum of 40 μg protein per well to yield a sufficient signal in the anti-phosphotyrosine blot. Also include a molecular weight marker.
15. Place the gel in the electrophoresis unit and add running buffer to fully submerge the gel.
16. Run the gel at 100 V until the blue line passes the stacking part of the gel, then increase the voltage to 120 V and run until the blue front is at the bottom of the gel. This will take about 2:45 to 3 h.
17. Transfer the gel onto a nitrocellulose or PVDF membrane (0.45 μm pore size). Pre-wet all the transfer material with the transfer buffer.
18. Stack the blotting sandwich as follows. At each step remove air bubbles using the roller provided with the transfer apparatus to ensure homogenous transfer. This is especially important when the gel is placed onto the membrane.
    (a) Transfer case (clear side).
    (b) Sponge.
    (c) Whatman paper.
    (d) Membrane.
    (e) Gel.

(f) Whatman paper.

(g) Sponge.

(h) Black side of the transfer case.

19. Place the transfer cassette into the transfer tank, making sure that the black sides of the cassette and tank are aligned. Fill the tank with cold transfer buffer until the cassette is fully submerged.
20. When using 1 mm gels, run the transfer at 200 mA for 1 h. When using 1.5 mm gels, run the transfer at 200 mA for 1.5 h (*see* also **Note 7**).
21. After transfer, place the membrane in a small container and add 3 % BSA-TBST blocking buffer so that the membrane is fully covered. Incubate for 1 h with gentle shaking at room temperature.
22. Replace the blocking buffer with 3 % BSA-TBST buffer containing a phospho-specific antibody or the 4G10 antibody (dilution 1/3,000) (*see* **Notes 8–10**).
23. Incubate overnight on an orbital shaker at 4 °C.
24. Wash three times with 1× TBST (5 min for each wash) at room temperature with gentle shaking.
25. Add an appropriate HRP-conjugated secondary antibody (e.g., HRP-conjugated anti-mouse secondary antibody when using 4G10 as primary antibody) diluted in TBST—5 % milk, and incubate for 1 h at room temperature with gentle shaking (*see* **Note 11**).
26. Wash three times with 1× TBST (5 min for each wash) at room temperature with gentle shaking.
27. Reveal the blots using the chemiluminescent (ECL) detection kit (*see* **Note 12**). Make sure to cover all the membrane with ECL solution.
28. Drain the ECL and wrap the membrane in thin plastic foil so that the film does not get wet during exposure.
29. Expose membrane to film for 1 min and develop film. Depending on the signal strength, expose more or less (ranging from a few seconds up to 20 min) (*see* **Note 13**).
30. Optionally, membranes can be stripped and re-probed with another antibody (*see* **Note 14**).
31. To control for equal loading, strip the membrane (*see* **Note 14**) and blot with a control antibody, usually against actin, tubulin, or GAPDH, all of which are very abundant proteins that can easily be detected even after a second strip. Alternatively, control blots can be performed with pan-antibodies to the protein that was probed with phospho-specific antibodies.

Select PTP inhibitors that dose-dependently augment the phosphorylation level of substrate proteins without affecting global tyrosine phosphorylation for further consideration. Confirm selected inhibitors by repeating this experiment. In order to quantify western blots, scan films in gray scale at a resolution of a least 600 dpi and analyze the scans using the ImageJ software (http://rsbweb.nih.gov/ij/index.html).

3.5.2 Time Course Experiment

To perform a time course after TCR triggering, follow these steps:

1. Distribute 25×10^6 cells each into two 1.5 mL microtubes in 500 μL 37 °C-prewarmed additive-free RPMI.
2. Add 1 μL vehicle (DMSO) or inhibitor (at optimal dose) and incubate cells for 30–45 min at 37 °C with gentle shaking (Thermomixer Comfort set at 37 °C and 400 rpm).
3. Distribute each sample into five microtubes (5×10^6 cells in 100 μL/microtube) and place tubes on ice.
4. Add 10 μg of purified anti-CD3 antibody to four tubes of each the vehicle- and inhibitor-treated cells and mix by pipetting up and down (one tube from each condition will be left unstimulated).
5. Incubate all samples on ice for 20 min (this step allows an equal binding of anti-CD3 without initiating the signaling cascade).
6. Add 500 μL of 37 °C-prewarmed complete medium to all tubes and incubate at 37 °C. For a stimulation time course, incubate one tube from each treatment condition (vehicle and inhibitor) for 5, 15, 30, or 60 min. Stop the activation reaction by adding 500 μL of ice cold PBS and centrifuge immediately at 2,000 × *g* for 5 min to pellet the cells.
7. Continue the experiment by following the protocol described above, starting at point 7.

The experiment should be repeated to confirm results. As described above, western blots can be quantified with the ImageJ software.

3.6 Ca^{2+} Flux Assay in Jurkat T Cells

Calcium is an important secondary intracellular messenger connecting the proximal TCR signaling pathways to NFAT-dependent gene transcription in T cells. Initiation of the TCR signaling complexes leads to the generation of inositol-1,4,5-trisphosphate (IP3), which binds to IP3 receptors present on the endoplasmic reticulum (ER) membrane. This leads to the release of calcium from the ER and subsequent increase of free cytosolic concentration of calcium, which binds and activates the Ca^{2+} sensor calmodulin (CaM). Activated CaM binds and modulates the activity of several downstream targets including the phosphatase calcineurin, leading to dephosphorylation and translocation of NFAT (reviewed in ref. 23). Measuring intracellular calcium fluxes in real-time after TCR stimulation can be used to evaluate and locate the effect of a test compound on the signaling pathways between the TCR and NFAT-regulated gene transcription. Moreover, the increased sensitivity of recently developed high-affinity calcium dyes (such as Fluo-3 and Fluo-4) allows for the detection of relatively small changes in calcium fluxes, making this technique attractive for drug screening. We recommend testing inhibitors at several doses, e.g., 0–0.3–1–3–10 μM

1. Harvest cells. Approximately 7×10^6 cells per reaction are needed.
2. Wash cells once in complete medium.
3. Resuspend cells in complete medium to a cell concentration of 7×10^6 cells/mL.
4. In the absence of direct light, add Fluo-4 AM to a final concentration of 1 μM.
5. Incubate cells at 37 °C in the dark for 45 min.
6. Wash cells twice with additive-free RPMI and resuspend in additive-free RPMI.
7. Add vehicle or inhibitors to cells.
8. Incubate cells at 37 °C in the dark for 30 min, centrifuge, and discard supernatant.
9. Resuspend cells with ice-cold RPMI (containing 2.5 % FBS as well as vehicle or inhibitor), to a cell concentration of 7×10^6 cells/mL and rest cells on ice for a further 30 min.
10. Set up FACS machine (FACSCanto II): SSC-A, FSC-A, FSC-H, and FITC channel. Use blue argon laser for excitation at 488 nm, and FITC channel at 518 nm to read emission (Ca^{2+}-bound Fluo-4 has maximum absorption at 494 nm and maximum emission at 516 nm).
11. Pre-equilibrate each tube with cells at 37 °C for 5 min (For convenience, set up a water-bath at 37 °C next to the flow cytometer).
12. Transfer tube to FACS machine, record baseline levels (1,000–5,000 events/s) for 1 min, add 5 μg/mL OKT3, and record for another 3 min. Optionally, 2 μM ionomycin can be added at this point, and signals recorded for an additional 1 min (*see* **Note 15**).
13. Analyze data using the FlowJo program. Use the FlowJo kinetics analysis, following their step-by-step descriptions (http://www.flowjo.com/home/tutorials/TN/kineticsweb2010.pdf).

3.7 Isolation of PBMC

Specific inhibitors that are non-toxic and efficacious in modulating TCR signaling in Jurkat T cells are further evaluated in human PBMC. T cells constitute about 40 % of PBMC, which can be relatively easily isolated either from buffy coats or whole blood from healthy volunteers. Alternatively, cryopreserved PBMC can be used (*see* **Notes 16** and **17**). The protocol below is applicable to both buffy coat and whole blood, each of which typically comes as a 50 mL sample.

1. Mix the 50 mL buffy coat (or blood) with an equal volume of additive-free RPMI at room temperature.
2. Isolate PBMC using the standard method of density gradient centrifugation on Ficoll-Paque PLUS, following the vendor's recommendations (http://www.stemcell.com/~/media/Technical%20Resources/1/F/F/B/3/29630_PIS_2_1_2.ashx).
3. After centrifugation of the buffy coat (blood)/ficoll mixture, carefully collect the white intermediate ring containing PBMC by using a disposable sterile plastic pipette, and transfer to a new 50 mL tube.
4. Add pre-warmed additive-free RPMI to a final volume of 50 mL.
5. Centrifuge at $160 \times g$ for 10 min at room temperature.
6. Remove medium by aspirating gently with a pipette and resuspend the cells in 25 mL additive-free RPMI.
7. Centrifuge at $160 \times g$ for 10 min at room temperature and carefully remove the supernatant by aspiration.
8. Resuspend the extracted PBMC in 1 mL additive-free RPMI. Count the cells using a hemacytometer after trypan blue exclusion to estimate the total number of extracted live cells. (Live cells do not incorporate trypan blue while dead cells do and thus, become blue. The percentage of cell viability is typically above 90 %.)

From a single buffy coat (50 mL), up to 7×10^8 cells can be isolated, whereas PBMC preparation from whole blood (50 mL) yields ~1×10^8 cells. At this point, PBMC are ready for any of the assays described below, or can be cryopreserved for later use (*see* **Note 18**).

3.8 Treatment of PBMC with PTP Inhibitors

Inhibitor treatment of PBMC is similar to the treatment of Jurkat T cells, described in Subheading 3.2.

1. Prepare 500× (or higher concentrated, depending on solubility) working solutions of inhibitors in DMSO. For dose–response studies prepare 1/3-log steps serial dilutions (highest final concentration ~50-times inhibitor IC_{50} value, but not exceeding 50 μM, *see* **Note 19**). We recommend four to five

inhibitor dilutions (e.g., 30–10–3–1–0.3–0 μM final inhibitor concentration), so that the lowest final concentration is equal to or lower than the inhibitor IC_{50} value against the recombinant PTP.

2. Distribute cells into sterile 1.5 mL microtubes at 10×10^6 cells in 500 μL of additive-free RPMI (*see* **Note 20**).
3. Add 1 μL vehicle (DMSO) or inhibitor solution and incubate cells at 37 °C for 30 min with gentle shaking. We typically use the Eppendorf Thermomixer Comfort, set at 37 °C and 400 rpm.
4. Dilute the treated cells in an equal volume (i.e., 500 μL) of RPMI containing 20 % FBS (final FBS concentration will be 10 %). Cells are now ready for testing.

3.9 Measuring T Cell Surface Expression of CD25 and CD69 on Stimulated PBMC

Both CD25 and CD69 (Leu-23) are surface expressed activation markers on T cells, and thus amenable to flow cytometry analysis. CD69 is a leukocyte surface receptor expressed very early (2–3 h) after TCR stimulation. Resting T cells do not express CD69 [24–27], making it an ideal marker for TCR-induced T cell activation, including the assessment of small molecules on TCR signaling [28]. CD25 is the ligand-specific IL-2 receptor α-chain (IL-2Rα), expressed only on activated T cells starting ~24 h after TCR stimulation [29]. Flow cytometry analysis of either CD69 or CD25 or both offers the possibility to simultaneously determine the effects of small-molecule inhibitors on different T cell subsets, including naïve, memory CD4 and CD8, and regulatory T (Treg) cells, without prior cell sorting (*see* **Note 21**). Since this assay is non-radioactive, it also offers the possibility to collect cell supernatants for subsequent evaluation of cytokine production (*see* Subheading 3.11), even if the collected volumes are very small.

1. Using a 48-well flat-bottom plate, seed, per well, 1×10^6 vehicle- or inhibitor-treated PBMC in 500 μL complete medium. For each inhibitor condition and vehicle, two sets of two wells are needed (for two time points and stimulated and unstimulated samples).
2. Add 1 μg purified anti-CD3 antibody to the stimulated samples and mix by pipetting up and down (*see* **Note 22**).
3. Incubate cells at 37 °C and 5 % CO_2.
4. Collect the first set of cells 6 h and the second set 24 h after stimulation by transferring cell suspensions into 4-mL tubes.
5. Centrifuge at $300 \times g$ for 10 min at room temperature.
6. Collect all the supernatants and aliquot them individually (50 μL each). Freeze at −80 °C until use for cytokine analysis (*see* Subheading 3.11) (*see* **Note 23**).

7. Resuspend cell pellets in 1 mL PBS supplemented with 2 % BSA (PBS/BSA). Mix by pipetting gently up and down to get a homogenous cell suspension, and centrifuge at 300 × *g* for 10 min at room temperature.
8. Meanwhile, prepare the staining antibody cocktail. Mix APC-CD3, PE-CD25, and FITC-CD69 (10 μL each per tube to be stained). Protect the antibody cocktails from light. (To investigate effects of inhibitors on specific T cell subsets, mix antibody cocktails accordingly.)
9. After centrifugation of cells, discard the supernatants and resuspend each cell pellet in 150 μL PBS/BSA by pipetting up and down.
10. Split each cell suspension equally into two tubes (50 μL in each), one for the non-stained control and one for the staining of $CD3^+$ T cells.
11. Add 50 μL PBS/BSA to the non-stained control samples. Add 50 μL of the anti-CD3 cocktail to the second set of tubes.
12. Incubate tubes on ice in the dark for 20 min.
13. Wash cells by adding 500 μL cold PBS and centrifuging at 300 × *g* for 10 min.
14. Discard supernatant and add 300 μL of 2 % PAF.
15. Cells are now ready to be analyzed by flow cytometry. As cells are fixed, they can also be analyzed the next day (store them at 4 °C in the dark, e.g., in the refrigerator) (*see* **Note 24**).
16. Flow cytometry can be performed with any flow cytometer equipped with a minimum of two lasers (*see* **Notes 25** and **26**).

Compounds that have no effect on CD25 and CD69 expression, compared to vehicle control, should be excluded from further consideration.

3.10 Measuring Cell Proliferation by ^3H Thymidine Incorporation in Stimulated PBMC

T cell activation results in the proliferation of primary T cells in culture, which can be measured directly by monitoring the rate of DNA synthesis. Metabolic incorporation of tritiated thymidine (^{3}H-TdR) into cellular DNA is the most commonly used method, because it is highly sensitive and allows detection of poorly proliferating cells such as primary T cells. In addition, the ^{3}H-TdR incorporation assay is time and cost effective and offers the possibility to screen the effects of several inhibitors, culture conditions, stimulations, etc. in a single 96-well plate.

1. Seed 2×10^5 vehicle- or inhibitor-treated PBMC in 200 μL complete medium per well (96-well round-bottom plate). Prepare two triplicate sets of wells, one for the stimulated, and one for the unstimulated samples.
2. Add 1 μg of soluble purified anti-CD3 antibody to the stimulated samples and mix gently by pipetting up and down.

3. Incubate the plates for 72 h at 37 °C and 5 % CO_2 (*see* **Note 27**).
4. 6 h before the end of the incubation, add 25 μL ^{3}H-TdR working solution into each well. Continue incubation for another 6 h.
5. Harvest cells onto a UniFilter plate using an Omnifilter-96 cell harvester.
6. Let the plate dry at room temperature for at least 24 h.
7. Attach the backing tape to the bottom of the dried plate and add 25 μL scintillation liquid buffer in all wells using a multichannel pipette.
8. Cover the plate with the MicroAmp optical adhesive film.
9. Place the plate into the scintillation counter and measure the radioactivity of the incorporated ^{3}H-TdR as counts per min (cpm).
10. Determine the effects of inhibitors on cell proliferation as follows:

    [%] = cpm (inhibitor treated)/cpm (vehicle treated) × 100

CAUTION: DISPOSE RADIOACTIVE WASTES INTO APPROPRIATE WASTE CONTAINERS.

Exclude compounds from further consideration that affect proliferation of unstimulated PBMC. Depending on the specific function of the PTP of interest in T cells, select compounds for further consideration accordingly. For instance, if the PTP plays an oncogenic role (like SHP2 (*PTPN11*) in various leukemias), select compounds that inhibit proliferation. If, on the other hand the, PTP inhibits T cell activation and is implicated in immunodeficiency, then select inhibitors that augment T cell activation/proliferation.

3.11 Cytokine Bead Array (CBA) Assays on PBMC

Multiplexed bead-based immunoassays such as CBA allow for reliable quantification of several soluble proteins simultaneously on cell supernatant volumes as low as 25–50 μL. The optimal number of cells needed to detect the analyte in the supernatant after anti-CD3 activation is 1×10^6 cells per well in a maximum volume of 500 μL of culture medium. Besides the possibility of multiplexing to quantify many cytokines simultaneously, CBA is also more time and cost effective than the conventional ELISA method. Several CBA kits are commercially available. However, for characterizing cytokine profiles in T cells, the preconfigured Thl/Th2/Th17 CBA kit is most commonly used. This kit contains reagents for the quantification of the human Thl/Th2/Th17 cytokines IL-2, IL-4, IL-6, IL-10, IL-17A, IFN-γ, and TNF-α, and is routinely used to evaluate the effects of immunotherapeutic agents [30]. Perform the CBA assay according to the instructions that come with this kit.

1. Prepare Th1/Th2/Th17 cytokine standard solutions according to the kit's instructions.
2. Proceed with serial dilutions of the standards as recommended by the vendor (nine standard dilutions).
3. Pool all the capture beads, as they are provided in individual bottles (A1–A7) (*see* **Note 28**).
4. Thaw the frozen supernatants (aliquots of 50 μL) from PBMC cultured for the evaluation of CD25 and CD69 expression levels (*see* Subheading 3.9). Prepare duplicate sets of two tubes for each supernatant, one with non-diluted sample, one with the sample diluted 1:1 with additive free RPMI (*see* **Note 29**).
5. Vortex the pooled beads and add 50 μL in each of the standard, sample, and negative control (RPMI) tubes.
6. Add 50 μL of the human Th1/Th2/Th17 PE detection reagent to all tubes and incubate in the dark at room temperature for 3 h.
7. Wash by adding 1 mL of the provided wash buffer and centrifuge at 200 × *g* for 5 min.
8. Carefully discard the supernatant and resuspend the pellets in 300 μL wash buffer.

Samples are now ready for immediate analysis by flow cytometry. Supported flow cytometers include BD Accuri, BD FACSAria, BD FACSArray, BD FACSCalibur, BD FACSCanto, BD FACSVerse, and BD LSRFortessa (*see* www.bdbiosciences.com/cbasetup). Analyze results using the FCAP Array program, a software dedicated to perform micro-bead based multiplexed analyses. Comprehensive step-by-step instructions are available online (http://www.bdbiosciences.com/documents/Analysis_of_data_from_CBA_using_FCAPArray.pdf).

3.12 Immunoblot Assays on Purified Primary T Cells

PBMC are a mixed cell population, only 40 % of which constitute T cells. While with flow cytometry, effects of inhibitors on T cells (or even subsets of T cells) can be separated and recorded, PBMC cannot be used to verify such effects in immunoblot assays, as tyrosine-phosphorylated proteins from non-T cells may mask the effect of the inhibitor. Therefore, primary human T cells are purified by negative selection from freshly isolated PBMC (*see* Subheading 3.7) using the StemSep Human T Cell Enrichment Kit. Follow the selection protocol step-by-step according to the vendor's instructions. Expect to recover >1×10^8 T cells from ~7×10^8 purified PBMC (from 50 mL buffy coat).

Perform the immunoblot experiments as described for Jurkat T cells in Subheading 3.5. The protocol only differs in terms of cell numbers and TCR stimulation. We recommend using 1×10^6

primary T cells per tube/condition instead of 5×10^6 cells as indicated for Jurkat T cells. Jurkat T cells are abundantly available, and cell lysates from 5×10^6 cells will allow performing several western blot experiments with different antibodies. However, isolating primary T cells is quite time and cost intensive, which is the reason why we recommend performing the assay with the minimal amount of cells that yield western blots with sufficient signal. If you are planning on probing with more than one phospho-antibody, blot with the more important one first, strip the membrane, and then blot with the second antibody (*see* **Note 14**). Strip the membrane again for blotting with a loading control antibody (*see* Subheading 3.5), or choose a loading control antibody to a protein with a different molecular weight, compared to the primary antibody, which will allow re-probing without stripping. If the signal for re-probing with the second phospho-antibody after stripping is not satisfactory, double the cell numbers and run two gels. For the activation of primary T cells, use Dynabeads Human T-Activator CD3/CD28, instead of anti-CD3 as described for Jurkat T cells. Add 2 μL of Dynabeads to all tubes except the non-stimulated samples and keep following the protocol described in Subheading 3.5.

We regard a dose-dependent effect of a PTP inhibitor at sub- and low-micromolar concentration on a direct substrate of the PTP (without affecting global tyrosine phosphorylation) as the "gold standard" in terms of efficacy of a PTP inhibitor in T cells. To confirm the specificity of the inhibitor, we recommend testing it in T cells where the target PTP has been acutely eliminated via RNA interference as described in the next section.

3.13 Testing Specificity of Inhibitors in Primary T Cells Using RNA Interference

To determine whether inhibitors specifically inhibit the PTP of interest in primary human T cells, purified T cells can be used, in which the target PTP has been acutely eliminated (residual expression <20 % of normal expression) by short interfering RNA (siRNA)- or small hairpin RNA (shRNA)-mediated knockdown. Signaling in T cells with severely reduced protein levels of the PTP of interest is expected to be similar to signaling in T cells treated with a specific inhibitor of the PTP. However, this may not always be the case. Inhibition of PTP activity by the inhibitor, or eliminating the protein altogether via RNAi, may have different outcomes due to a more complex function of the PTP that is not based on its phosphatase activity alone [31]. Nonetheless, the combined action of PTP inhibitor and PTP knockdown should not alter signaling beyond what is observed for each alone.

While siRNA duplexes are transfected into cells and allow downregulation of a specific PTP for a short period of time (days), shRNA are introduced into cells through expression vectors, allowing long-term and regulated delivery in single cells and whole organisms [32]. However, efficient transduction of specific shRNA with low residual expression rates of the target protein is difficult to achieve in primary T cells. Therefore, we recommend the use of siRNA duplexes to downregulate the PTP of interest. Prior to testing the efficacy of specific siRNA duplexes, we suggest determining optimal transfection conditions using a non-targeting fluorescently labeled siRNA such as Block-iT-FITC. This will allow evaluation of the transfection efficiency by a simple flow cytometry analysis based on FITC labeling. Next, siRNA duplexes that efficiently reduces the protein level of the target PTP need to be identified. Most vendors suggest a set of four different siRNA for the initial screening, which relies on the availability of well working antibodies for the protein of interest. Once at least two efficient siRNAs are found, time course evaluation of the protein or mRNA levels after siRNA transfection, using western blot or qRT-PCR, respectively, should be performed. Once optimal conditions for the specific PTP knockdown are established, immunoblot assays can be employed to test the specificity of a PTP inhibitor as described above (Subheadings 3.12 and 3.5).

Use the Amaxa Human T Cell Nucleofector Kit for unstimulated human T cells and the corresponding transfection program for human T cells following the vendor's instructions.

1. Purify human T cells by negative selection from freshly prepared PBMC (*see* Subheading 3.12).
2. Count cells and make aliquots of 5×10^6 cells per 1.5 mL sterile microtube. Prepare six tubes for each target gene: two tubes for the control non-targeting siRNA, and two tubes for each of the two specific siRNAs (the first tube is for vehicle treated cells; the second tube is for PTP inhibitor treated cells).
3. Pellet cells by centrifugation at $200 \times g$ for 10 min at room temperature. Discard supernatants.
4. Resuspend cells into 100 μL of Nucleofector solution (at room temperature).
5. Add siRNAs at the optimal concentration (as determined in the preliminary assays).
6. Transfer cells into the Nucleocuvette Vessels.
7. Place the cuvette into the retainer of the Nucleofector and start nucleofection by selecting the appropriate program.
8. After run completion, collect cells and add 500 μL of 37 °C pre-warmed additive-free RPMI.
9. Incubate for 2 h at 37 °C and 5 % CO_2.
10. Meanwhile, prepare an anti-CD3 coated flat-bottom 48-well plate. Add 50 μL of anti-CD3 antibody (20 μg/mL in sterile Milli-Q water) to each needed well and incubate for 1 h at 37 °C. Rinse the wells twice with sterile PBS before use.

11. Transfer cells to the anti-CD3 coated plate. Add 1 μg/mL (final concentration) of soluble anti-CD28 antibody in a total volume of 500 μL cell suspension.
12. Incubate cells at 37 °C and 5 % CO_2 for optimal knockdown time (as determined in the preliminary assays) minus 24 h.
13. 24 h prior to the assay, wash cells twice using 37 °C pre-warmed additive-free RPMI.
14. Plate cells into a new 48-well plate in a maximum of 500 μL complete medium.
15. Incubate at 37 °C and 5 % CO_2 for the remaining 24 h.

Cells are now ready for immunoblot assays. Proceed as described in Subheadings 3.12 and 3.5, respectively. Selective PTP inhibitors are expected to have a similar effect on phosphorylation levels of substrate proteins as observed in cells in which the PTP has been acutely eliminated by RNAi. However, the combination of knockdown and small-molecule inhibition should not result in greater effects than observed for each alone.

4 Notes

1. Preparation of paraformaldehyde (PAF). Weigh 4 g of PAF in beaker. Add 80 mL of 1× PBS. Cover with parafilm. Place the beaker on magnetic stirrer/heater (52 °C). Mix with moderate stirring until the solution becomes clear. (At the beginning, solution will appear milky.) Heating as well as a slight alkaline pH are necessary to dissolve PAF. Since the pH of PBS is around 7.4, PAF dissolves easily in the PBS. Adjust the final volume to 100 mL with PBS. Store the solution in small aliquots, depending on the usage, at −20 °C. Solution can also be stored at room temperature up to 1 week; longer storage at room temperature will lead to formation of formic acid in the solution.
2. Different combinations of fluorochrome-conjugated antibodies can be used depending on the available flow cytometer.
3. LAR II reagent should preferably be made fresh. Do not use LAR II reagent older than 2 weeks, because older reagent will give lower readings. Stop/Glo reagent comes prepared. Do not use Stop/Glo reagent older than 1 month. Add 50× substrate fresh before each use.
4. To perform an immunoprecipitation (IP) experiment, a minimum of 20×10^6 cells are needed. Lyse the cells in 1 mL RIPA buffer and collect supernatants in fresh 1.5 mL microtubes.

   To reduce nonspecific binding of proteins to beads, a pre-clearing step is required:

(a) Add 30 μL of bead slurry to the cell lysate (use A- or G-Sepharose beads, depending on the primary antibody used for IP. For a monoclonal antibody use G-Sepharose beads; for a polyclonal antibody use A-Sepharose beads).

(b) Incubate for 1 h at 4 °C with rotation of tubes.

(c) Centrifuge at 15,000 × *g* at 4 °C for 10 min.

(d) Collect supernatants and discard beads.

Perform the actual IP as follows:

(a) Add an optimal amount[1] of primary antibody to the cell lysates and incubate for 1 h at room temperature or overnight at 4 °C with gentle rotation. (Include one IP condition using an isotype control antibody as a control.)

(b) Add 40–50 μL of bead slurry to each sample and incubate for 2–3 h at 4 °C with gentle rotation.

(c) Centrifuge the tubes at 15,000 × *g* for 5 min. Remove the supernatants. Keep the supernatants (flow-through), as you may need them to evaluate the IP efficiency (*see* footnote 1).

(d) Wash the beads three times by adding 1 mL RIPA buffer and centrifuging at 15,000 × *g* for 5 min.

(e) Discard supernatants and add 2× SDS-PAGE sample buffer (v/v), pipet mix and boil the samples at 95 °C for 5 min. (Boiling not only denatures the protein, but also separates it from the beads.)

(f) Centrifuge at 15,000 × *g* for 5 min and collect supernatants. Do not discard beads immediately as they may be needed to check for quantitative protein release into the sample buffer.

(g) Separate the samples via SDS-PAGE and proceed with anti-phosphotyrosine western blotting.

5. We recommend the use of pervanadate (PV) as a positive control for PTP inhibition. Prepare a fresh 10 mM PV solution in additive-free RPMI as follows. First, incubate 10 μL of 200 mM Na_3VO_4 stock solution and 5 μL of 30 % H_2O_2 solution for 5 min at room temperature until the solution turns yellow due to PV formation. Then, add 185 μL RPMI and mix well (color will turn slightly pink). For the treatment of Jurkat cells, add

[1] The amount of antibody will depend on the abundance of protein in the sample, and the affinity of the protein to the antibody. Determine the optimal amount of primary antibody by running test IPs with several different concentrations of antibody. Therefore, separate each IP, together with the flow-through, by SDS-PAGE and reveal the levels of the protein by western blot analysis. An optimal antibody concentration will lead to a complete depletion of the protein in the flow-through.

10 μL PV solution to 1 mL cell suspension (20×10^6 cells/mL) in additive-free RPMI (final PV concentration will be 100 μM). Incubate cells while shaking at 500 rpm at 37 °C for 2–5 min using a thermomixer. Lyse cells immediately and proceed with SDS-PAGE and western blotting as described in Subheading 3.5 (Jurkat cells treated with PV do not need to be stimulated with anti-CD3).

6. The use of gradient gels allows a better separation of proteins with a greater range of molecular weights in a single gel. In addition, proteins with similar molecular weights are more likely to separate in gradient gels than in a linear gels [33]. Better separation of proteins allows better evaluation of the effects of a specific PTP inhibitor on the global tyrosine phosphorylation. When using gels with more than 12-wells, e.g., 15-well gels, we recommend using 1.5 mm gels, as smaller wells in the 1 mm gels will not allow loading enough protein.
7. Do not transfer longer than 1 or 1.5 h, respectively, as longer transfer may lead to the loss of low molecular weight proteins. Also, if the substrate of interest is a low molecular weight protein (MW < 20 kDa), use membranes with a smaller pore size (0.2 μm) and transfer for no longer than 30 min. If both, higher and lower molecular weight proteins are of interest, cut the gel at 20 kDa and transfer separately. Alternatively, use two membranes with different pore sizes stacked together. Low molecular weight proteins will go through the first membrane (0.45 μm) and be "trapped" in the second membrane (0.2 μm).
8. When using an anti-phosphotyrosine antibody, better results in blocking membranes are achieved by diluting the primary antibody in BSA instead of milk, as this significantly decreases the background levels.
9. Primary antibodies solutions can be re-used two or three times if stored at 4 °C. In this case, the BSA-TBST used to prepare the primary antibody solution should contain 0.01 % azide.
10. To save antibodies, adjust a small plastic bag to the size of the membranes + 1 cm and seal it using a plastic bags sealer. Usually, for a small membrane, the maximum volume of antibody solution used is ~4 mL.
11. The secondary antibody solution should not contain azide, and thus should be freshly prepared.
12. For detection of global tyrosine phosphorylation and phospho-specific western blots, such as phospho-ERK1/2, use the regular ECL detection kit. For phospho-specific antibodies against low abundance proteins, or if regular ECL does not yield a sufficient signal, use a more sensitive ECL such as ECL Plus (Pierce) or Western Lightning Plus ECL (Perkin Elmer).

13. If the signal is too strong, wait for 10 or 15 min for slight extinction and expose again to the film. Alternatively, stack two or three films together and expose briefly; the signal will be lower in the second or third film.
14. Membranes can be stripped once or even twice (depending on the amount of proteins initially loaded and on the abundance of the protein of interest in the sample). Stripping buffer; 50 mM Tris–HCl pH 6.8, 2 % SDS, 100 mM β-mercaptoethanol. Incubate membranes in stripping buffer (in sealed plastic bags with membranes fully submerged) in 70 °C water bath for 5–10 min. Rinse 4–5 times in 1× PBS + 0.1 % Tween. The membranes are now stripped and ready for re-blocking and incubation with a second primary antibody.
15. Ionomycin is a potent and selective calcium ionophore agent. It acts as a carrier of calcium that enhances the intracellular calcium concentration, leading to the activation of the calcium sensor CaM and subsequent calmodulin and NFAT activation. Shortcutting the proximal TCR signaling to activate NFAT is a commonly used positive control when evaluating the effects of agents that modulate TCR signaling pathways.
16. If using frozen PBMC, resuspend cells in RPMI with 10 % serum and incubate them overnight in a cell incubator. The next day, cells that were damaged by the freezing procedure will be dead, and the viable, intact cells can be counted from a culture aliquot. Percentage of recovered cells after freezing–thawing is about 70 %.
17. If working with whole blood, 50 mL can be collected in five vacutainer tubes (of 10-mL) containing anticoagulant and EDTA (potassium salt). Employing PBMC collected from healthy volunteers or patients offers the possibility to evaluate the effects of a PTP inhibitor on cells from a donor with a specific genotype. However, preparing PBMC from whole blood results in about seven times fewer cells, compared to preparation from buffy coat.
18. To cryopreserve PBMC, makes aliquots of 10^7 cells in RPMI containing 20 % FBS and 10 % DMSO (final concentrations), transfer into 2 mL cryovials, freeze at –80 °C for 24 h, and then store vials in liquid nitrogen.
19. A 500× solution will ensure a maximal final DMSO concentration of 0.2 %. In cases where low solubility of inhibitors requires higher DMSO concentrations, we have used up to 0.5 % DMSO. However, higher DMSO concentrations should be avoided and definitely not exceed 0.5 %, as higher concentrations will cause cytotoxicity.
20. Treatment of PBMC with PTP inhibitor should be done in the absence of serum in the culture medium, as serum proteins can deplete the inhibitor through nonspecific binding. Thus, cells

are treated with inhibitor (or vehicle control) for 30 min in additive-free RPMI, before the suspension is diluted twofold in RPMI containing 20 % FBS. Alternatively, cells may be rinsed after the inhibitor treatment and resuspended in RPMI supplemented by 10 % FBS.

21. If effects of inhibitors on specific T cell subsets are observed, results need to be confirmed on sorted T cells. We have successfully done these experiments after magnetic negative selection of $CD4^+$ T cells and $CD4^+CD45RA^-$ memory T cells. When using sorted T cells, cells are activated by coating a flat-bottom 96-well plate using anti-CD3 (20 μg/well in 50 μL PBS); soluble anti-CD28 antibody (1 μg/mL final concentration) is added to the cultured cells. Do not use CD3/CD28 cross-linked beads for flow cytometry evaluation of CD69 and CD25. Anti-CD25 and anti-CD69 antibodies can bind to the beads used for cross-linking, resulting in a false positive signal. Because using sorted T cells requires a larger number of cells and also is less cost effective than working with non-sorted T cells, we recommend this procedure only as a final step in the evaluation of a highly efficacious PTP inhibitor. Nonetheless, specific effects of PTP inhibitors on a particular T cell subset may have significant implications for future in vivo studies.
22. The appropriate concentration of anti-CD3 antibody should be determined prior to this assay. We have used several concentrations and found that, using the anti-CD3 antibody clone UCHT1, the lowest concentration that gives the highest proliferation index is 1 μg/mL, when activating PBMC. Other anti-CD3 antibodies can also be used. We found that the HIT3a clone yields similar results to the UCHT1 clone. However, the OKT3 clone required higher concentrations to obtain results similar to UCHT1 and HIT3a.
23. Do not re-freeze/re-use, as re-freezing will substantially degrade the cytokines in the medium.
24. We did not observe any differences between an immediate and a 24 h delayed analysis of cells. However, a further delayed analysis (beyond 24 h) leads to altered quality of dot plots due to loss of staining and increase of cellular debris.
25. Before running the acquisition on a flow cytometer, we usually calibrate the machine using the three-color BD CaliBRITE kit containing unlabeled beads, fluorescein isothiocyanate (FITC)-labeled beads, phycoerythrin (PE) labeled beads, and peridin chlorophyll protein (PerCP)-labeled beads. In addition, we add the allophycocyanin (APC)-labeled beads to setup all four colors used in the assay. This procedure allows determining the correct settings (compensation and detector/amplitude) when several colors are simultaneously used.

26. We suggest the following settings for the FACS Calibur II lasers (CellQuest software; BD Biosciences). However, these values are meant as an example/starting point, and need to be adjusted, depending on the specific equipment available.

    Compensations:

    - FL1—45 % of FL2.
    - FL2—25 % of FL1.
    - FL2—29 % of FL3.
    - FL3—12 % of FL2.
    - FL3—9 % of FL4.
    - FL4—23 % of FL3.

    Detector/amplitude:

    - FSC: E00 Linear mode
    - SSC: 489 Linear mode.
    - FL-1: 564 Log mode.
    - FL-2: 664 Log mode.
    - FL-3: 745 Log mode.
    - FL-5: 885 Log mode.
27. After 2 days, cell aggregates are formed upon anti-CD3 activation and are easily visible under optical microscope. Potential effects of inhibitors on PBMC proliferation can already be noticed at this stage, as the size of the cell aggregates is proportionally related to cell proliferation.
28. We recommend pooling only the required volumes of capture beads for the assay. Keep the remaining beads in the original packaging as provided by the vendor. Pooled beads cannot be re-used.
29. We recommend the use of one non-diluted and one half-diluted sample, in order to ensure that the mean fluorescence values fall into the range of the standard curve. Further dilutions are not required when working with cell supernatants, unless higher cell numbers than the ones recommended in this protocol (1×10^6) are used per sample.

Acknowledgment

This work was supported by NIH grants R03MH095532, R03MH084230, and R21CA132121 (to L.T.), by the Belgian National Funds for Scientific Research (FRS-FNRS), and by a grant from the University of Liège (Belgium) (to S.R.).

References

1. Mustelin T, Vang T, Bottini N (2005) Protein tyrosine phosphatases and the immune response. Nat Rev Immunol 5:43–57
2. Alonso A, Sasin J, Bottini N et al (2004) Protein tyrosine phosphatases in the human genome. Cell 117:699–711
3. Mustelin T, Alonso A, Bottini N et al (2004) Protein tyrosine phosphatases in T cell physiology. Mol Immunol 41:687–700
4. Mustelin T, Rahmouni S, Bottini N et al (2003) Role of protein tyrosine phosphatases in T cell activation. Immunol Rev 191:139–147
5. Goebel-Goody SM, Baum M, Paspalas CD et al (2012) Therapeutic implications for striatal-enriched protein tyrosine phosphatase (STEP) in neuropsychiatric disorders. Pharmacol Rev 64:65–87
6. Julien SG, Dube N, Hardy S et al (2011) Inside the human cancer tyrosine phosphatome. Nat Rev Cancer 11:35–49
7. Pulido R, Hooft van Huijsduijnen R (2008) Protein tyrosine phosphatases: dual-specificity phosphatases in health and disease. FEBS J 275:848–866
8. Rhee I, Veillette A (2012) Protein tyrosine phosphatases in lymphocyte activation and autoimmunity. Nat Immunol 13:439–447
9. Tautz L, Pellecchia M, Mustelin T (2006) Targeting the PTPome in human disease. Expert Opin Ther Targets 10:157–177
10. Tonks NK (2006) Protein tyrosine phosphatases: from genes, to function, to disease. Nat Rev Mol Cell Biol 7:833–846
11. Vang T, Miletic AV, Arimura Y et al (2008) Protein tyrosine phosphatases in autoimmunity. Annu Rev Immunol 26:29–55
12. Barr AJ (2010) Protein tyrosine phosphatases as drug targets: strategies and challenges of inhibitor development. Future Med Chem 2:1563–1576
13. Bialy L, Waldmann H (2005) Inhibitors of protein tyrosine phosphatases: next-generation drugs? Angew Chem Int Ed Engl 44:3814–3839
14. Vintonyak VV, Antonchick AP, Rauh D et al (2009) The therapeutic potential of phosphatase inhibitors. Curr Opin Chem Biol 13:272–283
15. Vang T, Liu WH, Delacroix L et al (2012) LYP inhibits T-cell activation when dissociated from CSK. Nat Chem Biol 8:437–446
16. Negro R, Gobessi S, Longo PG et al (2012) Overexpression of the autoimmunity-associated phosphatase PTPN22 promotes survival of antigen-stimulated CLL cells by selectively activating AKT. Blood 119: 6278–6287
17. Tautz L, Mustelin T (2007) Strategies for developing protein tyrosine phosphatase inhibitors. Methods 42:250–260
18. Abraham RT, Weiss A (2004) Jurkat T cells and development of the T-cell receptor signalling paradigm. Nat Rev Immunol 4:301–308
19. Mosmann T (1983) Rapid colorimetric assay for cellular growth and survival: application to proliferation and cytotoxicity assays. J Immunol Methods 65:55–63
20. Sherf BA, Navarro SL, Hannah RR (1996) Dual-luciferase™ reporter assay: an advanced co-reporter technology integrating firefly and renilla luciferase assays. Promega Notes 57:02
21. Sergienko E, Bobkova E, Vasile S et al (2010) Selective HePTP inhibitors: probe 1. Probe Reports from the Molecular Libraries Program, Bethesda (MD): National Center for Biotechnology Information (US)
22. Sergienko E, Xu J, Liu WH et al (2012) Inhibition of hematopoietic protein tyrosine phosphatase augments and prolongs ERK1/2 and p38 activation. ACS Chem Biol 7:367–377
23. Hogan PG, Lewis RS, Rao A (2010) Molecular basis of calcium signaling in lymphocytes: STIM and ORAI. Annu Rev Immunol 28:491–533
24. Cebrian M, Yague E, Rincon M et al (1988) Triggering of T cell proliferation through AIM, an activation inducer molecule expressed on activated human lymphocytes. J Exp Med 168:1621–1637
25. Cosulich ME, Rubartelli A, Risso A et al (1987) Functional characterization of an antigen involved in an early step of T-cell activation. Proc Natl Acad Sci USA 84:4205–4209
26. Hara T, Jung LK, Bjorndahl JM et al (1986) Human T cell activation. III. Rapid induction of a phosphorylated 28 kD/32 kD disulfide-linked early activation antigen (EA 1) by 12-o-tetradecanoyl phorbol-13-acetate, mitogens, and antigens. J Exp Med 164: 1988–2005
27. Reddy M, Eirikis E, Davis C et al (2004) Comparative analysis of lymphocyte activation marker expression and cytokine secretion profile in stimulated human peripheral blood mononuclear cell cultures: an in vitro model to monitor cellular immune function. J Immunol Methods 293:127–142
28. Porebski G, Gschwend-Zawodniak A, Pichler WJ (2011) In vitro diagnosis of T cell-mediated drug allergy. Clin Exp Allergy 41:461–470

29. Liao W, Lin JX, Leonard WJ (2011) IL-2 family cytokines: new insights into the complex roles of IL-2 as a broad regulator of T helper cell differentiation. Curr Opin Immunol 23:598–604
30. Morgan E, Varro R, Sepulveda H et al (2004) Cytometric bead array: a multiplexed assay platform with applications in various areas of biology. Clin Immunol 110:252–266
31. Weiss WA, Taylor SS, Shokat KM (2007) Recognizing and exploiting differences between RNAi and small-molecule inhibitors. Nat Chem Biol 3:739–744
32. Kim DH, Rossi JJ (2007) Strategies for silencing human disease using RNA interference. Nat Rev Genet 8:173–184
33. Walker JM (1984) Gradient SDS polyacrylamide gel electrophoresis. Methods Mol Biol 1:57–61

Chapter 16

Interactor-Guided Dephosphorylation by Protein Phosphatase-1

Shannah Boens, Kathelijne Szekér, Aleyde Van Eynde, and Mathieu Bollen

Abstract

Protein phosphatase-1 (PP1) is an essential enzyme for every eukaryotic cell and catalyzes more than half of all protein dephosphorylations at serine and threonine residues. The free catalytic subunit of PP1 shows little substrate selectivity but is tightly regulated in vivo by a large variety of structurally unrelated PP1-interacting proteins (PIPs). PIPs form highly specific dimeric or trimeric PP1 holoenzymes by acting as substrates, inhibitors, and/or substrate-specifiers. The surface of PP1 contains many binding sites for short PP1-docking motifs that are combined by PIPs to create a PP1-binding code that is universal, specific, degenerate, nonexclusive, and dynamic. These properties of the PP1-binding code can be used for the rational design of small molecules that disrupt subsets of PP1 holoenzymes and have a therapeutic potential.

Key words Protein phosphatase-1, Ser/Thr dephosphorylation, Substrate selection, Subcellular targeting

1 Introduction

The reversible phosphorylation of proteins is one of the most widespread posttranslational modifications, mediating responses to internal and external signals in a variety of cellular processes [1–3]. In eukaryotes 30–70 % of all proteins are phosphorylated on tyrosine (Tyr), threonine (Thr), and/or serine (Ser) residues [4–6]. Protein phosphorylation is catalyzed by protein kinases, which transfer the γ-phosphate of ATP to a hydroxyl side chain, resulting in the formation of a phosphate monoester. Protein phosphatases hydrolyze these phosphate monoesters and make protein phosphorylation a reversible modification [4].

Protein phosphatases are classified into four superfamilies based upon the structural conservation and mechanism of action of their catalytic domain [7–11]. The protein tyrosine phosphatase (PTP) superfamily includes the classical receptor and non-receptor

José Luis Millán (ed.), *Phosphatase Modulators*, Methods in Molecular Biology, vol. 1053, DOI 10.1007/978-1-62703-562-0_16, © Springer Science+Business Media, LLC 2013

protein tyrosine phosphatases, but also the dual-specificity protein phosphatases, which dephoshorylate Tyr as well as Ser/Thr residues. Some PTPs can also dephosphorylate non-protein substrates like phospholipids, RNA, and glycogen. A few protein Ser/Thr/Tyr phosphatases belong to the haloacid dehalogenase (HAD) superfamily of hydrolases. The metal-dependent protein phosphatase (PPM) superfamily contains monomeric protein Ser/Thr phosphatases that are activated by divalent cations. Finally, the phosphoprotein phosphatase (PPP) superfamily includes protein phosphatases 1–7, which also exclusively dephosphorylate Ser and Thr residues. PPPs have a structurally related catalytic domain and an identical catalytic mechanism [9, 12]. Biochemical studies revealed that PPPs catalyze more than 90 % of all dephosphorylations in eukaryotic cells.

Mammalian genomes only encode ~40 protein Ser/Thr phosphatases but more than 400 protein Ser/Thr kinases. This discrepancy can be explained by a different diversification strategy of these enzymes during evolution. Protein Ser/Thr kinases have mainly evolved by gene duplication and subsequent sequence diversification. In contrast, protein Ser/Thr phosphatases, in particular PP1 and PP2A, have mainly diversified by forming stable complexes with an ever increasing number of regulatory subunits. This explains why protein Ser/Thr kinases and phosphatases show the same diversity and substrate selectivity at the holoenzyme level, despite their hugely different number of catalytic subunits. Of all PPP superfamily members, PP1 has by far the most diverse protein interactome [9, 13]. Currently, about 200 structurally unrelated mammalian PP1 interacting proteins (PIPs) are known. These form stable holoenzyme complexes with one or several of the four PP1 isoforms, i.e., PP1α, β/δ, and the two splice variants γ1 and γ2. When bound to PP1, these PIPs interfere with the activity, subcellular localization and substrate specificity of the phosphatase. With such a diverse interactome it is not surprising that PP1 is the most abundant member of the PPP superfamily and catalyzes more than half of all protein dephosphorylation reactions. Indeed, recent quantitative proteome analyses of the human U2OS and HeLa cancer cell lines showed that the copy number of PP1 ranks in the hundreds of thousands and is, for example, three to six times higher than that of its closest relative PP2A [14, 15].

In this review we will address three important questions concerning PP1 holoenzymes: What is the structural basis of PIP–PP1 interactions? How do PIPs affect the activity and substrate specificity of PP1? Can the PIP–PP1 interaction code be exploited to rationally design small molecules that disrupt subsets of PP1 holoenzymes? We will illustrate the key points with data from the best-characterized PP1 holoenzymes.

2 The PIP–PP1 Interaction Code

The mammalian PP1 isoforms show about 90 % sequence identity and mainly differ in their extremities [13, 16, 17]. They fold into a compact globular domain that excludes, however, the C-terminal tail [18, 19]. The catalytic site of PP1 contains two metal ions and is located at the Y-shaped intersection of three putative substrate-binding clefts known as the hydrophobic, acidic, and C-terminal grooves. In addition, the surface of PP1 harbors numerous additional binding sites for PIPs, comprising both regulatory subunits and substrates. Often the binding of PIPs is mediated by short docking motifs [6, 20]. About ten distinct PP1 docking motifs have already been characterized in some detail (Table 1) and it seems likely that many more remain to be discovered. These binding motifs are often only found in one of the known PIPs but some of them are present in multiple PIPs. For example, the RVxF-type docking motif is found in about 90 % of all known PIPs and has been suggested to serve as an anchor for the initial recruitment of PP1, enabling the subsequent establishment of secondary PIP–PP1 interactions. In contrast, the IDoHa motif is only present in Inhibitor-2 and prevents access to the active site by binding to the hydrophobic and acid grooves of PP1. Some PP1 docking motifs are isoform-specific, explaining why some holoenzymes are only formed with one isoform of PP1. Thus, the AnkCap consists of ankyrin repeats that specifically enclose the C-terminus of PP1 β/δ [21]. Importantly, more than two thirds of all PIPs have a PP1-interaction domain that is predicted to be intrinsically disordered [22, 23]. It has been well established that these interaction domains adopt a defined fold upon binding to PP1. This structural feature is important because it enables multiple docking motifs to cover a large part of the PP1 surface [21, 22]. However, even after binding to PP1 some fragments of PP1-interaction domains can remain disordered. These disordered fragments are likely to mediate binding to other ligands or to interact with PP1 in a more dynamic manner that cannot be visualized using X-ray crystallography [24–26].

Most PIPs combine multiple docking motifs to form a unique and stable complex with PP1 (Fig. 1). This combinatorial control has led to the concept of a PP1-binding code [6, 27]. This interaction code is *universal* in that PP1 docking motifs and their binding sites are conserved from yeast to man. A nice illustration of this property is that the lethality of a lack of PP1 in budding yeast can be rescued by the expression of human PP1 [28]. The PP1 binding code is also *specific*: none of the PP1 docking motifs is known to mediate binding to closely related phosphatases of PP1. For example, the RVxF-type PP1 docking motif does not bind to PP3 (also known as calcineurin or PP2B), even though this phosphatase has a hydrophobic groove that is topologically equivalent to that of the

Table 1
The diversity of PP1 docking motifs

Acronym	Type	Site of interaction at the PP1 surface[a]	# Vertebrate PIPs
RVxF	5-Residue motif [KR]-[KR]-[V/I]-{FIMYDP}-[FW]	Opposite to the catalytic site, known as the RVxF-binding groove	Nearly 90 % of all PIPs
SILK	4-Residue motif [GS]-I-L-[RK]	Opposite to the catalytic site and adjacent to the RVxF-binding groove	7
MyPhoNe	8-Residue motif R-x-x-Q-[VIL]-[KR]-x-[YW]	Close to the hydrophobic substrate-binding groove	7
SpiDoC	8-Residue motif	The C-terminal substrate-binding groove	2
ΦΦ	2-Residue motif, hydrophobic residues	Opposite to the catalytic site and adjacent to the RVxF-binding groove	2
RNYF	4-Residue motif	ND	1
IDoHA	Inhibitory α-helix structure	The hydrophobic and acidic substrate-binding grooves	1
BiSTriP	Leucine-rich repeats	Triangular region between the C-terminal and hydrophobic substrate-binding grooves[b]	1
AnkCap	Ankyrin repeat domain	C-terminal tail of PP1β	1
NiHiP	α-Helix structure	Bottom surface	1

AnkCap Ankyrin repeat Cap, *BiSTriP* Bipartite docking site of the leucine-rich repeat protein SDS22 interacting with a Triangular region that is delineated by helices $\alpha_{4\text{-}6}$ of PP1, *IDoHA* Inhibitor-2 Docking site for the Hydrophobic and Acidic grooves, *MyPhoNE* Myosin Phosphatase N-terminal Element, *ND* not determined, *NiHiP* NIPP1 Helix that interacts with PP1, *SpiDoC* Spinophilin Docking site for the C-terminal groove. Based on [6, 20, 21, 30, 34, 54, 55]

[a]As compared to the position of the catalytic site that lies at the Y-shaped intersection of three putative substrate-binding grooves, namely, the hydrophobic, acidic, and C-terminal grooves

[b]Only mapped by mutagenesis experiments

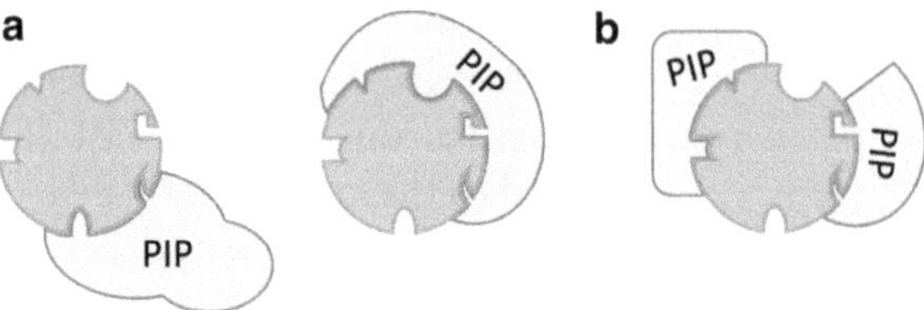

Fig. 1 Structure of PP1 holoenzymes. (**a**) A dimeric PP1 holoenzyme with multiple interaction sites. (**b**) A trimeric holoenzyme with two PIPs

RVxF-binding channel of PP1 [6, 29]. Another property of the PP1 code is that the binding motifs are *degenerate*. Thus, the consensus sequence of the RVxF-motif is $[K_{55}R_{34}][K_{28}R_{26}][V_{94}I_6]$ {FIMYDP}$[F_{83}W_{17}]$, where the subscript numbers indicate the percentage of residue occurrence in validated PIPs and the braced residues are excluded [20, 30, 31]. This degeneracy generates docking motifs with different affinities for PP1 and explains why the relative contribution of a specific docking motif to the overall PP1 binding affinity is PIP-dependent. For example, mutation of the RVxF-motif disrupts the interaction of some PIPs with PP1 but has no measurable effect on other PP1–PIP interactions. Yet another characteristic of the PP1-binding code is that it is *nonexclusive*: binding of one PIP does not necessarily exclude the binding of a second PIP as long as they each have at least one PP1 binding motif that is not common (Fig. 1b). If two PIPs share a PP1-binding motif the variant with the highest affinity occupies the corresponding binding groove of PP1 [32]. Trimeric PP1 holoenzymes often contain, in addition to the catalytic subunit, a regulatory and an inhibitory subunit [6, 33]. Finally, the PP1 binding code is highly *dynamic*. PIPs exist in a large molar excess and compete for binding to the limited number of PP1 molecules. This is well illustrated by the overexpression of PIPs, resulting in the competitive disruption of PP1 holoenzymes and the retargeting of PP1 to the cellular compartment that contains the overexpressed PIP. The dynamicity of the PIP–PP1 interaction is subject to regulation. For example, the phosphorylation of many PIPs on residues within or flanking the RVxF-motif decreases their affinity for PP1, reducing their ability to compete with other PIPs for PP1 binding [6, 27].

3 How PIPs Create Specific PP1 Holoenzymes

For most PIPs the physiological relevance of their binding to PP1 is not yet understood. However, the well-studied PIPs can be functionally classified into inhibitors, substrates or substrate-specifiers (Fig. 2). Here, we will discuss the properties of some of the best characterized PIPs and show how they direct the function of PP1.

3.1 Inhibitor-PIPs

Inhibitor-PIPs block access to the catalytic site and thereby indiscriminately prevent the dephosphorylation of all PP1 substrates. They can form a stable dimeric complex with PP1 but are also known to form trimeric complexes with PP1 and subsets of PIPs. The best characterized inhibitory PIPs are Inhibitor-2, Inhibitor-1, CPI-17, and their paralogs.

Inhibitor-2 has three major PP1-docking motifs. Two of these, the RVxF and SILK-motifs, only serve to anchor PP1, while the docking of the IDoHA α-helix motif to the acidic and hydrophobic grooves blocks access to the active site [25, 34]. Some structural and modeling data suggest that Inhibitor-2 also induces the release of a metal ion from the catalytic site [6, 24, 34]. The Inhibitor-2 induced loss of a metal ion is consistent with a large

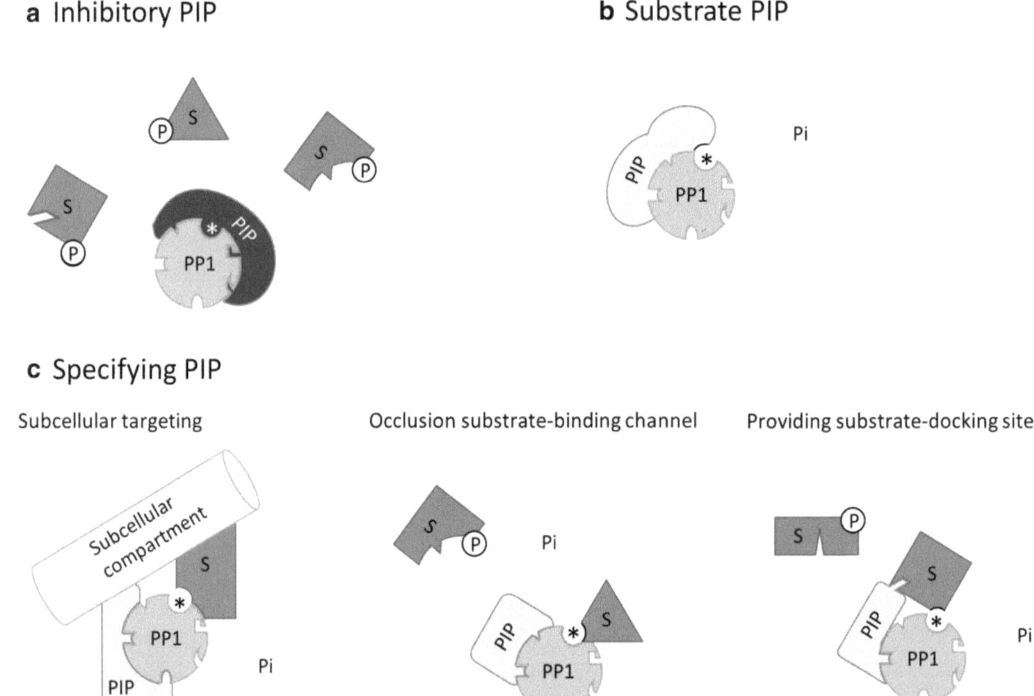

Fig. 2 Functional classication of PIPs. (**a**) Inhibitory-PIPs block access of substrates to the active site. (**b**) Substrate-PIPs are themselves substrates for associated PP1. (**c**) Substrate-specifying PIPs act by anchoring PP1 in a subcellular compartment that contains a subset of PP1 substrates (*left*), by occlusion of a substrate-binding channel (*middle*), or by serving as a substrate-recruitment platform (*right*)*, catalytic site of PP1; Pi, inorganic phosphate

body of biochemical data obtained in various laboratories showing that Inhibitor-2 gradually converts PP1 into an inactive, MgATP-dependent phosphatase [24, 35]. Reactivation of this complex depends on the phosphorylation of Inhibitor-2 by one of various protein kinases, including GSK3, and has been suggested to be associated with the reincorporation of a metal ion at the active site. Intriguingly, the phosphorylation of Inhibitor-2 is only transient because it is rapidly dephosphorylated by associated PP1. Hence, Inhibitor-2 is both a substrate and an inhibitor of PP1. Inhibitor-2 has also been shown to form trimeric complexes with PP1 and either a substrate (Nek2A, Aurora A) or a substrate-specifier (Spinophilin) [32, 36, 37]. In the latter complex, it is Spinophilin that occupies the RVxF-binding groove, at the expense of the weaker RVxF-motif of Inhibitor-2.

Inhibitor-1 and its paralogs (DARPP32, IPP5) have an RVxF-docking site. In addition, a Thr that is 23 residues C-terminal to the RVxF-motif functions as a pseudosubstrate when phosphorylated and inhibits the dephosphorylation of all PP1 substrates [12, 19].

It is not known what prevents the dephosphorylation of Inhibitor-1 by PP1. Inhibitor-1 has also been found to act as an inhibitory subunit of a trimeric complex with PP1 and the eIF2α-targeting subunit GADD34 [38].

CPI-17 and its paralogs (PHI1, KEPI, GBPI) are selective inhibitors of PP1 holoenzymes with the myosin-targeting subunit MYPT1 [39]. Unlike its paralogs, CPI-17 lacks an RVxF motif. Inhibition of PP1 depends on phosphorylation of CPI-17 at Thr38, resulting in a major conformational change and the exposure of a specific loop that is predicted to bind in the active site as a pseudosubstrate [33]. The holoenzyme specificity of CPI-17 stems from the fact that the PP1-MYPT1 complex, in contrast to other PP1 complexes, cannot dephosphorylate CPI-17 at Thr38. Hence, CPI-17 is both a substrate and an inhibitor of PP1, depending on the nature of the substrate-specifying PIP that is associated with PP1.

3.2 Substrate-PIPs

It has been well-established that some PIPs are themselves substrates for associated PP1. One group of substrate-PIPs consists of protein kinases, which currently represent 6 % of all validated vertebrate PIPs. For example, PP1 forms a complex with the centrosomal protein kinase Nek2A [40, 41]. Nek2A phosphorylates and inactivates PP1, whereas PP1 dephosphorylates and inactivates Nek2A. This double negative feedback loop has the properties of a bistable switch, with either PP1 or Nek2A fully activated. A further complexity arises from the binding of Inhibitor-2, which inhibits PP1 and promotes Nek2A activation. At the G2-M transition of the cell cycle, the balance is tilted in favor of Nek2A through phosphorylation of both PP1 and Inhibitor-2 by cyclin-dependent kinases (Cdks). In contrast, DNA-damage results in the dephosphorylation of PP1 and the dissociation of Inhibitor-2, which switches off signaling by Nek2A. A similarly tight regulation applies to a mitotic complex of Aurora A, PP1 and Inhibitor-2 [36].

Another well-established substrate-PIP is the Retinoblastoma protein (Rb) [42], which hampers the G1-S transition by the binding and inhibition of E2F transcription factors. Phosphorylation of Rb by Cdks disrupts the Rb–E2F interaction and allows cell-cycle progression. Following dephosphorylation by associated PP1, Rb is reconverted to an E2F inhibitor [42]. Intriguingly, the RVxF-type PP1 docking motif in the C-terminus of Rb overlaps with the binding site for Cdks, making the binding of PP1 and Cdks mutually exclusive [43]. This mechanism prevents the simultaneous phosphorylation and dephosphorylation of Rb.

3.3 Substrate-Specifiers

Substrate-specifiers act by (1) anchoring PP1 in a subcellular compartment that contains a subset of PP1 substrates, (2) occluding substrate-binding channels, and/or (3) providing additional binding sites for substrates. Subcellular anchoring PIPs limit the action radius of PP1 to local substrates. They include targeting PIPs for the plasma membrane (e.g., integrin αIIB), the endoplasmic reticulum (e.g., GADD34), glycogen particles (e.g., G_M-subunit), myofibrils (e.g., MYPT1), nucleoli (NOM1), centrosomes (e.g., Nek2A), kinetochores (e.g., KNL1), and chromosomes (e.g., Repo-Man). Some of these PIPs have multiple targeting domains which enables them to function as signal integrators. For example, MYPT1 has independent binding sites for actin-binding proteins (myosin, merlin, moesin), the Retinoblastoma protein and the mitotic kinase PLK1, which are all established substrates of MYPT1-associated PP1 [44]. The multi-targeting property of MYPT1 contributes to the integration of actomyosin cytoskeleton function and cell-cycle progression. Many PP1 interactors also interfere with substrate recruitment. More than half of all PIPs inhibit PP1 when glycogen phosphorylase *a* is used as substrate [30]. A recent structural analysis of the Spinophilin-PP1 complex showed that Spinophilin binds to the C-terminal groove of PP1 and thereby competitively disrupts the recruitment of glycogen phosphorylase *a* [22, 32]. The inhibition of the phosphorylase phosphatase activity of PP1 by NIPP1 involves a entirely different mechanism that depends on a polybasic stretch N-terminal to the PP1-docking RVxF motif [45, 46]. This polybasic stretch is not visible in the crystal of the NIPP1-PP1 complex, suggesting that it inhibits the phosphorylase phosphatase activity through dynamic interactions with acidic residues of PP1. The binding of some PIPs changes the electrostatic surface properties of PP1 quite dramatically [21, 32, 45] which is likely to interfere with substrate selection. In addition, some PIPs have additional docking sites for PP1 substrates. For example, the N-terminus of NIPP1 largely consists of a ForkHead Associated (FHA) domain that recruits a subset of substrates that are phosphorylated on phospho-TP dipeptide motifs and are dephosphorylated by NIPP1-associated PP1 in a regulated manner [47, 48]. Substrate selection can also involve the specific dephosphorylation of a single site of a substrate that is phosphorylated at multiple sites. A complex of PP1γ and Repo-Man selectively dephosphorylates histone H3 at Thr3 during prometaphase but does not dephosphorylate other histone H3 sites (e.g., Ser10, Ser28) which are, however, also dephosphorylated by PP1 at the mitotic exit [49]. The molecular basis of this site selectivity is not understood.

4 Therapeutic Potential of PP1

The properties of the PP1 interaction code suggests that it is possible to develop peptides or small molecules that bind to the surface of PP1 and selectively disrupt (subsets of) PP1 holoenzymes. Such tools can be used for functional studies (chemical genetics) but have also therapeutic potential to treat phosphorylation diseases [50]. One such PP1-disrupting peptide (PDP3) is a cell-permeable variant of the combined RVxF and ΦΦ motifs of NIPP1 [51]. This peptide was shown to generate a cellular pool of PDP3-associated PP1, which is fully active and dephosphorylates established mitotic substrates of PP1. These findings show that a major function of PIPs is to restrain PP1 and prevent the cellular accumulation of free PP1. The activation of PP1 using PDP3 or a small molecule with the same effect has also therapeutic potential, for example to sensitize tumors to protein kinase inhibitors that are already widely used for chemotherapy. The reverse approach, i.e., to develop molecules that bind to PIPs and prevent their interaction with PP1 is probably even more specific to target PP1 holoenzymes. However, this is definitely also a more challenging avenue since the interaction domain of most PIPs is intrinsically disordered and it is not trivial to rationally design small molecules that forces such domains to adopt a conformation that prevents their binding to PP1. The proof-of-concept for this approach has been delivered by the serendipitous finding that Guanabenz, an α2-adrenergic antagonist that is used to treat hypertension, binds to GADD34 and hampers its interaction with PP1. Since the PP1-GADD34 complex is involved in the regulation of ER-stress through dephosphorylation of the translation initiation factor eIF2α, Guanabenz analogues have potential to treat protein folding diseases [52, 53].

Acknowledgments

This work was supported by a "Flemish Concerted Research Action" (GOA 10/16) and the "Foundation against cancer."

References

1. Olsen JV, Blagoev B, Gnad F et al (2006) Global, in vivo, and site-specific phosphorylation dynamics in signaling networks. Cell 127:635–648
2. Mann M, Jensen ON (2003) Proteomic analysis of post-translational modifications. Nat Biotechnol 21:255–261
3. Manning G, Plowman GD, Hunter T et al (2002) Evolution of protein kinase signaling from yeast to man. Trends Biochem Sci 27:514–520
4. Ubersax JA, Ferrell JE Jr (2007) Mechanisms of specificity in protein phosphorylation. Nat Rev Mol Cell Biol 8:530–541
5. Cohen P (2000) The regulation of protein function by multisite phosphorylation-a 25 year update. Trends Biochem Sci 25:596–601
6. Heroes E, Lesage B, Gornemann J et al (2013) The PP1 binding code: a molecular-lego strategy that governs specificity. FEBS J 280(2):584–595
7. Moorhead GB, Trinkle-Mulcahy L, Ulke-Lemee A (2007) Emerging roles of nuclear protein phosphatases. Nat Rev Mol Cell Biol 8:234–244
8. Brautigan DL (2013) Protein Ser/Thr phosphatases—the ugly ducklings of cell signalling. FEBS J 280(2):324–345

9. Shi Y (2009) Serine/threonine phosphatases: mechanism through structure. Cell 139:468–484
10. Alonso A, Sasin J, Bottini N et al (2004) Protein tyrosine phosphatases in the human genome. Cell 117:699–711
11. Tonks NK (2012) Protein tyrosine phosphatases: from housekeeping enzymes to master-regulators of signal transduction. FEBS J 280(2):346–378
12. Barford D, Das AK, Egloff MP (1998) The structure and mechanism of protein phosphatases: insights into catalysis and regulation. Annu Rev Biophys Biomol Struct 27:133–164
13. Ceulemans H, Stalmans W, Bollen M (2002) Regulator-driven functional diversification of protein phosphatase-1 in eukaryotic evolution. Bioessays 24:371–381
14. Beck M, Schmidt A, Malmstroem J et al (2011) The quantitative proteome of a human cell line. Mol Syst Biol 7:549
15. Nagaraj N, Wisniewski JR, Geiger T et al (2011) Deep proteome and transcriptome mapping of a human cancer cell line. Mol Syst Biol 7:548
16. Cohen PT (2002) Protein phosphatase 1–targeted in many directions. J Cell Sci 115:241–256
17. Ceulemans H, Bollen M (2004) Functional diversity of protein phosphatase-1, a cellular economizer and reset button. Physiol Rev 84:1–39
18. Egloff MP, Cohen PT, Reinemer P et al (1995) Crystal structure of the catalytic subunit of human protein phosphatase 1 and its complex with tungstate. J Mol Biol 254:942–959
19. Goldberg J, Huang HB, Kwon YG et al (1995) Three-dimensional structure of the catalytic subunit of protein serine/threonine phosphatase-1. Nature 376:745–753
20. Bollen M, Peti W, Ragusa MJ et al (2010) The extended PP1 toolkit: designed to create specificity. Trends Biochem Sci 35:450–458
21. Terrak M, Kerff F, Langsetmo K et al (2004) Structural basis of protein phosphatase 1 regulation. Nature 429:780–784
22. Ragusa MJ, Dancheck B, Critton DA et al (2010) Spinophilin directs protein phosphatase 1 specificity by blocking substrate binding sites. Nat Struct Mol Biol 17:459–464
23. Dancheck B, Nairn AC, Peti W (2008) Detailed structural characterization of unbound protein phosphatase 1 inhibitors. Biochemistry 47:12346–12356
24. Cannon JF (2012) How phosphorylation activates the protein phosphatase-1 * inhibitor-2 complex. Biochim Biophys Acta 1834:71–86
25. Marsh JA, Dancheck B, Ragusa MJ et al (2010) Structural diversity in free and bound states of intrinsically disordered protein phosphatase 1 regulators. Structure 18:1094–1103
26. Peti W, Nairn AC, Page R (2012) Folding of intrinsically disordered protein phosphatase 1 regulatory proteins. Curr Phys Chem 2:107–114
27. Bollen M (2001) Combinatorial control of protein phosphatase-1. Trends Biochem Sci 26:426–431
28. Gibbons JA, Kozubowski L, Tatchell K et al (2007) Expression of human protein phosphatase-1 in Saccharomyces cerevisiae highlights the role of phosphatase isoforms in regulating eukaryotic functions. J Biol Chem 282:21838–21847
29. Yang J, Roe SM, Prickett TD et al (2007) The structure of Tap42/alpha4 reveals a tetratricopeptide repeat-like fold and provides insights into PP2A regulation. Biochemistry 46:8807–8815
30. Hendrickx A, Beullens M, Ceulemans H et al (2009) Docking motif-guided mapping of the interactome of protein phosphatase-1. Chem Biol 16:365–371
31. Meiselbach H, Sticht H, Enz R (2006) Structural analysis of the protein phosphatase 1 docking motif: molecular description of binding specificities identifies interacting proteins. Chem Biol 13:49–59
32. Dancheck B, Ragusa MJ, Allaire M et al (2011) Molecular investigations of the structure and function of the protein phosphatase 1-spinophilin-inhibitor 2 heterotrimeric complex. Biochemistry 50:1238–1246
33. Eto M, Kitazawa T, Matsuzawa F et al (2007) Phosphorylation-induced conformational switching of CPI-17 produces a potent myosin phosphatase inhibitor. Structure 15:1591–1602
34. Hurley TD, Yang J, Zhang L et al (2007) Structural basis for regulation of protein phosphatase 1 by inhibitor-2. J Biol Chem 282:28874–28883
35. Beullens M, Van Eynde A, Stalmans W et al (1992) The isolation of novel inhibitory polypeptides of protein phosphatase 1 from bovine thymus nuclei. J Biol Chem 267:16538–16544
36. Satinover DL, Leach CA, Stukenberg PT et al (2004) Activation of Aurora-A kinase by protein phosphatase inhibitor-2, a bifunctional signaling protein. Proc Natl Acad Sci USA 101:8625–8630
37. Eto M, Elliott E, Prickett TD et al (2002) Inhibitor-2 regulates protein phosphatase-1 complexed with NimA-related kinase to induce centrosome separation. J Biol Chem 277:44013–44020

38. Connor JH, Weiser DC, Li S et al (2001) Growth arrest and DNA damage-inducible protein GADD34 assembles a novel signaling complex containing protein phosphatase 1 and inhibitor 1. Mol Cell Biol 21:6841–6850
39. Eto M (2009) Regulation of cellular protein phosphatase-1 (PP1) by phosphorylation of the CPI-17 family, C-kinase-activated PP1 inhibitors. J Biol Chem 284:35273–35277
40. Helps NR, Luo X, Barker HM et al (2000) NIMA-related kinase 2 (Nek2), a cell-cycle-regulated protein kinase localized to centrosomes, is complexed to protein phosphatase 1. Biochem J 349:509–518
41. Mi J, Guo C, Brautigan DL et al (2007) Protein phosphatase-1alpha regulates centrosome splitting through Nek2. Cancer Res 67:1082–1089
42. Durfee T, Becherer K, Chen PL et al (1993) The retinoblastoma protein associates with the protein phosphatase type 1 catalytic subunit. Genes Dev 7:555–569
43. Hirschi A, Cecchini M, Steinhardt RC et al (2010) An overlapping kinase and phosphatase docking site regulates activity of the retinoblastoma protein. Nat Struct Mol Biol 17:1051–1057
44. Grassie ME, Moffat LD, Walsh MP et al (2011) The myosin phosphatase targeting protein (MYPT) family: a regulated mechanism for achieving substrate specificity of the catalytic subunit of protein phosphatase type 1delta. Arch Biochem Biophys 510:147–159
45. O'Connell N, Nichols SR, Heroes E et al (2012) The molecular basis for substrate specificity of the nuclear NIPP1:PP1 holoenzyme. Structure 20:1746–1756
46. Beullens M, Van Eynde A, Vulsteke V et al (1999) Molecular determinants of nuclear protein phosphatase-1 regulation by NIPP-1. J Biol Chem 274:14053–14061
47. Tanuma N, Kim SE, Beullens M et al (2008) Nuclear inhibitor of protein phosphatase-1 (NIPP1) directs protein phosphatase-1 (PP1) to dephosphorylate the U2 small nuclear ribonucleoprotein particle (snRNP) component, spliceosome-associated protein 155 (Sap155). J Biol Chem 283:35805–35814
48. Minnebo N, Görnemann J, O'Connell N et al (2013) NIPP1 maintains EZH2 phosphorylation and promoter occupancy at proliferation-related target genes. Nucleic Acids Res 41(2):842–854
49. Qian J, Lesage B, Beullens M et al (2011) PP1/Repo-man dephosphorylates mitotic histone H3 at T3 and regulates chromosomal aurora B targeting. Curr Biol 21:766–773
50. Bollen M, De Munter S, Kohn M (2012) Challenges and opportunities in the development of protein phosphatase-directed therapeutics. ACS Chem Biol 8(1):36–45
51. Chatterjee J, Beullens M, Sukackaite R et al (2012) Development of a peptide that selectively activates protein phosphatase-1 in living cells. Angew Chem Int Ed Engl 51:10054–10059
52. Tsaytler P, Harding HP, Ron D et al (2011) Selective inhibition of a regulatory subunit of protein phosphatase 1 restores proteostasis. Science 332:91–94
53. Fullwood MJ, Zhou W, Shenolikar S (2012) Targeting phosphorylation of eukaryotic initiation factor-2alpha to treat human disease. Prog Mol Biol Transl Sci 106:75–106
54. Llanos S, Royer C, Lu M et al (2011) Inhibitory member of the apoptosis-stimulating proteins of the p53 family (iASPP) interacts with protein phosphatase 1 via a non-canonical binding motif. J Biol Chem 286:43039–43044
55. Ceulemans H, Vulsteke V, De Maeyer M et al (2002) Binding of the concave surface of the Sds22 superhelix to the alpha 4/alpha 5/alpha 6-triangle of protein phosphatase-1. J Biol Chem 277:47331–47337

Chapter 17

Structure, Regulation, and Pharmacological Modulation of PP2A Phosphatases

Caroline Lambrecht, Dorien Haesen, Ward Sents, Elitsa Ivanova, and Veerle Janssens

Abstract

Protein phosphatases of the type 2A family (PP2A) represent a major fraction of cellular Ser/Thr phosphatase activity in any given human tissue. In this review, we describe how the holoenzymic nature of PP2A and the existence of several distinct PP2A composing subunits allow for the generation of multiple structurally and functionally different PP2A complexes, explaining why PP2A is involved in the regulation of so many diverse cell biological and physiological processes. Moreover, in human disease, most notably in several cancers and Alzheimer's Disease, PP2A expression and/or activity have been found significantly decreased, underscoring its important functions as a major tumor suppressor and tau phosphatase. Hence, several recent preclinical studies have demonstrated that pharmacological restoration of PP2A activity, as well as pharmacological PP2A inhibition, under certain conditions, may be of significant future therapeutic value.

Key words Ceramide, CIP2A, Forskolin, Fostriecin, FTY720, Norcantharidin, Polycations, Okadaic acid, PP2A, Protein phosphatase 2A, SET, Sphingolipids, Subunit, Tau, Tumor suppressor

1 PP2A Phosphatases: Multifunctional Signalling Enzymes with Complex Structure

Type 2A Protein Phosphatases (PP2A) belong to the PhosphoProtein Phosphatase (PPP) enzymes, which also comprise the Ser/Thr phosphatases PP1, PP2B (calcineurin), PP4, PP5, PP6, and PP7 [1]. Together with PP1, the ubiquitously expressed PP2A phosphatases constitute the bulk of Ser/Thr phosphatase activity in a given tissue or cell [1]. Structurally, PP2A phosphatases are holoenzymes, constituted of at least two subunits (reviewed in refs. 2, 3). The PP2A core enzyme consists of a catalytic subunit ($PP2A_C$) and a structural A subunit (A/PR65). This dimer ($PP2A_D$) can associate with a wide range of regulatory B-type subunits to form different PP2A heterotrimers ($PP2A_T$) (Fig. 1). Within these trimeric complexes the B-type subunits determine the catalytic activity, substrate specificity, and hence, the physiological

José Luis Millán (ed.), *Phosphatase Modulators*, Methods in Molecular Biology, vol. 1053,
DOI 10.1007/978-1-62703-562-0_17, © Springer Science+Business Media, LLC 2013

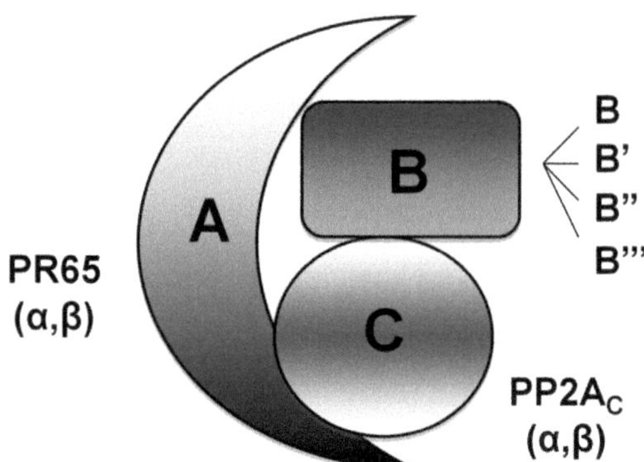

Fig. 1 Structure of PP2A. The majority of PP2A complexes within a given cell or tissue are heterotrimers, consisting of a catalytic C subunit (PP2A$_C$, existing as an α and a β isoform), a structural A subunit (PR65, existing as an α and a β isoform), and one of several regulatory B-type subunits. In humans, approximately 24 different B-type subunits have been described, which belong to four different families (B, B′, B″, and B″′). In addition, about 1/3 of PP2A can occur as a dimer of just one A and one C subunit (PP2A$_D$)

functions of the holoenzyme. Although the stoichiometry of PP2A within cells is poorly understood [4], there is evidence that approximately one-third of PP2A occurs as PP2A$_D$, while the majority of PP2A complexes are heterotrimers [2].

The catalytic subunit exists as an α and a β isoform, which are encoded by two distinct ubiquitously expressed genes: *PPP2CA* and *PPP2CB*, the latter being expressed tenfold lower due to a weaker promoter [2]. Interestingly, PP2A$_C$ appears to be synthesized as an inactive enzyme in order to prevent promiscuous/unrestricted activity of free PP2A$_C$ immediately after its translation [5]. Clearly, this necessitates a subsequent activation mechanism that is somehow coupled to the assembly of the entire holoenzyme [6]; it is within this holoenzyme context that the presence of a specific B-type subunit will restrict PP2A activity to its appropriate substrates. This biogenesis of active PP2A$_T$ is a complex step-by-step process that is controlled by several essential PP2A regulators and modulators (reviewed in ref. 4). In addition, specific posttranslational modifications of the highly conserved C-terminal tail of the C subunit can provide additional contacts with specific B-type subunits and thus contribute to or interfere with the assembly of particular holoenzymes (reviewed in ref. 7). Besides phosphorylation [8, 9], of particular interest in this context is PP2A$_C$ carboxymethylation, a conserved and reversible modification of the carboxy-terminus that is coordinated by the PP2A methyltransferase LCMT1 (Leucine Carboxyl Methyl Transferase 1, encoded by *LCMT1*) [10] and the methylesterase PME-1 (PP2A Methyl

Esterase-1, encoded by *PPME*) [11]. PME-1 yet fulfils another function as a binding and stabilizing protein of inactive $PP2A_C$ [6, 11, 12]. The PP2A activating protein PTPA (Protein phosphatase Two A Phosphatase Activator, encoded by *PPP2R4*) is required to refold $PP2A_C$ into an active conformation [5, 12–14], which allows efficient methylation by LCMT1 and subsequent B-type subunit binding [15].

The structural A subunit also exists in two isoforms, PR65/Aα and PR65/Aβ, encoded by two different genes, *PPP2R1A* and *PPP2R1B*. In a given tissue, the expression level of PR65/Aα is much higher than that of PR65/Aβ [2]. Both isoforms are composed of 15 HEAT repeats, which have a major role in binding B-type and C subunits. The B-type subunits bind in a mutually exclusive manner to repeats 1–10 and the C subunit to repeats 11–15. From the crystal structures of free PR65/Aα [16], $PP2A_D$ [17] and several trimeric PP2A holoenzymes [18–21], it became clear that the A subunit is a highly flexible molecule. Both binding of the C and the B subunits induce conformational changes, allowing proper access of the C subunit to the substrate [22], which is recruited into the complex by the B-type subunit.

The B-type subunits are expressed in a tissue-specific manner and mediate the localization and the substrate specificity of the PP2A holoenzyme (Table 1). Four families of human B-type subunits have been identified to date.

The PR55/B55 or B subunits are encoded by four different genes (*PPP2R2A*, *PPP2R2B*, *PPP2R2C*, and *PPP2R2D*), and encompass different splice variants (isoforms α, β, β_2, γ, and δ). Both Bα, and Bδ are widely expressed, while Bβ, $B\beta_2$, and Bγ are highly enriched in brain [23, 24, 28, 29, 50–52]. The B subunit family has been implicated in the regulation of many cell biological processes and signalling pathways, including mitosis (reviewed in refs. 53, 54), apoptosis (reviewed in ref. 55), extracellular signal-regulated kinase signalling (reviewed in ref. 3), transforming growth factor-β signalling [56], DNA-damage signalling [57], and tau dephosphorylation (reviewed in ref. 58).

The PR61/B56 or B′ subunits represent the largest B-type subunit family (isoforms α to ε) and are encoded by five different genes (*PPP2R5A*, *PPP2R5B*, *PPP2R5C*, *PPP2R5D*, and *PPP2R5E*), some of which are alternatively spliced or give rise to alternatively translated transcripts. B′β and B′δ have high expression in brain, while B′α and B′γ isoforms are particularly abundant in heart [38, 40, 45, 46, 59–63]. All B′ isoforms, except B′γ1, are phosphoproteins [38]. These holoenzymes have targets in mitosis/meiosis (reviewed in refs. 53, 54), apoptosis (reviewed in ref. 55), development (reviewed in ref. 64) and dopaminergic signalling (reviewed in ref. 65). Importantly, they also regulate Akt, Wnt, and c-Myc signalling, and are therefore considered as the main tumor suppressive PP2A subunits (reviewed in refs. 2, 3, 64, 66–69).

Table 1
Human PP2A B-type subunits, their tissue distribution and subcellular localization

Gene	Proteins	Aliases	Subcellular localization	Tissue distribution (highest expression)	References
PPP2R2A	Bα	PR55α, B55α	Cytoskeleton (vimentin, microtubules), cytoplasm, nucleus, plasma membrane, Golgi, ER	Widespread (testis)	[23–27]
PPP2R2B	Bβ1	PR55β, B55β	Cytosol	Testis, brain	[23–25]
	Bβ2	B55β2	Mitochondria	Brain	[28]
PPP2R2C	Bγ	PR55γ, B55γ	Cytoskeleton (microtubules, intermediate filaments)	Brain	[24, 25, 29]
PPP2R2D	Bδ	PR55δ, B55δ	Cytosol	Widespread (testis)	[24]
PPP2R3A	B″α1	PR130	Nucleus, cytosol, plasma membrane, Golgi, centrosome	Widespread (kidney, heart)	[30–33]
	B″α2	PR72	Nucleus, cytosol	Skeletal muscle, heart	
PPP2R3B	B″β1	PR70 (PR48)	Cytosol (nucleus)	Unknown	[32, 34, 35]
	B″β2	PR70	Unknown	Skeletal muscle, heart	
PPP2R3C	B″γ	G5PR	Cytosol, nucleus, centrosome	Widespread (kidney, testis)	[36, 37]
PPP2R5A	B′α	PR61α, B56α	Cytoplasm, kinetochore/centromere, mitochondria	Widespread (heart, muscle)	[37–39]
PPP2R5B	B′β	PR61β, B56β	Cytoplasm, kinetochore/centromere	Highly enriched in brain	[37, 38, 40]

PPP2R5C	B′γ1	PR61γ1, B56γ1	Nucleus (speckles) > cytoplasm, focal adhesions, kinetochore/ centromere, Golgi	Widespread (heart, muscle)	[37, 38, 41–44]
	B′γ2	PR61γ2, B56γ2	Nucleus > cytoplasm, kinetochore/ centromere, Golgi	Widespread (heart, muscle)	
	B′γ3	PR61γ3, B56γ3	Nucleus > cytoplasm, kinetochore/ centromere	Widespread (heart, muscle)	
PPP2R5D	B′δ1	PR61δ, B56δ, 74-kDa/B″δ	Nucleus > cytoplasm, kinetochore/ centromere	Widespread (brain)	[37, 38, 40, 45]
	B′δ2		Unknown		
	B′δ3		Unknown		
PPP2R5E	B′ε	B56ε, PR61ε	Cytoplasm > nucleus, kinetochore/ centromere	Widespread (brain)	[37, 38, 40, 46]
	B56ε-s		Cytoplasm		
STRN	striatin	B‴/PR93	Caveolae	Highly enriched in brain	[47–49]
STRN3	SG2NA	B‴/PR110	Nucleus, Golgi, caveolae	Brain, skeletal muscle	
STRN4	zinedin		Golgi, caveolae	Highly enriched in brain	

Human PR72/B″ subunits are encoded by three different genes (*PPP2R3A*, *PPP2R3B*, and *PPP2R3C*), the first two of which give rise to two alternative splice variants. All B″ subunits are ubiquitously expressed, except B″α2 and B″β2, which are almost exclusively found in heart and skeletal muscle [30, 32, 34–36, 70, 71]. The B″ subunits are EF-hand containing Ca^{2+}-binding proteins [72] with established functions in noncanonical Wnt signalling (reviewed in ref. 3), Epidermal growth factor (EGF) signalling [33], DNA synthesis [2, 73], pocket protein dephosphorylation (reviewed in ref. 74), and neuronal signalling (reviewed in ref. 65).

PR110/B‴ (striatin), PR93/B‴ (SG2NA), and zinedin represent the fourth B-type family and are encoded by three genes (*STRN*, *STRN3* and *STRN4*), which are all highly expressed in brain [47, 75, 76]. They are calmodulin-binding proteins with reported roles in Hippo signalling [77], Golgi polarization [78], and cytokinesis/abscission [79].

Together, the combinatorial assembly of all these diverse subunits gives rise to over 90 different PP2A complexes, all with potentially different physiological functions. Most of these functions have been elucidated in cellular models, while only a limited number of in vivo studies have been performed in mammalians. While genetic ablation of Cβ in mice results in no overt phenotypes [80], knockout mice of Cα [81] or Aα [82] are embryonic lethal, confirming PP2A phosphatases serve very essential functions. In contrast, genetic ablation of a single B-type subunit (PR61/B′δ) in mice is viable and gives rise to a number of discrete neurological phenotypes [83, 84], clearly demonstrating that despite significant homology within a given B-type subunit family, individual B-type subunits serve nonredundant functions in vivo. The future generation of additional B-type subunit KO mouse models should further provide insights into these physiological roles.

2 Regulation of Cellular PP2A Activity

2.1 Second Messengers and B-Type Subunit Phosphorylation

In support of the view that protein phosphatases are not at all “lazy housekeeping genes” that merely serve to passively counteract the “highly regulated protein kinases”, it has been demonstrated that PP2A activity is responsive to specific extracellular stimuli or stresses. Through the generation of second messengers, such as cAMP, Ca^{2+} ions, and ceramides, these stimuli can indeed either directly or indirectly (through activation of protein kinases) regulate PP2A activity.

Ca^{2+} ions, generated for instance upon glutaminergic neuronal stimulation, directly bind to the EF-hand motifs of the B″ subunits, resulting in increased affinity for $PP2A_D$ and increased phosphatase activity [72, 73, 85]. **Ceramide**, a sphingolipid mediating the biological effects of several cytokines, hormones and growth factors, has been found to increase PP2A activity in several ways: directly (as demonstrated in PP2A phosphatase assays in vitro) [86] and indirectly, by blocking binding of SET (a cellular PP2A inhibitor, see further) [87] or by inhibiting PP2A demethylation [88]. Several **protein kinases** affect PP2A activity through phosphorylation of the B-type subunits, notably of the B′ family. The double-stranded-RNA-dependent protein kinase (PKR) was shown to interact with and phosphorylate PR61/B′α, both in vitro and in vivo, resulting in increased PP2A activity [89]. Extracellular signal-regulated kinase (ERK) phosphorylates the conserved Ser337 residue of PR61/B′γ (and/or β), leading to dissociation of PR61/B′ from the holoenzyme and PP2A inactivation. This event is regulated by the immediate early gene product IEX-1 (immediate early response gene X-1) and is part of a negative-feedback loop in ERK-mediated signalling [90]. DNA-damage triggers ATM-dependent phosphorylation of PR61/B′γ2, γ3 and δ, leading to an increase in PR61/B′-PP2A complexes and direction of PP2A phosphatase activity toward p53 [91]. Likewise, DNA-damage-induced Chk1 activity results in Ser37 phosphorylation of PR61/B′δ, leading to increased $PP2A_{T61}$ assembly and subsequent dephosphorylation of Cdc25 [92]. PP2A is also responsive to elevation of intracellular cAMP levels, which lead to PP2A activation through Protein Kinase A (PKA)-dependent phosphorylation of PR61/B′δ on Ser566 [93–95]. More recently, evidence was presented that this site may additionally be a direct target of Protein Kinase C (PKC) [96].

2.2 Cellular PP2A Inhibitors: An Emerging Group

PP2A catalytic activity can be directly inhibited by an emerging set of specific cellular PP2A inhibitory proteins, which may either target all or very specific PP2A holoenzymes (reviewed in ref. 97).

ANP32a, also called PP2A Inhibitor 1, belongs to a family of at least 9 members, of which only the a and e isoforms show PP2A inhibitory ability [98]. ANP32a and ANP32e are phosphoproteins: a tyrosine phosphorylation on an unknown site dissociates them from PP2A allowing increased PP2A activity. They are localized both in the cytoplasm and the nucleus. ANP32a has multiple cellular functions and it is currently unclear to what extent inhibition of PP2A contributes to all of these functions [97].

SET, also called PP2A Inhibitor 2, is a multifunctional phosphoprotein that can be found both in the nucleus and the cytoplasm. Like ANP32a, its potent PP2A inhibitory activity was discovered using classical biochemical methods [99]. Two splice variants of SET exist, each with a different N-terminal region. The PP2A inhibitory activity resides within a 95 amino-acid region immediately C-terminally of these different N terminal regions, explaining why both isoforms can inhibit PP2A. It has been documented that SET inhibits $PP2A_T$, $PP2A_D$ as well as monomeric $PP2A_C$ in vitro, suggesting that inhibition occurs through binding with the C subunit. Like ANP32a, SET is a phosphoprotein and Serine 9 and Serine 24 have been identified as potential PKC sites of functional importance [97]. However, how these phosphorylations regulate SET's PP2A inhibitory activity remains controversial [100–103].

CIP2A, or Cancerous Inhibitor of PP2A, was originally identified as a novel co-precipitating partner of the PP2A A subunit [104]. At the moment it is mechanistically unclear how CIP2A might inhibit PP2A, but there are major indications only a very limited set of cellular PP2A substrates is affected, most notably c-Myc, Akt and the Death-Associated Protein Kinase (reviewed in ref. 55).

cAMP-regulated phosphoproteins, ARPP16 and ARPP19, and the closely related ENSA (α-endosulfine) have come into the picture as potential mitotic PP2A inhibitors [105, 106]. These proteins require phosphorylation on a Serine residue (Ser67 in ENSA, Ser62 in ARPP19) by the Greatwall kinase in order to bind a specific PP2A complex and inhibit its activity: once phosphorylated, they indeed strongly bind PR55/Bδ, but no other B-type subunits, $PP2A_D$ or monomeric $PP2A_C$.

TIP, type 2A interacting protein or TIPRL1, does not only interact with $PP2A_C$, but also with the catalytic subunits of the PP2A-like phosphatases, PP4 and PP6 [4]. The PP2A inhibitory activity resides in the C-terminal part, which is lacking in the shorter splice variant TIPRL2 that does not inhibit PP2A. TIP is functionally involved in DNA damage and repair signalling, where it regulates PP2A enzymes that oppose ATM/ATR-dependent phosphorylation events [107].

3 PP2A In Human Disease

Given the highly diverse physiological functions of PP2A, it should come of little surprise that deregulated PP2A activity and/or expression is found associated with human disease.

PP2A is now firmly established as a genuine tumor suppressor (reviewed in refs. 66–69) and is functionally impaired or genetically altered in several human solid cancers and leukemias. Currently discovered PP2A alterations include genetic mutations in *PPP2R1A* and *PPP2R1B* (reviewed in refs. 3, 66, 69), reduced expression of A and specific B-type subunits (for instance PR55/Bα in breast and lung cancer [57, 108]; PR61/B′γ in metastatic melanoma [109]; PR65/A, PR55/Bα, PR61/B′α, γ, and δ in c-KIT$^+$ AML [110]), and increased expression of PP2A inhibitors, most notably CIP2A and SET. Increased CIP2A expression is found in an impressive number of diverse human cancers and correlates with cancer aggressiveness and poor prognosis (reviewed in ref. 97).

Increased SET expression is mainly found in leukemias, such as CML, Philadelphia-chromosome-positive ALL, T-cell ALL, B-cell CLL, and AML [111–114].

PP2A deregulation is also found in non-neoplastic diseases, such as Alzheimer's disease (AD) (reviewed in ref. 58). AD onset is facilitated by hyperphosphorylation of the microtubule-associated protein Tau, which results in increased Tau aggregation, formation of neurofibrillary tangles, microtubule disruption and eventually neurodegeneration. PP2A is the major Tau phosphatase in brain that accounts for over 70 % of Tau dephosphorylation [115]. In postmortem AD brains, several abnormalities testifying from impaired PP2A function have been found, including reduced expression of specific PP2A subunits, reduced $PP2A_C$ carboxymethylation, increased expression of SET and ANP32a, SET cleavage and mislocalization, and increased $PP2A_C$ phosphorylation (reviewed in ref. 58). Thus, it seems that decreased PP2A activity may significantly contribute to AD pathology by inducing Tau hyperphosphorylation.

4 Pharmacological Modulation of PP2A

Recent findings have suggested that pharmacological modulation of PP2A in cellular and in vivo models of human cancers or AD is a feasible approach, identifying PP2A as an exciting novel druggable target for these diseases (reviewed in refs. 112, 116–118). In addition, there are a number of other diseases, such as Parkinson's disease and diabetes, in which there is currently little clinical evidence for PP2A deregulation, but in which modulation of specific PP2A complexes might still be of future therapeutic interest (reviewed in refs. 119, 120).

4.1 PP2A Activating Compounds: Opportunities to Restore PP2A Activity in Disease?

Activation of PP2A by small molecules offers therapeutic possibilities to treat AD, other neurodegenerative disorders and cancer. As mentioned before, PP2A is well-established as a genuine tumor suppressor and an important tau phosphatase. A logical therapeutic strategy should therefore aim at (re) activating impaired PP2A function to counteract cancer cell survival or tau hyperphosphorylation-induced neurodegeneration. Evidently, in order to rationalize the choice of a particular therapeutic strategy, detailed information about the mechanism(s) of pathological PP2A deregulation in these conditions is mandatory, and this may significantly vary between different pathological states or cancer types. Only then, one can specifically target the deregulated PP2A activity and avoid the unspecific or undesired side effects, which may result from (over)stimulating general PP2A activity.

Historically, the first compounds that were found to directly modulate PP2A activity in vitro were **polycations**, such as polylysine and protamine; hence, PP2A was originally named PolyCation Stimulated (PCS) phosphatase. The cation sensitivity of different Ser/Thr phosphatases is one of the biochemical tools that can be used to discriminate between the in vitro activities of PP1 (inhibited by polycations), PP2B (Ca^{2+}/calmodulin-dependent), PP2C (Mg^{2+}-dependent) and PP2A (activated by polycations) when measured on phosphoprotein substrates [121, 122]. Because of their highly basic nature, histones such as Histone H1, were found to mimic this stimulatory property of polycations, but the physiological relevance of this phenomenon is still unclear. The sensitivity of PP2A to polycations is both substrate- and holoenzyme-dependent: when using a phosphoprotein substrate such as phosphorylase-*a*, the free catalytic subunit is being stimulated the least, while the trimeric forms can be stimulated to a larger extent than $PP2A_D$ [123]. Especially $PP2A_{T\ B''}\alpha_2$ can be highly stimulated by very low polycation concentrations [123], although it was shown later on, that a partially proteolysed B″/PR72 subunit ("PR45"), resulting from limited calpain-mediated cleavage of PR72, was in fact responsible for this peculiar property [124]. Although mechanistically still elusive, an interaction of polycations both with the phosphoprotein substrate and the enzyme is required to facilitate PP2A-mediated dephosphorylation [125, 126]. That may also explain why polycations are without any effect towards phosphopeptide PP2A substrates [127]. To date, it remains unclear whether the obvious effects of polycations on PP2A activity in vitro may become of future therapeutic value.

As explained before, PP2A is one of the prime targets of **ceramide**. Through the use of ceramide derivatives, it has been demonstrated that several of the unique structural features of ceramide are strictly required for allosteric activation of PP2A, including the amide group, the primary and secondary hydroxyl groups, and the *trans* configuration of the double bond of the sphingoid backbone [128]. Ceramide activation of PP2A is also affected by the acyl chain length (C10 > C6 > C2) [86], the solvent conditions [129] and the nature of the PP2A complex itself [86, 128, 129].

Other sphingoids, such as **sphingosine** and its derivatives, share structural features with ceramides and also act as allosteric PP2A activators in vitro [86, 87, 128]. In addition, sphingosine, *N,N′*-dimethyl sphingosine (DMS) and phytosphingosine, but not dihydrosphingosine or sphingosine 1-phosphate, directly bind to ANP32a, thereby relieving the inhibitory action of ANP32a on PP2A [130]. This may be another mechanism explaining the pro-apoptotic and PP2A-activating effects of sphingosine in cells [131]. Another pertinent example is **FTY720 (Fingolimod)**, a synthetic sphingosine analogue and known immunosuppressant that is currently approved for clinical use in relapsing multiple sclerosis patients. In vitro, FTY720 can activate phosphorylase-*a* activity of a PP2A trimer with PR55/B subunit [132]. In cells, FTY720 becomes biologically activated upon phosphorylation by sphingosine kinase, which turns it into a high-affinity agonist of sphingosine-1-phosphate receptors [133]. However, this modification seems dispensable for its PP2A activating ability and the resulting anti-proliferative and pro-apoptotic effects of FTY720 in various cancer cell lines. Depending on the cancer cell type, these effects involve PP2A-mediated modulation of Akt, ERK, p70S6K, JAK2, STAT5, and cyclin D1 activity, but not Bcl-2 phosphorylation [110, 132, 134–136]. Besides a direct allosteric effect on PP2A, the mechanism of PP2A activation by FTY720 may involve relieve of the inhibitory function of SET, because FTY720-induced apoptosis is particularly efficient in leukemias characterized by increased SET expression [134]. Indicative for its potential future clinical use, FTY720 treatment in vivo, in mice models of several leukemia types, significantly prolonged animal survival, without any obvious side effects on normal hematopoiesis [110, 134–136].

Alternative strategies to restore PP2A function in conditions where its inactivation results from overexpression of cellular PP2A inhibitors (such as SET, CIP2A, and ANP32a) and that aim at interfering with the inhibitor–PP2A interaction itself (as far as such an interaction is biochemically sufficiently described), involve the use of interfering peptides. **COG112 and COG449**, two apolipoprotein E mimetic peptides, have been shown to interfere with the SET–PP2A interaction [137]. They induce mitochondria-dependent apoptosis and have potent anti-proliferative effects in models of glioma [138], breast adenocarcinoma [138], B-cell CLL, and non-Hodgkin lymphoma [139], without considerable effects on noncancerous cells. In models of AD, they reduce tau phosphorylation, behavioral deficits, plaques, and tangles [140, 141]. The therapeutic effect of the AD drug **memantine**, a low affinity antagonist to glutamate NMDA receptors, has partially also been contributed to decreasing SET-induced inhibition of PP2A activity [142]. Targeting the PP2A–CIP2A interaction may result in similar future favorable therapeutic outcome in several cancer types with increased CIP2A expression (reviewed in ref. 55), but whether this is feasible, awaits the elucidation of the mechanism by which CIP2A inhibits PP2A.

Forskolin, or other elevators of intracellular cAMP such as **rolipram** (a cAMP phosphodiesterase inhibitor) can induce PP2A activation [111, 143, 144]. Although most of their cellular effects depend on augmented PKA and adenylate cyclase activities, the mechanism by which forskolin activates PP2A does not require an increase in PKA activity [143]. This is best proven by the persisting PP2A stimulatory effect of 1,9-dideoxy-forskolin [111], a forskolin derivative that does not result in adenylate cyclase activation and increased cAMP production. In cellular [111, 113, 144] and in vivo [111] models of CLL, CML, and AML, rolipram, forskolin, and 1,9-dideoxy-forskolin show marked PP2A-dependent induction of apoptosis and inhibition of tumorigenesis.

Xylulose-5-phosphate, heparin, and sodium selenate are all anionic PP2A activators. **Heparin**, a highly sulphated, and thus negatively charged, glycosaminoglycan polymer, stimulates PP2A activity in cells and on phosphorylase-*a* in vitro [122, 145], possibly by dissociating the B subunit [146]. **Xylulose-5-phosphate** is a sugar phosphate that activates a specific PP2A form isolated from high-carbohydrate-fed rat liver (trimer with PR55/Bδ subunit) to dephosphorylate some substrates involved in carbohydrate metabolism (Fru-6-P,2-kinase: Fru-2,6-bisphosphatase and the transcription factor ChREBP) [147, 148]. **Sodium selenate**, but not selenite or other selenium derivatives, can activate $PP2A_D$ in vitro when measured on pNPP or a phosphoThr peptide [149]. In addition, it reduces tau phosphorylation, neurodegeneration, behavioral and memory deficits in cellular and mouse models of AD [149, 150] by a mechanism that may involve enhanced binding of PP2A to tau [150].

Certain naturally occurring chemopreventatives or antioxidants, such as **dithiolethione** and **dihydroxy phenylethanol (DPE)** have also been shown to increase PP2A activity in cells [151, 152]. Effects of **tocopherols** (vitamin E and its derivatives, such as alpha-tocopheryl succinate) on PP2A are complex: they influence PP2A activity directly [153] or indirectly through a ceramide-dependent mechanism [154]. **Eicosanoyl-5-hydroxytryptamide** (EHT) shares similar properties to tocopherols and activates PP2A by inhibiting PME-1-mediated $PP2A_C$ demethylation in vitro, in hippocampal neurons and in a mouse model of PD [155].

A summary of the major pharmacological PP2A activators is given in Table 2.

4.2 The Therapeutic Potential of PP2A Inhibitory Compounds

In addition to PP2A reactivating strategies, it has been convincingly shown that inhibition of PP2A can also lead to cancer cell death, most likely as a consequence of improper cell cycle progression and sustained checkpoint activation, eventually resulting in mitotic catastrophe and apoptotic cell death (reviewed in refs. 55, 117). This reflects the critical roles of PP2A in the control and regulation of proper cell cycle progression [53, 54]. Another rationale behind the therapeutic use of PP2A inhibitors originates from the role of PP2A in DNA damage and repair signalling pathways. In this case, PP2A inhibition in combination with a DNA damaging agent [156] or DNA repair inhibitor [57] significantly sensitizes cancer cells to chemotherapy-induced cell death. Currently, the major drawback of such PP2A-inhibitory strategies is definitely their lack of specificity, in the sense that both healthy as well as diseased cells are affected, resulting in too much cytotoxicity. This problem might be overcome in future, in part by improving drug

Table 2
PP2A activating compounds

Compound	Activating effect observed	References
Ceramides	In vitro, in cells	[86–88, 128, 129]
Sphingosine / *N*,*N*′-dimethyl sphingosine (DMS) / phytosphingosine	In vitro, in cells	[86, 87, 128, 130, 131]
FTY720 and analogues	In vitro, in cells, in vivo	[110, 132, 134–136]
Memantine	In vitro, in cells	[142]
Forskolin / 1,9-dideoxy forskolin	In cells, in vivo	[111, 113, 143]
Rolipram	In cells	[144]
Sodium selenate	In vitro, in cells, in vivo	[149, 150]
ApoE peptide derivates	In vitro, in cells, in vivo	[137–141]
Dithiolethione	In cells	[151]
Dihydroxy phenylethanol (DPE)	In vitro, in cells	[152]
α-Tocopherols	In vitro, in cells	[153, 154]
Eicosanoyl-5-hydroxytryptamide (EHT)	In vitro, in cells, in vivo	[155]
Xylulose-5-phosphate	In vitro, in cells, in vivo	[147, 148]
Polycations, polyamines	In vitro	[122, 123]
Heparine	In vitro, in cells	[122, 145]
Carnosic acid	In cells	[161]
Taurolidine	In cells	[162]
1,8-Naphthyridines	In cells	[163]
Troglitazone	In cells	[164]

Indicative for their potential mechanism-of-action (direct vs. indirect), the conditions in which their PP2A activating effects have been observed (in vitro, in cells, in vivo) are included (*cfr.* main text for more details)

delivery strategies, in part by designing PP2A inhibitors that do not target the catalytic subunit (and hence all PP2A complexes) but only very specific PP2A holoenzymes or PP2A-substrate interactions.

There are several naturally occurring toxins that can inhibit Ser/Thr protein phosphatases in general, and PP2A in particular (summarized in Table 3). All of these inhibitors target the phosphatase catalytic subunit and cannot distinguish between different holoenzyme complexes. Among these, **Okadaic acid** (OA), a polyether fatty acid produced by marine dinoflagellates, inhibits PP2A most potently. OA acts through direct $PP2A_C$ binding via a hydro-

Table 3
PP2A inhibitory compounds and their estimated IC_{50} values (in nM)

Compound	IC_{50} PP2A	IC_{50} PP1	References
Okadaic acid (OA)	0.02–0.5	10–200	[117, 165]
Acanthifocilin (ACA), OA-analogue	1	20	[117, 165]
OA-methyl ester	900	>25.10^4	[117, 165]
OA-9-anthryldiazo-methyl ester (ADAM)	900	>25.10^4	[117, 165]
Acanthifocilin methyl ester (ACAM)	5000	>25.10^4	[117, 165]
Dinophysistoxin-1	0.6	55	[165]
Tautomycin	0.94–32	0.23–22	[117]
Calyculin-A	0.5–1	2–2.8	[117, 166]
Microcystin-LR	0.04–2.0	0.1–6.0	[167]
7-Desmethyl-microcystin-RR	1	3	[167]
Nodularin-V	0.03–1.0	0.5–3	[167]
Cantharidin (CA)	40–194	473–1,100	[168, 169]
Norcantharidin	500	2,000	[168, 169]
Cantharidic acid	53	562	[168, 169]
Palasonin (CA-analogue)	120	656	[168, 169]
Endothall (CA-analogue)	970	5,000	[168, 169]
Fostriecin	1.1–1.5	61.10^3–83.10^3	[162, 170]
Cytostatin	22–36	>1.10^5	[162, 170]
Phoslactomycins	3,700	>1.10^6	[162, 170]
Rubratoxin A	170	>2.10^5	[163]

For comparison, the IC_{50} values for inhibition of PP1 (in nM) are included. It is important to notice that the IC_{50} values are in fact relative values, since—depending on the mode of inhibition—they can be function of the PP2A enzyme and/or substrate concentration

phobic cage that is not conserved in other Ser/Thr phosphatases, explaining why OA has such a high affinity for PP2A (reviewed in ref. 117). Several OA analogues with slightly different properties than OA have been developed (Table 3).

Fostriecin is another phosphatase inhibiting compound that binds the catalytic subunit of Ser/Thr phosphatases and triggers premature entry in mitosis. It has a low toxicity and a high specificity for PP2A and the PP2A-like phosphatase PP4, but unfortunately, it is not very stable (reviewed in refs. 55, 117, 157). Investigators tried to overcome these issues by developing fostriecin analogues such as **phoslactomycins** and **cytostatins** [55, 117, 158]. **Cytostatin** exhibits a potent anti-metastatic effects in tumor cells; it has however a poor stability, limiting its clinical use. **Rubratoxin** is far more stable than cytostatin and fostriecin, has comparable selective PP2A inhibitory activity and shows anti-metastatic effects in mice, but unfortunately, is still too hepatotoxic to justify clinical use [55, 159].

Cantharidin is a natural PP2A inhibitory compound, secreted by blister beetles and the Spanish fly. Despite its potential as a less selective inhibitor of PP2A (it equally well inhibits PP1), its clinical use is prohibited due to high nephrotoxicity. However, its demethylated analogue, **Norcantharidin**, is less toxic and has anticancer properties in many tumor cell types (reviewed in refs. 55, 117, 160). Several other Cantharidin analogues have been described, some of which are more selective for PP2A (Table 3).

Microcystin-LR (and derivatives), Calyculin A, Tautomycin, and Nodularin are nonspecific PP2A inhibitors, which retain significant potential to inhibit PP1 as well (Table 3).

Thus, despite the existence of several different classes of PP2A inhibitors, some of which are selective for PP2A (vs. PP1), none of them can be considered as 100 % specific. Moreover, when used in vivo or in cells as a means to discriminate between the activities of different Ser/Thr phosphatases, one has to realize their effects are not only dependent on the specificity of the inhibition itself, but also on the permeability of the cells (or tissues) for these inhibitors, which can significantly differ for distinct cell types and inhibitors [2].

5 Concluding Remarks

Despite increasing evidence that pharmacological modulation of PP2A is feasible, with obvious therapeutic benefits in treating cancer or neurodegenerative diseases, PP2A remains a challenging therapeutic target, not the least because of its complex structure and our incomplete understanding of the (patho)physiological functions and regulation of specific PP2A holoenzymes in normal vs. diseased cells. Together with structural data, this knowledge should definitely enable us in the near future to modulate phosphatase activity of just a selective group of PP2A holoenzymes, specific PP2A–regulator interactions or PP2A–substrate interactions, in order to further improve the already promising strategies that we have described in this review.

Acknowledgments

C.L. and D.H. are supported by an I.W.T. fellowship of the Flemish Agency for Innovation by Science and Technology. V.J.'s research is funded by the Research Foundation-Flanders (G.0582.11N) and the Interuniversity Attraction Poles Programme initiated by the Belgian Science Policy Office (IAP P7/13).

References

1. Moorhead GB, De Wever V, Templeton G et al (2009) Evolution of protein phosphatases in plants and animals. Biochem J 417:401–409
2. Janssens V, Goris J (2001) Protein phosphatase 2A: a highly regulated family of serine/threonine phosphatases implicated in cell growth and signalling. Biochem J 353:417–439
3. Eichhorn PJ, Creyghton MP, Bernards R (2009) Protein phosphatase 2A regulatory subunits and cancer. Biochim Biophys Acta 1795:1–15
4. Sents W, Ivanova E, Lambrecht C et al (2013) The biogenesis of active protein phosphatase 2A holoenzymes: a tightly regulated process creating phosphatase specificity. FEBS J 280:644–661
5. Fellner T, Lackner DH, Hombauer H et al (2003) A novel and essential mechanism determining specificity and activity of protein phosphatase 2A (PP2A) in vivo. Genes Dev 17:2138–2150
6. Hombauer H, Weismann D, Mudrak I et al (2007) Generation of active protein phosphatase 2A is coupled to holoenzyme assembly. PLoS Biol 5:e155
7. Janssens V, Longin S, Goris J (2008) PP2A holoenzyme assembly: in cauda venenum (the sting is in the tail). Trends Biochem Sci 33:113–121
8. Chen J, Martin BL, Brautigan DL (1992) Regulation of protein serine-threonine phosphatase type-2A by tyrosine phosphorylation. Science 257:1261–1264
9. Schmitz MH, Held M, Janssens V et al (2010) Live-cell imaging RNAi screen identifies PP2A-B55alpha and importin-beta1 as key mitotic exit regulators in human cells. Nat Cell Biol 12:886–893
10. De Baere I, Derua R, Janssens V et al (1999) Purification of porcine brain protein phosphatase 2A leucine carboxyl methyltransferase and cloning of the human homologue. Biochemistry 38:16539–16547
11. Ogris E, Du X, Nelson KC et al (1999) A protein phosphatase methylesterase (PME-1) is one of several novel proteins stably associating with two inactive mutants of protein phosphatase 2A. J Biol Chem 274:14382–14391
12. Longin S, Jordens J, Martens E et al (2004) An inactive protein phosphatase 2A population is associated with methylesterase and can be re-activated by the phosphotyrosyl phosphatase activator. Biochem J 380:111–119
13. Jordens J, Janssens V, Longin S et al (2006) The protein phosphatase 2A phosphatase activator is a novel peptidyl-prolyl cis/trans-isomerase. J Biol Chem 281:6349–6357
14. Leulliot N, Vicentini G, Jordens J et al (2006) Crystal structure of the PP2A phosphatase activeator: implications for its PP2A-specific PPIase activity. Mol Cell 23:413–424
15. Stanevich V, Jiang L, Satyshur KA et al (2011) The structural basis for tight control of PP2A methylation and function by LCMT-1. Mol Cell 41:331–342
16. Groves MR, Hanlon N, Turowski P et al (1999) The structure of the protein phosphatase 2A PR65/A subunit reveals the conformation of its 15 tandemly repeated HEAT motifs. Cell 96:99–110
17. Xing Y, Xu Y, Chen Y et al (2006) Structure of protein phosphatase 2A core enzyme bound to tumor-inducing toxins. Cell 127:341–353
18. Cho US, Xu W (2006) Crystal structure of a protein phosphatase 2A heterotrimeric holoenzyme. Nature 445:53–57
19. Xu Y, Xing Y, Chen Y et al (2006) Structure of the protein phosphatase 2A holoenzyme. Cell 127:1239–1251
20. Xu Y, Chen Y, Zhang P et al (2008) Structure of a protein phosphatase 2A holoenzyme: insights into B55-mediated Tau dephosphorylation. Mol Cell 31:873–885
21. Xu Z, Cetin B, Anger M et al (2009) Structure and function of the PP2A-shugoshin interaction. Mol Cell 35:426–441
22. Grinthal A, Adamovic I, Weiner B et al (2010) PR65, the HEAT-repeat scaffold of phosphatase PP2A, is an elastic connector that links force and catalysis. Proc Natl Acad Sci USA 107:2467–2472

23. Mayer RE, Hendrix P, Cron P et al (1991) Structure of the 55-kDa regulatory subunit of protein phosphatase 2A: evidence for a neuronal-specific isoform. Biochemistry 30: 3589–3597
24. Strack S, Chang D, Zaucha JA et al (1999) Cloning and characterization of B delta, a novel regulatory subunit of protein phosphatase 2A. FEBS Lett 460:462–466
25. Strack S, Zaucha JA, Ebner FF et al (1998) Brain protein phosphatase 2A: developmental regulation and distinct cellular and subcellular localization by B subunits. J Comp Neurol 392:515–527
26. Sontag E, Nunbhakdi-Craig V, Bloom GS et al (1995) A novel pool of protein phosphatase 2A is associated with microtubules and is regulated during the cell cycle. J Cell Biol 128:1131–1144
27. Turowski P, Myles T, Hemmings BA et al (1999) Vimentin dephosphorylation by protein phosphatase 2A is modulated by the targeting subunit B55. Mol Biol Cell 10:1997–2015
28. Dagda K, Zaucha JA, Wadzinski BE et al (2003) A developmentally regulated, neuron-specific splice variant of the variable subunit Bbeta targets protein phosphatase 2A to mitochondria and modulates apoptosis. J Biol Chem 278:24976–24985
29. Zolnierowicz S, Csortos C, Bondor J et al (1994) Diversity in the regulatory B-subunits of protein phosphatase 2A: identification of a novel isoform highly expressed in brain. Biochemistry 33:11858–11867
30. Hendrix P, Mayer-Jackel RE, Cron P et al (1993) Structure and expression of a 72-kDa regulatory subunit of protein phosphatase 2A. Evidence for different size forms produced by alternative splicing. J Biol Chem 268:15267–15276
31. Takahashi M, Shibata H, Shimakawa M et al (1999) Characterization of a novel giant scaffolding protein, CG-NAP, that anchors multiple signaling enzymes to centrosome and the golgi apparatus. J Biol Chem 274:17267–17274
32. Zwaenepoel K, Louis JV, Goris J et al (2008) Diversity in genomic organisation, developmental regulation and distribution of the murine PR72/B″ subunits of protein phosphatase 2A. BMC Genomics 9:393
33. Zwaenepoel K, Goris J, Erneux C et al (2010) Protein phosphatase 2A PR130/B″α_1 subunit binds to the SH2 domain-containing inositol polyphosphate 5-phosphatase 2 and prevents epidermal growth factor (EGF) induced EGF receptor degradation sustaining EGF mediated signaling. FASEB J 24:538–547
34. Schiebel K, Meder J, Rump A et al (2000) Elevated DNA sequence diversity in the genomic region of the phosphatase PPP2R3L gene in the human pseudoautosomal region. Cytogenet Cell Genet 91:224–230
35. Yan Z, Fedorov SA, Mumby MC et al (2000) PR48, a novel regulatory subunit of protein phosphatase 2A, interacts with Cdc6 and modulates DNA replication in human cells. Mol Cell Biol 20:1021–1029
36. Kono Y, Maeda K, Kuwahara K et al (2002) MCM3-binding GANP DNA-primase is associated with a novel phosphatase component G5PR. Genes Cells 7:821–834
37. Foley EA, Maldonado M, Kapoor TM (2011) Formation of stable attachments between kinetochores and microtubules depends on the B56-PP2A phosphatase. Nat Cell Biol 13:1265–1271
38. McCright B, Rivers AM, Audlin S et al (1996) The B56 family of protein phosphatase 2A (PP2A) regulatory subunits encodes differentiation-induced phosphoproteins that target PP2A to both nucleus and cytoplasm. J Biol Chem 271:22081–22089
39. Ruvolo PP, Clark W, Mumby M et al (2002) A functional role for the B56 alpha-subunit of protein phosphatase 2A in ceramide-mediated regulation of Bcl2 phosphorylation status and function. J Biol Chem 277:22847–22852
40. Martens E, Stevens I, Janssens V et al (2004) Genomic organisation, chromosomal localisation tissue distribution and developmental regulation of the PR61/B′ regulatory subunits of protein phosphatase 2A in mice. J Mol Biol 336:971–986
41. Ito A, Kataoka TR, Watanabe M et al (2000) A truncated isoform of the PP2A B56 subunit promotes cell motility through paxillin phosphorylation. EMBO J 19:562–571
42. Gigena MS, Ito A, Nojima H et al (2005) A B56 regulatory subunit of protein phosphatase 2A localizes to nuclear speckles in cardiomyocytes. Am J Physiol Heart Circ Physiol 289:H285–H294
43. Ito A, Koma Y, Sohda M et al (2003) Localization of the PP2A B56gamma regulatory subunit at the Golgi complex: possible role in vesicle transport and migration. Am J Pathol 162:479–489
44. Lee TY, Lai TY, Lin SC et al (2010) The B56gamma3 regulatory subunit of protein phosphatase 2A (PP2A) regulates S phase-specific nuclear accumulation of PP2A and the G1 to S transition. J Biol Chem 285:21567–21580
45. Tanabe O, Nagase T, Murakami T et al (1996) Molecular cloning of a 74-kDa regulatory subunit (B″ or delta) of human protein phosphatase 2A. FEBS Lett 379:107–111

46. Jin Z, Shi J, Saraf A et al (2009) The 48-kDa alternative translation isoform of PP2A:B56epsilon is required for Wnt signaling during midbrain-hindbrain boundary formation. J Biol Chem 284:7190–7200
47. Castets F, Rakitina T, Gaillard S et al (2000) Zinedin, SG2NA, and striatin are calmodulin-binding, WD repeat proteins principally expressed in the brain. J Biol Chem 275: 19970–19977
48. Baillat G, Moqrich A, Castets F et al (2001) Molecular cloning and characterization of phocein, a protein found from the Golgi complex to dendritic spines. Mol Biol Cell 12:663–673
49. Gaillard S, Bartoli M, Castets F et al (2001) Striatin, a calmodulin-dependent scaffolding protein, directly binds caveolin-1. FEBS Lett 508:49–52
50. Healy AM, Zolnierowicz S, Stapleton AE et al (1991) CDC55, a Saccharomyces cerevisiae gene involved in cellular morphogenesis: identification, characterization, and homology to the B subunit of mammalian type 2A protein phosphatase. Mol Cell Biol 11: 5767–5780
51. Pallas DC, Weller W, Jaspers S et al (1992) The third subunit of protein phosphatase 2A (PP2A), a 55-kilodalton protein which is apparently substituted for by T antigens in complexes with the 36- and 63-kilodalton PP2A subunits, bears little resemblance to T antigens. J Virol 66:886–893
52. Schmidt K, Kins S, Schild A et al (2002) Diversity, developmental regulation and distribution of murine PR55/B subunits of protein phosphatase 2A. Eur J Neurosci 16:2039–2048
53. Wurzenberger C, Gehrlich DW (2011) Phosphatases: providing safe passage through mitotic exit. Nat Rev Mol Cell Biol 12:469–482
54. Barr FA, Elliott PR, Gruneberg U (2011) Protein phosphatases and the regulation of mitosis. J Cell Sci 124:2323–2334
55. Janssens V, Rebollo A (2012) The role and therapeutic potential of Ser/Thr phosphatase PP2A in apoptotic signaling networks in human cancer cells. Curr Mol Med 12:268–287
56. Batut J, Schmierer B, Cao J et al (2008) Two highly related regulatory subunits of PP2A exert opposite effects on TGF-beta/Activin/Nodal signaling. Development 135: 2927–2937
57. Kalev P, Simicek M, Vazquez I et al (2012) Loss of PPP2R2A inhibits homologous recombination DNA repair and predicts tumor sensitivity to PARP inhibition. Cancer Res 72(24):6414–6424
58. Torrent L, Ferrer I (2012) PP2A and Alzheimer disease. Curr Alzheimer Res 9:248–256
59. McCright B, Virshup DM (1995) Identification of a new family of protein phosphatase 2A regulatory subunits. J Biol Chem 270:26123–26128
60. Csortos C, Zolnierowicz S, Bako E et al (1996) High complexity in the expression of the B′ subunit of protein phosphatase 2A0. Evidence for the existence of at least seven novel isoforms. J Biol Chem 271:2578–2588
61. Tehrani MA, Mumby MC, Kamibayashi C (1996) Identification of a novel protein phosphatase 2A regulatory subunit highly expressed in muscle. J Biol Chem 271:5164–5170
62. Zolnierowicz S, Van Hoof C, Andjelkovic N et al (1996) The variable subunit associated with protein phosphatase $2A_0$ defines a novel multimember family of regulatory subunits. Biochem J 317:187–194
63. Tanabe O, Gomez GA, Nishito Y et al (1997) Molecular heterogeneity of the cDNA encoding a 74-kDa regulatory subunit (B″ or δ) of human protein phosphatase 2A. FEBS Lett 408:52–56
64. Yang J, Phiel C (2010) Functions of B56-containing PP2As in major developmental and cancer pathways. Life Sci 87:659–666
65. Walaas SI, Hemmings HC, Jr Greengard P et al (2011) Beyond the dopamine receptor: regulation and roles of serine/threonine protein phosphatases. Front Neuroanat 5:50
66. Janssens V, Goris J, Van Hoof C (2005) PP2A the expected tumor suppressor. Curr Opin Genet Dev 15:34–41
67. Arroyo JD, Hahn WC (2005) Involvement of PP2A in viral and cellular transformation. Oncogene 24:7746–7755
68. Arnold HK, Sears RC (2008) A tumor suppressor role for PP2A B56α through negative regulation of c-Myc and other key oncoproteins. Cancer Metastasis Rev 27:147–158
69. Westermarck J, Hahn WC (2008) Multiple pathways regulated by the tumor suppressor PP2A in transformation. Trends Mol Med 14:152–160
70. Voorhoeve PM, Hijmans EM, Bernards R (1999) Functional interaction between a novel protein phosphatase 2A regulatory subunit, PR59, and the retinoblastoma-related p107 protein. Oncogene 18:515–524
71. Stevens I, Janssens V, Martens E et al (2003) Identification and characterization of B″-subunits of protein phosphatase 2A in

Xenopus laevis oocytes and adult tissues. Eur J Biochem 270:376–387

72. Janssens V, Jordens J, Stevens I et al (2003) Identification and functional analysis of two Ca^{2+}-binding EF-hand motifs in the B″/PR72 subunit of protein phosphatase 2A. J Biol Chem 278:10697–10706
73. Davis AJ, Yan Z, Martinez B et al (2008) Protein phosphatase 2A is targeted to cell division control protein 6 by calcium-binding regulatory subunit. J Biol Chem 283:16104–16114
74. Kolupaeva V, Janssens V (2013) PP1 and PP2A phosphatases: cooperating partners in modulating retinoblastoma protein activation. FEBS J 280:627–643
75. Moreno CS, Park S, Nelson K et al (2000) WD40 repeat proteins striatin and S/G(2) nuclear autoantigen are members of a novel family of calmodulin-binding proteins that associate with protein phosphatase 2A. J Biol Chem 275:5257–5263
76. Goudreault M, D'Ambrosio LM, Kean MJ et al (2009) A PP2A phosphatase high density interaction network identifies a novel striatin-interacting phosphatase and kinase complex linked to the cerebral cavernous malformation 3 (CCM3) protein. Mol Cell Proteomics 8:157–171
77. Ribeiro PS, Josué F, Wepf A et al (2010) Combined functional genomic and proteomic approaches indentify a PP2A complex as a negative regulator of Hippo signalling. Mol Cell 39:521–534
78. Kean MJ, Ceccarelli DF, Goudreault M et al (2011) Structure-function analysis of core STRIPAK proteins: a signaling complex implicated in Golgi polarization. J Biol Chem 286:25065–25075
79. Hyodo T, Ito S, Hasegawa H et al (2012) Misshapen-like kinase 1 (MINK1) is a novel component of striatin-interacting phosphatase and kinase (STRIPAK) and is required for the completion of cytokinesis. J Biol Chem 287:25019–25029
80. Gu P, Qi X, Zhou Y et al (2012) Generation of *Ppp2Ca* and *Ppp2Cb* conditional null alleles in mouse. Genesis 50:429–436
81. Götz J, Probst A, Ehler E et al (1998) Delayed embryonic lethality in mice lacking protein phosphatase 2A catalytic subunit Calpha. Proc Natl Acad Sci USA 95:12370–12375
82. Ruediger R, Ruiz J, Walter G (2011) Human cancer-associated mutations in the Aα subunit of protein phosphatase 2A increase lung cancer incidence in Aα knock-in and knockout mice. Mol Cell Biol 31:3832–3844
83. Louis JV, Martens E, Borghgraef P et al (2011) Mice lacking phosphatase PP2A subunit PR61/B′delta (*Ppp2r5d*) develop spatially restricted tauopathy by deregulation of CDK5 and GSK3beta. Proc Natl Acad Sci USA 108:6957–6962
84. Tadmouri A, Kiyonaka S, Barbado M et al (2012) Cacnb4 directly couples electrical activity to gene expression, a process defective in juvenile epilepsy. EMBO J 31:3730–3744
85. Ahn JH, Sung JY, McAvoy T et al (2007) The B″/PR72 subunit mediates Ca^{2+}-dependent dephosphorylation of DARPP-32 by protein phosphatase 2A. Proc Natl Acad Sci USA 104:9876–9881
86. Dobrowsky RT, Kamibayashi C, Mumby MC et al (1993) Ceramide activates heterotrimeric protein phosphatase 2A. J Biol Chem 268:15523–15530
87. Mukhopadhyay A, Saddoughi SA, Song P et al (2009) Direct interaction between the inhibitor 2 and ceramide via sphingolipid-protein binding is involved in the regulation of protein phosphatase 2A activity and signaling. FASEB J 23:751–763
88. Chen CL, Lin CF, Chiang CW (2006) Lithium inhibits ceramide- and etoposide-induced protein phosphatase 2A methylation, Bcl-2 dephosphorylation, caspase-2 activation, and apoptosis. Mol Pharmacol 70:510–517
89. Xu Z, Williams BRG (2000) The B56α regulatory subunit of protein phosphatase 2A is a target for regulation by double-stranded RNA-dependent protein kinase PKR. Mol Cell Biol 20:5285–5299
90. Letourneux C, Rocher G, Porteu F (2006) B56-containing PP2A dephosphorylate ERK and their activity is controlled by the early gene IEX-1 and ERK. EMBO J 25:727–738
91. Shouse GP, Nobumori Y, Panowicz MJ et al (2011) ATM-mediated phosphorylation activates the tumor-suppressive function of B56γ-PP2A. Oncogene 30:3755–3765
92. Margolis SS, Perry JA, Forester CM et al (2006) Role for the PP2A/B56delta phosphatase in regulating 14-3-3 release from Cdc25 to control mitosis. Cell 127:759–773
93. Usui H, Inoue R, Tanabe O et al (1998) Activation of protein phosphatase 2A by cAMP-dependent protein kinase-catalyzed phosphorylation of the 74-kDa B′δ regulatory subunit in vitro and identification of the phosphorylation sites. FEBS Lett 430:312–316
94. Ahn JH, McAvoy T, Rakhilin SV et al (2007) Protein kinase A activates protein phosphatase 2A by phosphorylation of the B56delta subunit. Proc Natl Acad Sci USA 104:2979–2984
95. Dodge-Kafka KL, Bauman A, Mayer N et al (2010) cAMP-stimulated protein phosphatase 2A activity associated with muscle A

kinase-anchoring protein (mAKAP) signaling complexes inhibits the phosphorylation and activity of the cAMP-specific phosphodiesterase PDE4D3. J Biol Chem 285: 11078–11086

96. Ahn JH, Kim Y, Kim HS et al (2011) Protein kinase C-dependent dephosphorylation of tyrosine hydroxylase requires the B56δ heterotrimeric form of protein phosphatase 2A. PLoS One 6:e26292
97. Haesen D, Sents W, Ivanova E et al (2012) Cellular inhibitors of protein phosphatase PP2A in cancer. Biomed Res (India) 23:SI197–SI211
98. Matilla A, Radrizzani M (2005) The Anp32 family of proteins containing leucine-rich repeats. Cerebellum 4:7–18
99. Li M, Makkinje A, Damuni Z (1996) The myeloid leukemia-associated protein SET is a potent inhibitor of protein phosphatase 2A. J Biol Chem 271:11059–11062
100. Adachi Y, Pavlakis GN, Copeland TD (1994) Identification of in vivo phosphorylation sites of SET, a nuclear phosphoprotein encoded by the translocation breakpoint in acute undifferentiated leukemia. FEBS Lett 340:231–235
101. Nowak SJ, Pai CY, Corces VG (2003) Protein phosphatase 2A activity affects histone H3 phosphorylation and transcription in Drosophila melanogaster. Mol Cell Biol 23:6129–6138
102. ten Klooster JP, Leeuwen I, Scheres N et al (2007) Rac1-induced cell migration requires membrane recruitment of the nuclear oncogene SET. EMBO J 26:336–345
103. Arnaud L, Chen S, Liu F et al (2011) Mechanism of inhibition of PP2A activity and abnormal hyperphosphorylation of tau by I2(PP2A)/SET. FEBS Lett 585:2653–2659
104. Junttila MR, Puustinen P, Niemelä M et al (2007) CIP2A inhibits PP2A in human malignancies. Cell 130:51–62
105. Gharbi-Ayachi A, Labbé JC, Burgess A et al (2010) The substrate of Greatwall kinase, Arpp19, controls mitosis by inhibiting protein phosphatase 2A. Science 330:1673–1677
106. Mochida S, Maslen SL, Skehel M et al (2010) Greatwall phosphorylates an inhibitor of protein phosphatase 2A that is essential for mitosis. Science 330:1670–1673
107. McConnell JL, Gomez RJ, McCorvey LRA et al (2007) Identification of a PP2A interacting protein that functions as a negative regulator of phosphatase activity in the ATM/ATR signaling pathway. Oncogene 26:6021–6030
108. Curtis C, Shah SP, Chin SF et al (2012) The genomic and transcriptomic architecture of 2,000 breast tumours reveals novel subgroups. Nature 486:346–352
109. Deichmann M, Polychronidis M, Wacker J et al (2001) The protein phosphatase 2A subunit Bgamma gene is identified to be differentially expressed in malignant melanomas by subtractive suppression hybridization. Melanoma Res 11:577–585
110. Roberts KG, Smith AM, McDougall F et al (2010) Essential requirement for PP2A inhibition by the oncogenic receptor c-KIT suggests PP2A reactivation as a strategy to treat c-KIT+ cancers. Cancer Res 70:5438–5447
111. Neviani P, Santhanam R, Trotta R et al (2005) The tumor suppressor PP2A is functionally inactivated in blast crisis CML through the inhibitory activity of the BCR/ABL-regulated SET protein. Cancer Cell 8:355–368
112. Perrotti D, Neviani P (2008) Protein phosphatase 2A (PP2A), a drugable tumor suppressor in Ph1(+) leukemias. Cancer Metastasis Rev 27:159–168
113. Cristóbal I, Garcia-Orti L, Cirauqui C et al (2011) PP2A impaired activity is a common event in acute myeloid leukemia and its activation by forskolin has a potent anti-leukemic effect. Leukemia 25:606–614
114. Cristóbal I, Garcia-Orti L, Cirauqui C et al (2012) Overexpression of SET is a recurrent event associated with poor outcome and contributes to protein phosphatase 2A inhibition in acute myeloid leukemia. Haematologica 97:543–550
115. Liu F, Grundke-Iqbal I, Iqbal K et al (2005) Contributions of protein phosphatases PP1, PP2A, PP2B and PP5 to the regulation of tau phosphorylation. Eur J Neurosci 22:1942–1950
116. McConnell JL, Wadzinski BE (2009) Targeting protein serine/threonine phosphatases for drug development. Mol Pharmacol 75:1249–1261
117. Kalev P, Sablina AA (2011) Protein phosphatase 2A as a potential target for anticancer therapy. Anticancer Agents Med Chem 11:38–46
118. Voronkov M, Braithwaite SP, Stock JB (2011) Phosphoprotein phosphatase 2A: a novel druggable target for Alzheimer's disease. Future Med Chem 3:821–833
119. Braithwaite SP, Voronkov M, Stock JB et al (2012) Targeting phosphatases as the next generation of disease modifying therapeutics for Parkinson's disease. Neurochem Int 61:899–906
120. Kowluru A, Matti A (2012) Hyperactivation of protein phosphatase 2A in models of glucolipotoxicity and diabetes: potential mecha-

nisms and functional consequences. Biochem Pharmacol 84:591–597
121. Ingebritsen TS, Cohen P (1983) The protein phosphatases involved in cellular regulation. 1. Classification and substrate specificities. Eur J Biochem 132:255–261
122. Pelech S, Cohen P (1985) The protein phosphatases involved in cellular regulation. 1. Modulation of protein phosphatases-1 and 2A by histone H1, protamine, polylysine and heparin. Eur J Biochem 148:245–251
123. Waelkens E, Goris J, Merlevede W (1987) Purification and properties of polycation-stimulated phosphorylase phosphatases from rabbit skeletal muscle. J Biol Chem 262:1049–1059
124. Janssens V, Derua R, Zwaenepoel K et al (2009) Specific regulation of protein phosphatase 2A PR72/B″ subunits by calpain. Biochem Biophys Res Commun 386:676–681
125. King AJ, Andjelkovic N, Hemmings BA et al (1994) The phospho-opsin phosphatase from bovine rod outer segments. An insight into the mechanism of stimulation of type-2A protein phosphatase activity by protamine. Eur J Biochem 225:383–394
126. Cheng Q, Erickson AK, Wang ZX et al (1996) Stimulation of phosphorylase phosphatase activity of protein phosphatase 2A1 by protamine is ionic strength dependent and involves interaction of protamine with both substrate and enzyme. Biochemistry 35:15593–15600
127. Agostinis P, Goris J, Waelkens E et al (1987) Dephosphorylation of phosphoproteins and synthetic phosphopeptides. Study of the specificity of the polycation-stimulated and MgATP-dependent phosphorylase phosphatases. J Biol Chem 262:1060–1064
128. Chalfant CE, Szulc Z, Roddy P et al (2004) The structural requirements for ceramide activation of serine-threonine protein phosphatases. J Lipid Res 45:496–506
129. Chalfant CE, Kishikawa K, Mumby MC et al (1999) Long chain ceramides activate protein phosphatase-1 and protein phosphatase-2A. Activation is stereospecific and regulated by phosphatidic acid. J Biol Chem 274:20313–20317
130. Habrukowich C, Han DK, Le A et al (2010) Sphingosine interaction with acidic leucine-rich nuclear phosphoprotein-32A (ANP32A) regulates PP2A activity and cyclooxygenase (COX)-2 expression in human endothelial cells. J Biol Chem 285:26825–26831
131. Yoon CH, Kim MJ, Park MT et al (2009) Activation of p38 mitogen-activated protein kinase is required for death receptor-independent caspase-8 activation and cell death in response to sphingosine. Mol Cancer Res 7:361–370
132. Matsuoka Y, Nagahara Y, Ikekita M et al (2003) A novel immunosuppressive agent FTY720 induced Akt dephosphorylation in leukemia cells. Br J Pharmacol 138:1303–1312
133. Nagaoka Y, Otsuki K, Fujita T et al (2008) Effects of phosphorylation of immunomodulatory agent FTY720 (fingolimod) on antiproliferative activity against breast and colon cancer cells. Biol Pharm Bull 31:1177–1181
134. Neviani P, Santhanam R, Oaks JJ et al (2007) FTY720, a new alternative for treating blast crisis chronic myelogenous leukemia and Philadelphia chromosome-positive acute lymphocytic leukemia. J Clin Invest 117:2408–2421
135. Liu Q, Zhao X, Frissora F et al (2008) FTY720 demonstrates promising preclinical activity for chronic lymphocytic leukemia and lymphoblastic leukemia/lymphoma. Blood 111:275–284
136. Liu Q, Alinari L, Chen CS et al (2010) FTY720 shows promising in vitro and in vivo preclinical activity by downmodulating Cyclin D1 and phospho-Akt in mantle cell lymphoma. Clin Cancer Res 16:3182–3192
137. Christensen DJ, Ohkubo N, Oddo J et al (2011) Apolipoprotein E and peptide mimetics modulate inflammation by binding the SET protein and activating protein phosphatase 2A. J Immunol 186:2535–2542
138. Switzer CH, Cheng RY, Vitek TM et al (2011) Targeting SET/I(2)PP2A oncoprotein functions as a multi-pathway strategy for cancer therapy. Oncogene 30:2504–2513
139. Christensen DJ, Chen Y, Oddo J et al (2011) SET oncoprotein overexpression in B-cell chronic lymphocytic leukemia and non-Hodgkin lymphoma: a predictor of aggressive disease and a new treatment target. Blood 118:4150–4158
140. Vitek MP, Christensen DJ, Wilcock D et al (2012) APOE-mimetic peptides reduce behavioral deficits, plaques and tangles in Alzheimer's disease transgenics. Neurodegener Dis 10:122–126
141. Ghosal, K., Stathopoulos, A., Thomas, D. et al. (2012) The apolipoprotein-E-mimetic cog112 protects amyloid precursor protein intracellular domain-overexpressing animals from alzheimer's disease-like pathological features. Neurodegener Dis PMID= 22965147
142. Chohan MO, Khatoon S, Iqbal IG et al (2006) Involvement of I2PP2A in the abnormal hyperphosphorylation of tau and its reversal by Memantine. FEBS Lett 580:3973–3979
143. Feschenko MS, Stevenson E, Nairn AC et al (2002) A novel cAMP-stimulated pathway in

protein phosphatase 2A activation. J Pharmacol Exp Ther 302:111–118
144. Moon EY, Lerner A (2003) PDE4 inhibitors activate a mitochondrial apoptotic pathway in chronic lymphocytic leukemia cells that is regulated by protein phosphatase 2A. Blood 101:4122–4130
145. Mishra-Gorur K, Singer HA, Castellot JJ Jr (2002) Heparin inhibits phosphorylation and autonomous activity of Ca^{2+}/calmodulin-dependent protein kinase II in vascular smooth muscle cells. Am J Pathol 161:1893–1901
146. Kamibayashi C, Estes R, Slaughter C et al (1991) Subunit interactions control protein phosphatase 2A. Effects of limited proteolysis, N-ethylmaleimide, and heparin on the interaction of the B subunit. J Biol Chem 266:13251–13260
147. Nishimura M, Uyeda K (1995) Purification and characterization of a novel xylulose 5-phosphate-activated protein phosphatase catalyzing dephosphorylation of fructose-6-phosphate,2-kinase:fructose-2,6-bisphosphatase. J Biol Chem 270:26341–26346
148. Kabashima T, Kawaguchi T, Wadzinski BE et al (2003) Xylulose 5-phosphate mediates glucose-induced lipogenesis by xylulose 5-phosphate-activated protein phosphatase in rat liver. Proc Natl Acad Sci USA 100:5107–5112
149. Corcoran NM, Martin D, Hutter-Paier B et al (2010) Sodium selenate specifically activates PP2A phosphatase, dephosphorylates tau and reverses memory deficits in an Alzheimer's disease model. J Clin Neurosci 17:1025–1033
150. van Eersel J, Ke YD, Liu X et al (2010) Sodium selenate mitigates tau pathology, neurodegeneration, and functional deficits in Alzheimer's disease models. Proc Natl Acad Sci USA 107:13888–13893
151. Switzer CH, Ridnour LA, Cheng RY et al (2009) Dithiolethione compounds inhibit Akt signaling in human breast and lung cancer cells by increasing PP2A activity. Oncogene 28:3837–3846
152. Guichard C, Pedruzzi E, Fay M et al (2006) Dihydroxyphenylethanol induces apoptosis by activating serine/threonine protein phosphatase PP2A and promotes the endoplasmic reticulum stress response in human colon carcinoma cells. Carcinogenesis 27:1812–1827
153. Ricciarelli R, Tasinato A, Clément S (1998) alpha-Tocopherol specifically inactivates cellular protein kinase C alpha by changing its phosphorylation state. Biochem J 334:243–249
154. Neuzil J, Weber T, Schröder A et al (2001) Induction of cancer cell apoptosis by alpha-tocopheryl succinate: molecular pathways and structural requirements. FASEB J 15:403–415
155. Lee KW, Chen W, Junn E et al (2011) Enhanced phosphatase activity attenuates α-synucleinopathy in a mouse model. J Neurosci 31:6963–6971
156. Lu J, Kovach JS, Johnson F et al (2009) Inhibition of serine/threonine phosphatase PP2A enhances cancer chemotherapy by blocking DNA damage induced defense mechanisms. Proc Natl Acad Sci USA 106:11697–11702
157. Lewy DS, Gauss CM, Soenen DR et al (2002) Fostriecin: chemistry and biology. Curr Med Chem 9:2005–2032
158. Swingle MR, Amable L, Lawhorn BG et al (2009) Structure-activity relationship studies of fostriecin, cytostatin, and key analogs, with PP1, PP2A, PP5, and(beta12-beta13)-chimeras (PP1/PP2A and PP5/PP2A), provide further insight into the inhibitory actions of fostriecin family inhibitors. J Pharmacol Exp Ther 331:45–53
159. Wada S, Usami I, Umezawa Y et al (2010) Rubratoxin A specifically and potently inhibits protein phosphatase 2A and suppresses cancer metastasis. Cancer Sci 101:743–750
160. Liu D, Chen Z (2009) The effects of cantharidin and cantharidin derivates on tumour cells. Anticancer Agents Med Chem 9:392–396
161. Kar S, Palit S, Ball WB et al (2012) Carnosic acid modulates Akt/IKK/NF-κB signaling by PP2A and induces intrinsic and extrinsic pathway mediated apoptosis in human prostate carcinoma PC-3 cells. Apoptosis 17:735–747
162. Aceto N, Bertino P, Barbone D et al (2009) Taurolidine and oxidative stress: a rationale for local treatment of mesothelioma. Eur Respir J 34:1399–1407
163. de Los Ríos C, Egea J, Marco-Contelles J et al (2010) Synthesis, inhibitory activity of cholinesterases, and neuroprotective profile of novel 1,8-naphthyridine derivatives. J Med Chem 53:5129–5143
164. Cho DH, Choi YJ, Jo SA et al (2006) Troglitazone acutely inhibits protein synthesis in endothelial cells via a novel mechanism involving protein phosphatase 2A-dependent p70 S6 kinase inhibition. Am J Physiol Cell Physiol 291:C317–C326
165. Holmes CF, Luu HA, Carrier F et al (1990) Inhibition of protein phosphatases-1 and -2A with acanthifolicin. Comparison with diarrhetic shellfish toxins and identification of a region on okadaic acid important for phosphatase inhibition. FEBS Lett 270:216–218
166. Ishihara H, Martin BL, Brautigan DL et al (1989) Calyculin A and okadaic acid:

inhibitors of protein phosphatase activity. Biochem Biophys Res Commun 159:871–877

167. Eriksson JE, Toivola D, Meriluoto JA et al (1990) Hepatocyte deformation induced by cyanobacterial toxins reflects inhibition of protein phosphatases. Biochem Biophys Res Commun 173:1347–1353
168. Li YM, Mackintosh C, Casida JE (1993) Protein phosphatase 2A and its [3H]cantharidin/[3H]endothall thioanhydride binding site. Inhibitor specificity of cantharidin and ATP analogues. Biochem Pharmacol 46: 1435–1443
169. Sakoff JA, McCluskey A (2004) Protein phosphatase inhibition: structure based design. Towards new therapeutic agents. Curr Pharm Des 10:1139–1159
170. Kawada M, Kawatsu M, Masuda T et al (2003) Specific inhibitors of protein phosphatase 2A inhibit tumor metastasis through augmentation of natural killer cells. Int Immunopharmacol 3:179–188

Index

José Luis Millán (ed.), *Phosphatase Modulators*, Methods in Molecular Biology, vol. 1053,
DOI 10.1007/978-1-62703-562-0,

B

C

D

J

K

L

M

N

O

P

R

S

T

U

V

W

X

Z

MIX
Papier aus verantwortungsvollen Quellen
Paper from responsible sources
FSC® C105338

If you have any concerns about our products,
you can contact us on
ProductSafety@springernature.com

In case Publisher is established outside the EU,
the EU authorized representative is:
Springer Nature Customer Service Center GmbH
Europaplatz 3, 69115 Heidelberg, Germany

Printed by Libri Plureos GmbH
in Hamburg, Germany